HVAC LICENSING
EXAM STUDY GUIDE

About the Authors

Rex Miller, Professor Emeritus of Industrial Technology at the State University College at Buffalo, has taught technical courses at all levels for over 40 years. He is the author or co-author of more than 100 textbooks for vocational schools and industrial-arts programs, including McGraw-Hill's bestselling *Carpentry & Construction* and *Electrician's Pocket Manual*.

Mark R. Miller, Professor of Industrial Technology at the University of Texas in Tyler, has taught technical courses at all levels for over 20 years. He is the author or co-author of more than 30 textbooks for trade and technical programs, including *Audel Refrigeration: Home and Commercial* and *Audel Air Conditioning: Home and Commercial*.

HVAC LICENSING EXAM STUDY GUIDE

Rex Miller
Mark R. Miller

McGraw-Hill

New York Chicago San Francisco Lisbon London Madrid
Mexico City Milan New Delhi San Juan Seoul
Singapore Sydney Toronto

697
MIL

Copyright © 2007 by The McGraw-Hill Companies, Inc. All rights reserved.
Printed in the United States of America. Except as permitted under the United
States Copyright Act of 1976, no part of this publication may be reproduced or
distributed in any form or by any means, or stored in a data base or retrieval
system, without the prior written permission of the publisher.

1 2 3 4 5 6 7 8 9 0 QPD/QPD 0 1 3 2 1 0 9 8 7

ISBN-13: 978-0-07-148640-8
ISBN-10: 0-07-148640-2

*The sponsoring editor for this book was Larry S. Hager and the production supervisor was
Pamela A. Pelton. It was set in Baskerville by Lone Wolf Enterprises, Ltd. The art director
for the cover was Anthony Landi.*

Printed and bound by Quebecor/Dubuque.

This book is printed on acid-free paper.

McGraw-Hill books are available at special quantity discounts to use as premi-
ums and sales promotions, or for use in corporate training programs. For more
information, please write to the Director of Special Sales, McGraw-Hill
Professional, Two Penn Plaza, New York, NY 10121-2298. Or contact your local
bookstore.

This book is dedicated to

Patricia Ann Navara Miller

wife and mother

Contents

$$\text{PART A} \quad \textbf{HEAT}$$

---

* Space limitations preclude our including text material covering all the questions in each chapter. Outside sources should be consulted as you examine the whole field of air conditioning and refrigeration.

PART E **CODES AND CERTIFICATIONS**

Preface

This textbook has been prepared to aid in instructional programs in high schools, technical schools, trade schools, and community colleges. Adult evening classes and apprenticeship programs may also find it useful. This book provides over 800 questions, multiple-choice, true-false and some scrambled words to aid in name recognition. This book should be especially helpful to those studying on their own or in formal classes for the certification exams offered by the ARI, NATE, and others.

The book has been organized for quick recognition and for ease in locating material desired. A glossary has been provided for those who like to learn more than required of them. The scrambled words can be a mind teaser and an educational tool expanding your vocabulary and stretching your brain power.

Once the answers have been filled in, the book still has value in aiding those who want to review or dig deeper into this field of knowledge.

Note how additional information has been provided for each chapter for an extended or, in some cases, deeper coverage of the material. This book does not guarantee you will pass any test by studying its contents. It does have the ability to extend your knowledge of your field and expand your thinking power so that you will be better equipped to learn what is necessary to become a master of your specialty and trade.

Rex Miller
Mark R. Miller

Acknowledgments

No author works without being influenced and aided by others. Every book reflects this fact. This book is no exception. A number of people cooperated in providing technical data and illustrations. For this we are grateful.

We would like to thank those organizations that so generously contributed information and illustrations. The following have been particularly helpful:

ARI
ASHRAE
Carrier Air-Conditioning Co.
Dwyer Instruments, Inc.
Johnson Controls, Inc.
Kelvinator Company
Lennox Industries, Inc.
Marley Company
NATE
National Refrigerants, Inc.
Packless Industries Co.
Rheem Manufacturing Co.
Sears, Roebuck & Co.
Tecumseh Products, Co.
Thermal Engineering, Inc.
Tuttle and Bailey, Div. of Allied Thermal Corp.
Virginia Chemical Co.
Weksler Instrument Co.

Introduction

Techniques for Studying and Test-Taking

PREPARING FOR THE EXAM

1. **Make a study schedule.** Assign yourself a period of time each day to devote to preparation for your exam. A regular time is best, but the important thing is daily study.

2. **Study alone.** You will concentrate better when you work by yourself. Keep a list of questions you find puzzling and points you are unsure of to talk over with a friend who is preparing for the same exam. Plan to exchange ideas at a joint review session just before the test.

3. **Eliminate distractions.** Choose a quiet, well-lit spot as far as possible from telephone, television, and family activities. Try to arrange not to be interrupted.

4. **Begin at the beginning.** Read. Underline points that you consider significant. Make marginal notes. Flag the pages that you think are especially important with little Post-it™ Notes.

5. **Concentrate on the information and instruction chapters.** Study the Code Definitions, the Glossary of air-conditioning and refrigeration terms, and the Decoding Words of equipment and usage. Learn the language of the field. Focus on the technique of eliminating wrong answers. This information is important to answering all multiple-choice questions.

6. **Answer the practice questions chapter by chapter.** Take note of your weaknesses; use all available textbooks to brush up.

7. **Try the previous exams if available.** When you believe that you are well prepared, move on to these exams. If possible, answer an entire exam in one sitting. If you must divide your time, divide it into no more than two sessions per exam.

    • When you do take the practice exams, treat them with respect. Consider each as a dress rehearsal for the real thing. Time yourself accurately, and do not peek at the correct answers.

    • Remember, you are taking these for practice; they will not be scored; they do not count. So learn from them.

IMPORTANT: Do not memorize questions and answers. Any question that has been released will not be used again. You may run into questions that are very similar, but you will not be tested with the original ones. The included questions will give you good practice, but they will not be the same as any of the questions on your exam.

HOW TO TAKE AN EXAM

Get to the examination room about 10 minutes ahead of time. You'll get a better start when you are accustomed to the room. If the room is too cold, too warm, or not well ventilated, call these conditions to the attention of the person in charge.

Make sure that you read the instructions carefully. In many cases, test takers lose points because they misread some important part of the directions. (An example would be reading the incorrect choice instead of the correct choice.)

Don't be afraid to guess. The best policy is, of course, to pace yourself so that you can read and consider each question. Sometimes this does not work. Most exam scores are based only on the number of questions answered correctly. This means that a wild guess is better than a blank space. There is no penalty for a wrong answer, and you just might guess right. If you see that time is about to run out, mark all the remaining spaces with the same answer. According to the law of averages, some will be right.

However, you have bought this book for practice in answering questions. Part of your preparation is learning to pace yourself so that you need not answer randomly at the end. Far better than a wild guess is an educated guess. You make this kind of guess not when you are pressed for time but are not sure of the correct answer. Usually, one or two of the choices are obviously wrong. Eliminate the obviously wrong answers and try to reason among those remaining. Then, if necessary, guess from the smaller field. The odds of choosing a right answer increase if you guess from a field of two instead of from a field of four. When you make an educated guess or a wild guess in the course of the exam, you might want to make a note next to the question number in the test booklet. Then, if there is time, you can go back for a second look.

Reason your way through multiple-choice questions very carefully and methodically.

MULTIPLE-CHOICE TEST-TAKING TIPS

Here are a few examples that we can "walk through" together:

1.   On the job, your supervisor gives you a hurried set of directions. As you start your assigned task, you realize you are not quite clear on the directions given to you. The best action to take would be to:

   (a) continue with your work, hoping to remember the directions

   (b) ask a co-worker in a similar position what he or she would do

   (c) ask your supervisor to repeat or clarify certain directions

   (d) go on to another assignment

In this question you are given four possible answers to the problem described. Though the four choices are all possible actions, it is up to you to choose the best course of action in this particular situation.

Choice (a) will likely lead to a poor result; given that you do not recall or understand the directions, you would not be able to perform the assigned task properly. Keep choice (a) in the back of your mind until you have examined the other alternatives. It could be the best of the four choices given.

Choice (b) is also a possible course of action, but is it the best? Consider that the co-worker you consult has not heard the directions. How could he or she know? Perhaps his or her degree of incompetence is greater than yours in this area. Of choices (a) and (b), the better of the two is still choice (a).

Choice (c) is an acceptable course of action. Your supervisor will welcome your questions and will not lose respect for you. At this point, you should hold choice (c) as the best answer and eliminate choice (a).

The course of action in choice (d) is decidedly incorrect because the job at hand would not be completed. Going on to something else does not clear up the problem; it simply postpones your having to make a necessary decision.

After careful consideration of all choices given, choice (c) stands out as the best possible course of action. You should select choice (c) as your answer.

Every question is written about a fact or an accepted concept. The question above indicates the concept that, in general, most supervisory personnel appreciate subordinates questioning directions that may not have been fully understood. This type of clarification precludes subsequent errors. On the other hand, many subordinates are reluctant to ask questions for fear that their lack of understanding will detract from their supervisor's evaluation of their abilities.

The supervisor, therefore, has the responsibility of issuing orders and directions in such a way that subordinates will not be discouraged from asking questions. This is the concept on which the sample question was based.

Of course, if you were familiar with this concept, you would have no trouble answering the question. However, if you were not familiar with it, the method outlined here of eliminating incorrect choices and selecting the correct one should prove successful for you.

We have now seen how important it is to identify the concept and the key phrase of the question. Equally or perhaps even more important is identifying and analyzing the keyword or the qualifying word in a question. This word is usually an adjective or adverb. Some of the most common key words are:

most	least	best	highest		
lowest	always	never	sometimes		
most likely	greatest	smallest	tallest		
average	easiest	most nearly	maximum		
minimum	only	chiefly	mainly	but	or

Identifying these keywords is usually half the battle in understanding and, consequently, answering all types of exam questions.

Now we will use the elimination method on some additional questions.

2. On the first day you report for work after being appointed as an AC mechanic's helper, you are assigned to routine duties that seem to you to be very petty in scope. You should:

(a) perform your assignment perfunctorily while conserving your energies for more important work in the future

(b) explain to your superior that you are capable of greater responsibility

(c) consider these duties an opportunity to become thoroughly familiar with the workplace

(d) try to get someone to take care of your assignment until you have become thoroughly acquainted with your new associates

Once again we are confronted with four possible answers from which we are to select the best one.

Choice (a) will not lead to getting your assigned work done in the best possible manner in the shortest possible time. This would be your responsibility as a newly appointed AC mechanic's helper, and the likelihood of getting to do more important work in the future following the approach stated in this choice is remote. However, since this is only choice (a), we must hold it aside because it may turn out to be the best of the four choices given.

Choice (b) is better than choice (a) because your superior may not be familiar with your capabilities at this point. We therefore should drop choice (a) and retain choice (b) because, once again, it may be the best of the four choices.

The question clearly states that you are newly appointed. Therefore, would it not be wise to perform whatever duties you are assigned in the best possible manner? In this way, you would not only use the opportunity to become acquainted with procedures but also to demonstrate your abilities.

Choice (c) contains a course of action that will benefit you and the location in which you are working because it will get needed work done. At this point, we drop choice (b) and retain choice (c) because it is by far the better of the two.

The course of action in choice (d) is not likely to get the assignment completed, and it will not enhance your image to your fellow AC mechanic's helpers.

Choice (c), when compared to choice (d), is far better and therefore should be selected as the best choice.

Now let us take a question that appeared on a police-officer examination:

3.   An off-duty police officer in civilian clothes riding in the rear of a bus notices two teenage boys tampering with the rear emergency door. The most appropriate action for the officer to take is to:

   (a) tell the boys to discontinue their tampering, pointing out the dangers to life that their actions may create

   (b) report the boys' actions to the bus operator and let the bus operator take whatever action is deemed best

   (c) signal the bus operator to stop, show the boys the officer's badge, and then order them off the bus

   (d) show the boys the officer's badge, order them to stop their actions, and take down their names and addresses

Before considering the answers to this question, we must accept that it is a well-known fact that a police officer is always on duty to uphold the law even though he or she may be technically off duty.

In choice (a), the course of action taken by the police officer will probably serve to educate the boys and get them to stop their unlawful activity. Since this is only the first choice, we will hold it aside.

In choice (b), we must realize that the authority of the bus operator in this instance is limited. He can ask the boys to stop tampering with the door, but that is all. The police officer can go beyond that point. Therefore, we drop choice (b) and continue to hold choice (a).

Choice (c) as a course of action will not have a lasting effect. What is to stop the boys from boarding the next bus and continuing their unlawful action? We therefore drop choice (c) and continue to hold choice (a).

Choice (d) may have some beneficial effect, but it would not deter the boys from continuing their actions in the future.

When we compare choice (a) with choice (d), we find that choice (a) is the better one overall, and therefore it is the correct answer.

The next question illustrates a type of question that has gained popularity in recent examinations and that requires a two-step evaluation.

First, the reader must evaluate the condition in the question as being "desirable" or "undesirable." Once the determination has been made, we are then left with making a selection from two choices instead of the usual four.

4. A visitor to an office in a city agency tells one of the aides that he has an appointment with the supervisor, who is expected shortly. The visitor asks for permission to wait in the supervisor's private office, which is unoccupied at the moment. For the office aide to allow the visitor to do so would be:

   (a) desirable; the visitor would be less likely to disturb the other employees or to be disturbed by them

   (b) undesirable; it is not courteous to permit a visitor to be left alone in an office

   (c) desirable; the supervisor may wish to speak to the visitor in private

   (d) undesirable; the supervisor may have left confidential papers on the desk

First of all, we must evaluate the course of action on the part of the office aide of permitting the visitor to wait in the supervisor's office as being very undesirable. There is nothing said of the nature of the visit; it may be for a purpose that is not friendly or congenial. There may be papers on the supervisor's desk that he or she does not want the visitor to see or to have knowledge of. Therefore, at this point, we have to decide between choices (b) and (d).

This is definitely not a question of courtesy. Although all visitors should be treated with courtesy, permitting the visitor to wait in the supervisor's office is not the only possible act of courtesy. Another comfortable place could be found for the visitor to wait.

Choice (d) contains the exact reason for evaluating this course of action as being undesirable, and when we compare it with choice (b), choice (d) is far better.

TRUE-FALSE QUESTIONS AND ANSWERS

Keep in mind that tests or exams are also learning tools. They make you learn the assigned material so that you don't have to refer to sources other than your own brain's memory.

There are a number of types of questions utilized in everyday teaching and learning. The advantage of the multiple-choice type of question is that it makes you think and then utilize your reasoning power to arrive at an educated guess. That is, of course, if you didn't know the answer outright from previous experience or studying.

Next, there is the true-false type of question. It is easy to answer either true (T) or false (F), you are either right or wrong. You have a 50 percent chance of being right or wrong when you guess. One of the major reasons this type of test is used is its quick right or wrong answer. It makes you think fast and recall the material you recently read or studied and quickly focuses your learning on the necessary information.

Both types are easy to check and grade. They also make the instructor's role an easier one.

Another type of question and answer test is "fill-in the blank." This requires you to think in regards to the meaning of the sentence with the missing word. There are, however, many clues in the question or statement before you. This type of test can also be used with the blank filled in by four possible answers. This type then resembles the multiple-choice type of test and serves the same purpose.

There are certifying agencies organized and operating to aid you in obtaining the skills and knowledge to perform correctly in your chosen field. By requiring you to submit to a written exam on the material, you have an incentive to study hard and organize the material you have committed to memory. Passing the exam from one of the agencies makes it easier for them to hire you to do the work and know that you can do it with some degree of skill and perfection.

The inspector relies on you to do the assigned task properly and safely. The inspector has to be thoroughly familiar with all the contract documents, including the plans with all changes, specifications, and contract submittals such as shop drawings.

Inspectors have different responsibilities and authorities, depending on the organizational setup, and size and scope of the project. Each inspector should be clear on the answers to the many questions presented during a day in the field.

Inspectors have the task of examining a finished job and informing the specialist as to how his work meets or fails to meet the code requirements.

A STRATEGY FOR TEST DAY

On the exam day assigned to you, allow the test itself to be the main attraction of the day. Do not squeeze it in between other activities. Arrive rested, relaxed, and on time. In fact, plan to arrive a little bit early. Leave plenty of time for traffic tie-ups or other complications that might upset you and interfere with your test performance.

Here is a breakdown of what occurs on examination day and tips on starting off on the right foot and preparing to start your exam:

1.   In the test room the examiner will hand out forms for you to fill out and will give you the instructions that you must follow in taking the examination. Note that you must follow instructions exactly.

2.   The examiner will tell you how to fill in the blanks on the forms.

3.   Exam time limits and timing signals will be explained.

4.   Be sure to ask questions if you do not understand any of the examiner's instructions. You need to be sure that you know exactly what to do.

5.   Fill in the grids on the forms carefully and accurately. Filling in the wrong blank may lead to loss of veterans' credits to which you may be entitled or to an incorrect address for your test results.

6.   Do not begin the exam until you are told to begin.

7.   Stop as soon as the examiner tells you to stop.

8.   Do not turn pages until you are told to do so.

9.   Do not go back to parts you have already completed.

10. Any infraction of the rules is considered cheating. If you cheat, your test paper will not be scored, and you will not be eligible for appointment.

11. Once the signal has been given and you begin the exam, read every word of every question.

12. Be alert for exclusionary words that might affect your answer: words such as "not" "most," and "least."

MARKING YOUR ANSWERS

Read all the choices before you mark your answer. It is statistically true that most errors are made when the last choice is the correct answer. Too many people mark the first answer that seems correct without reading through all the choices to find out which answer is best.

Be sure to read the suggestions below now and review them before you take the actual exam. Once you are familiar with the suggestions, you will feel more comfortable with the exam itself and find them useful when you are marking your answer choices.

1. Mark your answers by completely blackening the answer space of your choice.

2. Mark only ONE answer for each question, even if you think that more than one answer is correct. You must choose only one. If you mark more than one answer, the scoring machine will consider you wrong even if one of your answers is correct.

3. If you change your mind, erase completely. Leave no doubt as to which answer you have chosen.

4. If you do any figuring on the test booklet or on scratch paper, be sure to mark your answer on the answer sheet.

5. Check often to be sure that the question number matches the answer space number and that you have not skipped a space by mistake. If you do skip a space, you must erase all the answers after the skip and answer all the questions again in the right places.

6. Answer every question in order, but do not spend too much time on any one question. If a question seems to be "impossible," do not take it as a personal challenge. Guess and move on. Remember that your task is to answer correctly as many questions as possible. You must apportion your time so as to give yourself a fair chance to read and answer all the questions. If you guess at an answer, mark the question in the test booklet so that you can find it again easily if time allows.

7. Guess intelligently if you can. If you do not know the answer to a question, eliminate the answers that you know are wrong and guess from among the remaining choices. If you have no idea whatsoever of the answer to a question, guess anyway. Choose an answer other than the first.

   The first choice is generally the correct answer less often than the other choices. If your answer is a guess, either an educated guess or a wild one, mark the question in the question booklet so that you can give it a second try if time permits.

8. If you happen to finish before time is up, check to be sure that each question is answered in the right space and that there is only one answer for each question. Return to the difficult questions that you marked in the booklet and try them again. There is no bonus for finishing early so use all your time to perfect your exam paper.

With the combination of techniques for studying and test taking as well as the self-instructional course and sample examinations in this book, you are given the tools you need to score high on your exam.

HVAC LICENSING EXAM STUDY GUIDE

Part A

HEAT

Chapter 1
HEAT SOURCES

STEAM AND HOT-WATER HEATING

Design and installation questions regarding steam and hot-water heating are asked in state licensing examinations at times reference is made to steam and hot-water heating piping systems. Therefore, steam and hot-water heating systems will briefly be discussed.

Steam Heating

Steam is an effective heating medium. It is adaptable to almost any type of building. One of the simplest steam-heating systems is the one-pipe gravity type. Installations of this type are usually limited to moderate-size buildings where the radiators can be positioned at least 24 inches above the water level in the boiler. This type of installation is simple to operate, and the initial cost is low.

Some inherent disadvantages of the one-pipe gravity steam heating systems are as follows:

1. Large piping and radiator valves are required to allow the condensate to return against the resistance offered by the steam flow.

2. Steam and condensate flow in opposite directions with a possibility of water hammer.

3. Air valves are required. Failure of these valves to always open to allow the escape of air may result in slow heat buildup and excessive fuel consumption.

4. Comfortable room temperatures are difficult to maintain unless the radiator valves are regulated by opening and closing. Automatic control of steam from the boiler may result in fluctuating room temperatures.

The two-pipe gravity steam heating system was developed to overcome the difficulty encountered when steam and condensate flow against each other in the same pipe. This system has all the disadvantages (to some degree) listed for the one-pipe system.

Additional disadvantages are as follows:

1. The midsections of the radiators may become air-bound if they are not water-sealed. This happens during the warm-up period when steam fills the radiator nearest the boiler and flows through and enters the return piping.

2. Installation of a valve at each end of all radiators is required so that the steam may be shut off. If this is not done, steam may be present in both the return and supply lines.

3. The returns from each radiator must be separately connected into a wet return header or water-sealed in some manner.

In a vapor system, thermostatic traps are used at each radiator and at the ends of the steam mains. The radiator inlet valves used in this system are of the graduated or orifice type. The steam pressure necessary is very low: often less than one pound. This type of installation can be used in buildings where 24 inches or more can be provided between the boiler water level and the end of the return line.

Advantages of a vapor steam-heating system include:

1. Air cannot enter the system as it is closed. Thus, a moderate vacuum is created by the condensation producing steam at lower temperatures.

2. An even, quiet circulation of steam is provided with no air binding or noisy water hammer.

3. Room temperatures can be closely and automatically regulated by thermostatic controls.

4. Radiator air valves are not required.

Disadvantages of a vapor steam-heating system include:

1. Comparatively large pipe sizes are necessary.

2. Only low steam pressure is possible.

3. The condensate must return to the boiler by gravity.

The condensate must return to the boiler by gravity in this type of heating system and may back up in the vertical return pipe when there is excess steam pressure in the boiler. As a result, an air eliminator must be installed well above the level of the water in the boiler, yet low enough for it to close before the return water is of sufficiently high level to enter the return main. Close control of the boiler pressure is required in this type of system.

A return-trap heating system closely resembles the vapor system except that the return trap provides a positive return of condensate to the boiler. This type can be used in all but the largest buildings if the equivalent direct radiation (EDR) capacity is not greater than the return-trap capacity.

Advantages of a return-trap steam-heating system include the following:

1. The pipe sizes may be smaller because of the higher steam pressures.

2. Return of condensate to the boiler is rapid and positive.

3. The system responds easily to thermostatic control.

4. Steam distribution may be balanced by the use of orifice valves.

There are some disadvantages to the return-trap heating system:

1. Steam circulation depends almost entirely on boiler pressure.

2. Sufficient head room above the boiler must be provided for piping installation.

3. Physical limitations to the size and capacity of return traps exist, which limit the boiler capacity.

If a heating system is limited by the boiler water-level height, by boiler capacity, or because a return trap cannot be used, a condensate-return system may be installed. In this system a condensation pump is installed to return the condensate to the boiler.

Condensate-return systems have the advantage of allowing the return lines to be located below the water level in the boiler and also allowing high steam pressures. However, large drip traps and piping are required for variable-vacuum and vacuum-return-line systems.

Vacuum-return-line systems are similar to condensate-return-type installations except that a vacuum pump is installed to provide a low vacuum in the return line to return the condensate to the boiler.

Advantages of a vacuum-return-line steam heating system include:

1. There is positive return of the condensate to the boiler.

2. Air is removed from the steam mechanically, resulting in a rapid circulation of the steam.

3. Smaller pipe sizes may be used because of the greater pressure differential between the supply and return lines.

The vacuum-air-line system is a variation of the one-pipe steam-heating system. Radiator air vents are replaced with air valves, the outlets of which are connected to a return air line. A vacuum pump is included to exhaust the system of air. Air-line valves are of the thermostatic type.

Advantages of this system include:

1. Steam circulation is more rapid.

2. Radiators heat efficiently at lower pressures.

3. Air vents are not required.

There are some disadvantages:

1. The steam may be noisy at times, as it and condensate flow in the same pipe.

2. Piping and radiator valves must be oversized to accommodate the flow of both steam and condensate.

Two-pipe medium- and high-pressure systems (with ranges of 25 to 125 psi) are used for space heating and for steam-process equipment such as water heaters, dryers, kettles, etc. The two-pipe high-pressure steam heating system is used for space heating and employs tube radiation, fan vector radiation, unit heaters, and fan units with blast coils. High-pressure thermostatic traps or inverted bucket traps are generally used in this type of installation to handle the condensate and to vent the air in the system.

Advantages of two-pipe medium-and high-pressure steam heating systems include:

1. Since the return water can be lifted into the return mains, condensate return lines can be elevated.

2. High-pressure or boiler-feed condensate pumps return the condensate directly to the boiler.

3. Smaller radiators or heat-exchange units and smaller pipe sizes may be used.

4. If the system is used for both heating and steam processing, a simplified piping system may be used with the supply mains being utilized for both the heating and processing equipment. A common condensate system may be installed.

MULTIPLE-CHOICE EXAM

1. When inspecting an installation of heating equipment, make sure there is enough room for the proper functioning and _____ of all the equipment.
    a. testing
    b. maintenance
    c. cleaning
    d. installation

2. Operation and maintenance instructions are usually posted on the _____ upon completion of the installation of heating equipment.
    a. equipment
    b. ceiling
    c. wall
    d. door

3. When checking a boiler installation to pass examination by a code inspector you should make sure the installation is correctly done in respect to the:
    a. ICC Building Code
    b. National Electrical Code
    c. ASME Code
    d. all of the above

4. When checking an installation of heating equipment so it will pass the code inspection, it is best to make sure that:
    a. the joints are tightly sealed
    b. the air is clean
    c. the water is clean
    d. the mortar is dry

5. The heat content of water is usually measured in:
    a. Btu
    b. degrees
    c. Celsius
    d. Fahrenheit

6. The five classes of steam heating systems are high pressure, medium pressure, low pressure, vapor and _____ systems.
    a. vacuum
    b. air
    c. suction
    d. blower

7. A steam trap must be properly sized to handle the full load of:
    a. steam
    b. hot water
    c. cold water
    d. condensate

8. Water hammering is a phenomenon that occurs when _____ remaining in the pipe flashes into steam.

    a. condensate   b. cold water

    c. hot water   d. warm water

9. Whenever a steam system is servicing an area whose outdoor temperature will drop below _____ the designer must make provisions to prevent freezing.

    a. 32 degrees F   b. 35 degrees F

    c. 40 degrees F   d. 10 degrees F

10. Which of the following is NOT a recommendation for minimizing freezing problems in steam systems?

    a. installing a strainer before all heating units

    b. oversize traps

    c. keeping condensate lines as short as possible

    d. where possible, not using overhead return

11. Unit heaters can be classified by:

    a. type of air mover (either propeller or centrifugal blower)

    b. airflow configuration (vertical or horizontal air delivery)

    c. heating medium (steam, hot water, gas, oil, or electricity)

    d. all of these

12. The disadvantage of the centrifugal blower fan unit heater is:

    a. it has a generally quieter operation than a propeller type

    b. it is less efficient than other types

    c. it is ideal for use in spot welding areas

    d. it has air discharge nozzles

13. Gas-heating unit heaters must conform to local codes and requirements for type and volume of gas burned and pressure _____ allowed in the line.

    a. drop   b. relief

    c. build-up   d. increase

14. Electric-heating unit heaters must conform to the local codes and the most often used

    a. National Electrical Code   b. International Code Council

    c. National Plumbing Code   d. Electrical Engineers Code

15. An important point to consider with every heating medium is the change in _____ _____ temperature.

    a. entering air   b. exiting air

    c. inside air   d. outside air

16. Appliances are classified as Category I and Category II. Category I appliances operate with a non-positive vent connection pressure and with a flue gas temperature of at least _____ degrees F above dew point.

    a. 180   b. 140

    c. 212   d. 32

17. Category II appliances operate with a non-positive vent connection pressure and with a flue gas temperature less than _____ degrees F.

    a. 180   b. 140

    c. 212   d. 32

18. Hydronic piping sizes should be sized for the demand of the system. These systems are installed according to the _____ .

    a. International Plumbing Code

    b. International Code Council

    c. International Fuel Gas Code

    d. International Mechanical Code

19. According to code, hydronic pipe shall not be made of:

    a. CPVC   b. copper

    c. brass   d. cast iron

20. Hydronic piping should be insulated to the thickness required by the.

    a. International Energy Conservation Code

    b. International Mechanical Code

    c. International Building Code

    d. National Builder's Code

21.   Heat exchangers were introduced in 1930 and since have gained wide acceptance. The heat exchanger is described as a viable means of transferring heat between liquids such as water to water and steam to water. Typical applications include:

   a.   cooling tower and water cooling

   b.   free-cycling cooling

   c.   highrise pressure interceptor applications

   d.   all the above

22.   The advantages of plate heat exchangers over other types of heat exchangers, such as shell-and-tube exchangers, include:

   a.   compact heat exchange

   b.   greater transfer efficiency

   c.   flexible design

   d.   all of these

   e.   none of these

23.   During normal operation, the heat exchanger must also be protected against large solids and _____ surges.

   a.   electrical   b.   pressure

   c.   water   d.   liquid

24.   Depending upon the nature of the transfer fluids and the application, performance of the heat exchanger may _____ over time.

   a.   upgrade   b.   degrade

   c.   improve   d.   explode

25.   Onsite cleaning of heat exchangers for encrusted scales and sedimentation include:

   a.   hot water

   b.   nitric, sulfuric, citric, or phosphoric acid

   c.   complexing agents such as EDTA or NTA

   d.   all of these

   e.   none of these

26. An alternative to external plate cleaning is on-site cleaning or cleaning in place (CIP). CIP cleaning is achieved by:

    a. chemical or partial dissolving of the deposit through chelation or chemical dissolution

    b. decomposition of the deposit containing sand, fiber, or other foreign particles

    c. reducing the adhesive force between the deposit and the plate

    d. killing the biological material present in the mass

    e. none of these

    f. all of these

27. In most cases, external leaks are sure signs of gasket failure. The most common cause for gasket failure is:

    a. gasket blowout   b. crushing

    c. old gaskets   d. all of these

    e. none of these

28. Radiant panel heating utilizes the following type of piping:

    a. copper   b. cross-linked polyethylene

    c. PEX   d. all of these

29. Radiant floor heat has been used for:

    a. 100 years   b. 500 years

    c. 50 years   d. centuries

30. Typical hydronic radiant flooring (HRF) can be used in:

    a. warehouses, garages, small commercial buildings

    b. small commercial buildings

    c. manufacturing facilities

    d. all of these

31. A well-insulated building is the best key to designing a successful HRF system. Which of the following is the best answer?

    a. The poorer the insulation, the better

    b. The hotter the heat source, the better

    c. A concrete HRF slab should be constructed below grade

    d. The earth itself acts as an insulative barrier

32. The HRF system heats by means of:

    a. radiation   b. forced hot air

    c. moving cold air   d. sunshine

33. Mean radiant temperature (MRT) is a term used to describe the collective effect on an occupant of all the surrounding _____ in an indoor environment.

    a. ceilings   b. surfaces

    c. walls   d. floors

34. HRT systems provide a high level of _____ by reducing drafts and creating even air-temperature distribution.

    a. concern   b. comfort

    c. air circulation   d. none of these

35. Structures with high ceilings can:

    a. cool rapidly with an HRF system

    b. obtain no benefit from an HRF system

    c. benefit greatly from an HRF system

    d. none of the above

36. The human body, as a surface, will either radiate its _____ to cooler surrounding surfaces or absorb _____ from warmer surfaces.

    a. energy   b. heat

    c. temperature   d. all of these

    e. none of these

37. MRT is a term used to describe the collective effect on an occupant of all the surrounding surfaces in an indoor environment. MRT means:

    a. mean radiant temperature

    b. more radiant temperatures

    c. more rapid temperature changes

    d. mean rapid temperature

38. All hydronic heating systems are susceptible to _____ entering the system through numerous sources such as threaded fittings, air vents, and gas-permeable materials.

    a. air   b. oxygen

    c. dust   d. gases

39. Which of the listed materials is NOT used in radiant-floor-heating installations?

    a. copper   b. rubber

    c. plastics   d. iron

40. Fuel-burning appliances must have the following characteristics and dimensions:

    a. must be accessible without moving permanent structures

    b. must have a 30-inch platform working space in front of the control side of the appliance with one exception: room heaters need only 18 inches of working space

    c. not less than 1 inch clear air space from combustible materials

    d. all must have a label

    e. none of these

    f. all of these

41. A hydronic system with a low water temperature has several advantages when viewed from the perspective of good solar-collector design. Which of the following is NOT a good solar-collector design?

    a. a low required water temperature in the hydronic system will increase the range of outdoor temperatures at which the solar system will be operative

    b. a low temperature in the hydronic system will allow operation of the solar collector at a lower mean liquid temperature and will increase the collection efficiency

    c. a low temperature in the hydronic system will allow operation of the solar collector at a higher mean liquid temperature and will increase the collection efficiency

42. Liquid solar-heating systems are generally _____ efficient than air-based ones.

    a. less   b. more

    c. much less   d. much more

43. All-water solar collector systems incorporate a _____ feature to prevent freezing damage.

    a. drain   b. dumping

    c. drainback   d. pumping

44. Leaks are a serious threat to the long-term operation of any solar heating system. To minimize the occurrence of leaks, the following should be used:

    a. welded or sweated joints should be used whenever possible

    b. glycol systems should contain an automatic makeup valve

    c. pumps should have gland seals

    d. none of these

45. What is heat tracing?

    a. circulating water pipes on external cooling towers

    b. prevention of condensation buildup in air lines

    c. the supplemental heating of fluid system piping through extraneous means

    d. exterior pipes carrying oil or fuel

46. For some piping networks, _____ heat tracing is the only practical method.

    a. hot water   b. steam

    c. electric   d. condensate

47. Heat-tracing duty can be continuous or _____.

    a. irregular   b. intermittent

    c. once in a while   d. never

48. Special heat-conducting _____ tapes are available where heat is to be conducted from the tracing to the rest of the circumference of the pipe.

    a. aluminum   b. copper

    c. plastic   d. iron

49. A typical heat-tracing application requires that the pipe be kept at _____ temperature above ambient conditions.

    a. the same   b. some

    c. the exact   d. approximately

50. In snow melting, designs for the system differ slightly depending on whether mineral-insulated cable or embedded wires are used. Embedded wires can be laid out in single strands or as _____ mats.

    a. prefabricated   b. interwoven

    c. solid   d. crested

TRUE-FALSE EXAM

1. In the installation of heating equipment, the inspector must make sure that each piece of material and each item of equipment have been approved well in advance of their need.

   True   False

2. The inspector must examine pressure boilers for conformance with the ASME Code.

   True   False

3. The inspector must require that refractories be kept wet.

   True   False

4. An inspector must be sure that all settings are constructed with provision for expansion and contraction of both refractories and pressure parts.

   True   False

5. An inspector must make sure that an uptake damper has been set for correct location, bearing material, and freedom of operation when cold.

   True   False

6. An inspector has to check for carbon dioxide in flue gas.

   True   False

7. Fuel-burning appliances must have a 30-inch platform or working space in front of the control side of the device.

   True   False

8. The exception to No.7 above is that room heaters need only 18 inches of working space.

   True   False

9. A Category IV appliance must operate with a positive vent pressure and a flue gas temperature of less than 140 degrees F above its dew point.

   True   False

10. The following type of code-approved appliance protection is 3 1/2-inch masonry without air space.

    True   False

11.   For code-approved appliance protection use a 1-inch glass fiber or mineral wool sandwiched between two sheets of 0.024-inch-thick metal with a ventilating air space.

   True   False

12.   If mineral wool batts are used, they must have a density of 8 pounds/foot^3 with a minimum melting point of 1,500 degrees F (816 degrees C).

   True   False

13.   In clothes-dryer ducts, the ducts must be galvanized steel or copper if passing through any fire-rated assembly.

   True   False

14.   Clothes-dryer ducts cannot exceed 25 feet in length and a minimum of 4 inches in diameter.

   True   False

15.   Gas dryers cannot be located in a room with other fuel-burning appliances.

   True   False

MULTIPLE-CHOICE ANSWERS

1. B	11. D	21. D	31. D	41. C
2. C	12. B	22. D	32. A	42. B
3. C	13. A	23. B	33. B	43. C
4. A	14. A	24. B	34. B	44. A
5. A	15. A	25. D	35. C	45. C
6. A	16. B	26. F	36. B	46. C
7. D	17. B	27. D	37. A	47. B
8. A	18. A	28. D	38. B	48. A
9. B	19. D	29. D	39. D	49. B
10. B	20. A	30. D	40. F	50. A

TRUE-FALSE ANSWERS

1. T	5. F	9. T	13. F
2. T	6. T	10. T	14. T
3. F	7. T	11. T	15. T
4. T	8. T	12. T	

Chapter 2
HEATING SYSTEMS

HOT-WATER HEATING

Hot-water heating systems transmit only sensible heat to radiators, as distinguished from steam systems that work principally by the latent heat of evaporation. The result is that the radiator temperature of a steam system is relatively high compared to that of a hot-water system. In a hot-water system, latent heat is not given off to a great degree, so more heating surface is required.

Advantages of hot-water heating include:

1. Temperatures may widely vary, so it is more flexible than low-pressure (above atmospheric) steam systems.

2. The radiators will remain warm for a considerable time after the heat-generating fire has gone out; thus the system is a reservoir for storing heat.

Disadvantages include:

1. There is a danger of freezing when not in use.

2. More or larger heating surfaces (radiators) are required than with steam systems.

There are actually two types of hot-water systems, depending on how heated water flows: thermal and forced circulation.

The word "thermal" refers to systems that depend on the difference in the weight of water per unit of volume at different temperatures to form the motive force that results in circulation. This type is rightfully called a gravity hot-water system. The difference in the density or weight of hot and cold water causes natural circulation throughout the system. This circulation is necessary in order for the water to carry the heat from the boiler to the radiators.

In the forced-circulation type of hot-water system a pump is used to force the water through the piping. Thus, the flow is entirely independent of the difference in water temperature.

Gravity hot-water systems are used mainly in small buildings such as homes and small business places.

Advantages of this type of system include:

1. Ease of operation.

2. Low installation costs.

3. Low maintenance costs.

Disadvantages include:

1. Possible water damage in case of leaks.

2. Rapid temperature changes result in a slow response from the system.

3. Properly balancing the flow of water to radiators is sometimes difficult.

4. Nonattendance when the heat-generating unit fails may result in a freeze-up.

5. Flow depends on gravity, and as a result larger pipe sizes are required for good operation.

Forced-hot-water heating systems require a pump that forces the water through the piping system. Limitations of flow, dependent on water-temperature differences, do not exist in this type of system. It may be of either the one- or two-pipe variety. In two-pipe systems, either direct or reversed returns and up-feed or down-feed mains may be used. The path of the water from the boiler into and through the radiator and back again to the boiler is almost the same length for each of the radiators in the system. It is common to use one-pipe forced-circulation systems for small and medium-sized buildings when hot water is used as the heating medium.

Advantages of forced circulation include:

1. There is rapid response to temperature changes.

2. Smaller pipe sizes may be installed.

3. Room temperatures can be automatically controlled if either the burner or the flow of water is thermostatically controlled.

4. There is less danger of water freeze and damage.

Disadvantages include:

1. All high points must be vented.

Radiant-panel heating is the method of heating a room by raising the temperature of one or more of its interior surfaces (floor, walls, or ceiling) instead of heating the air.

One of the most common methods of achieving radiant heating is by the installation of specially constructed pipe coils or lengths of tubing in the floor, walls, or ceiling. These coils generally consist of small-bore wrought-iron, steel, brass, or copper pipe, usually with an inside diameter of 3/8 to 1 inch. Every consideration should be given to complete building insulation when radiant panel heating is used.

Air venting is necessary to the proper control of any panel hot-water heating system. Collection of air in either the circuit pipe or pipe coils results in a shortage of heat. Because of the continuous slope of the coil connections, it may be sufficient to install automatic vents at the top of the return riser only, omitting such vents on the supply riser.

The following are some advantages of radiant-panel heating:

1. Radiant-panel heating eliminates radiators and grills, thus providing more floor space and resulting in better furniture arrangements and wall decorations.

2. There is less streaking of walls and ceiling due to lower velocities of air currents.

3. It provides warm floors in homes with no basement.

4. It simplifies interior architectural and engineering building designs.

5. A well-designed and installed radiant-panel heating system provides low operating and maintenance costs.

Hot-water radiant panels can be installed in nearly any type of building, with or without a basement or excavated section. Conventional hot-water boilers are used. Units of this type are available in compact types that fit into small spaces and are fired by gas or oil. Additions can be installed at any time, provided that the limitations of the boiler unit and circulating pump are not exceeded.

Radiant panels should never be used with steam; too many complications arise. In addition, domestic hot water should not be taken from the system for use in bathrooms or kitchens.

Provision must be made for draining the system if the need should arise. Care must be taken in design and installation to ensure that the system can be completely drained, with no water pockets that will hold water and result in damage in the event of a freeze.

The ceiling is the most satisfactory location for radiant-heating panels. A combination grid and continuous coil are used in a large installation where the heat requirements make it necessary to install several coils.

Ceiling panels should not be installed above plywood, composition board, or other insulating types of ceiling material. Surfaces of this type have an undesirable insulating effect that diminishes full heat output of the panel.

Radiant panels are often installed in floors. When this is done, the best arrangement is to place the coils in the concrete floor slab. Good results are obtained when the pipe or tubing is placed at least 2 inches below the floor surface or deeper if a heavy traffic load is anticipated. Allow a minimum of two weeks for the concrete to set before applying heat, and then apply heat gradually. Floor covering of terrazzo, linoleum, tile, or carpeting can be installed. If the water temperature is kept below the prescribed maximum of 85 degrees F, no damage will result to rugs, varnish, polish, or other materials. A typical piping diagram for a radiant floor panel avoids low places in the coil.

Heat loss to the ground from a floor panel laid directly on the ground can be expected. This loss is estimated to be from 10 to 20 percent of the heat provided for the room. Heat loss from outside slab edges can be greater than this amount. Slab edges should receive from 1/2 inch to 2 inches of waterproof insulation.

It is not customary to install radiant wall panels except to provide supplementary heat where ceiling and/or floor panels do not provide a required degree of comfort. Wall panels are occasionally installed in bathrooms where higher than normal heat temperatures may be desired. In a typical radiant wall-panel installation diagram it must be remembered that circulating pumps for use with radiant heating should have a higher head rating than for convector systems of the same capacity. This requirement exists because the coil pressure drop is considerably higher than the drop in a radiator or convector.

Certain fittings or devices are essential to the proper operation of a boiler. The piping diagram is a typical steam boiler with the control and indicating devices that are essential to its proper operation.

One of the most important features is the safety valve. Every boiler installation must have a safety valve installed to protect the boiler itself and the building occupants in case of malfunction. These valves are adjusted to open and relieve the internal pressure should it rise above a safe predetermined level. Numerous valves, gauges, and safety devices are found on all boilers.

The level of the water in a boiler must be maintained between certain limits; otherwise, serious damage to the boiler and building may result, as well as possible injury to the building occupants. Various safety devices are incorporated to protect against this possibility. Water gauges are provided as a means of visually checking the level. More sophisticated gauges employ floats that actuate a whistle or other alarm when the water level drops to a dangerous point.

Pressure-relief valves and fusible plugs are also used to protect the boiler in case of malfunction. Both of these devices will relieve dangerous high pressure under certain conditions.

Injectors are used to supply water against the high pressure existing within a boiler. This is done by means of the jet principle. Steam loops are often provided to return condensate to a boiler. These devices are entirely automatic and have only one moving part, a check valve at the bottom of the drop leg.

Heating pumps are usually used in steam heating systems to improve efficiency. Two principal types are available—condensation pumps and vacuum pumps. The type of heating system, cost, and individual requirements dictate which of these pumps must be used.

This heating section is a brief summary of steam and hot-water heating principles and in general contains the information necessary to answer questions pertaining to steam and hot-water heating that may be found in plumbing code examinations.

HOT-WATER BASEBOARD HEATING

A hydronic hot-water heating system circulates hot water to every room through baseboard panels. "Hydronic" is another term for forced hot-water heating. The hot water gives off its heat energy by utilizing fins attached to the tubing or channel carrying the hot water.

Baseboard panels mounted around the outer perimeter of the home provide a curtain of warmth that surrounds those inside. Radiant heat rays warm the room surfaces. Rising currents of convected warm air block out drafts and cold. Walls stay warm and cold spots are eliminated.

A boiler provides water between 120 and 210 degrees F. The water is pumped through the piping in the baseboard. Today's boilers are very efficient and very small in size. Most units are the size of an automatic washing machine; some are as small as suitcases and can be hung on the wall, depending on the size of the home being heated.

Circulating Pump

Booster pumps are used to circulate hot water through the pipes. These pumps are designed to handle a wide range of pumping capacities. They will vary in size from small booster pumps with a 5-gallon-per-minute capacity to those capable of handling thousands of gallons per minute.

Piping Arrangements

There are two different piping arrangements utilized by the hydronic hot-water system: the series loop and the one-pipe system, which is utilized in zoning. The zone-controlled system has two circulators that are attached to a single boiler, and separate thermostats are used to control the zones. There are, of course, other methods, but these two are among the most commonly used in home heating. The plumbing requirements are minimal, usually employing copper tubing with soldered joints.

One-pipe systems may be operated on either forced or gravity circulation. Care must be taken to design and install the system with the characteristic temperature drop found in the heat-emitting units farthest from the boiler in mind. The design of gravity circulation and one-pipe systems must be planned very carefully to allow for heat load and losses in the system. The advantage of the one-pipe system lies in its ability to allow one or two of the heat-emitting units to be shut off without interfering with the flow of hot water to the other units. This is not the case in the series-loop system, where the units are connected in series and form a part of the supply line. The piping varies so that one allows the cutoff and the other does not. It is obvious that the series-loop system is less expensive to install because it has a very simple piping arrangement.

Hot Water for Other Purposes

It is possible to use the boiler of a hydronic heating system to supply heat for such purposes as snow melting, a swimming pool, domestic hot water for household use, and other purposes. Separate circuits are created for each of these purposes, which are controlled by their own thermostats. Each is designed to tap into the main heating circuit from which it receives its supply of hot water. Hot water for household use, for instance, can be obtained by means of a heat exchanger or special coil inserted into the boiler. Note that the supply water does not come in contact with water being heated by the boiler for the baseboard units and other purposes.

One of the disadvantages of the hydronic system is its slow recovery time. If an outside door is opened during cold weather for any period of time, it takes a considerable length of time for the room to once again come up to a comfortable temperature. There is also noise made by the piping heating up and expanding and popping; it becomes rather annoying at night when you are in a quiet room and not too sleepy.

The main advantage, however, is its economical operation. The type of fuel used determines the expense. The boilers can be electrically heated, heated with natural gas, or heated with oil.

MULTIPLE-CHOICE EXAM

1. Hot-water heating systems are classified as low, medium and _____ temperature.

    a. boiling   b. high
    c. sub-cool   d. near boiling

2. Low-temperature water systems operate within the pressure and temperature limits of ASHRAE's Boiler Construction Code for low-temperature heating boilers. That means that they have a maximum allowable pressure for such boilers of 160 pounds/inch^2 and a maximum temperature of _____ degrees F.

    a. 250   b. 100
    c. 200   d. 75

3. Low-temperature water systems are generally used for space heating in single homes, residential buildings, and most commercial and _____ buildings such as office structures.

    a. college   b. industrial
    c. manufacturing   d. hotel

4. All hot-water systems require some type of _____ to overcome friction losses of the flowing water.

    a. heating   b. cooling
    c. suctioning   d. pumping

5. Manufacturers of radiators and other terminal units provide data on _____ losses through their equipment.

    a. high   b. low
    c. friction   d. resistance

6. Hot-water system piping must be sized to carry _____ _____ desired amount of heating water throughout the system.

    a. the minimum   b. the least
    c. the maximum   d. the greater

7. Infrared heating takes advantage of the fact that _____ can be used to convey heat.

    a. wind   b. air
    c. oil   d. light

8. Infrared heaters offer the user total _____ in heating a building.

a. use
b. control
c. flexibility
d. energy

9. A person feels comfortable or uncomfortable based on the _____ temperature, humidity, air flow, and degree of activity.

a. surrounding
b. outside
c. inside
d. external

10. Infrared units may use either _____ gas or propane.

a. methane
b. natural
c. ethanol
d. high-pressure

11. To achieve total heating with infrared systems, calculate the normal heat loss of the building or room to be heated and supply an _____ number of watts of infrared heat to replace what is being lost.

a. established
b. estimated
c. equal
d. equivalent

12. Gas-fired infrared units have many of the same advantages as _____ powered units.

a. water-
b. wind-
c. electronically-
d. electrically-

13. Infrared units should be mounted at as low a distance or height as possible. The lowest recommended mounting height is _____ feet to prevent too high a watt density on the upper part of one's body.

a. 40
b. 30
c. 20
d. 10

14. Electrical heating systems are either centralized or _____.

a. free-standing
b. closed-ended
c. open-ended
d. decentralized

15. A centralized system may be:

a. a heated-water system
b. an electric boiler
c. a heat pump
d. all of the above

16.   Radiant units may be:

   a.   quartz-tube elements with reflectors

   b.   quartz lamps with reflector

   c.   heat lamps

   d.   valance (cove) units

   e.   none of these

   f.   all of these

17.   Unit heaters are available in three types; which of these is not one of them?

   a.   cabinet heaters

   b.   horizontal projection heaters

   c.   vertical projection heaters

   d.   baseboard heaters

18.   Heat pumps use electricity more _____ than resistance heaters.

   a.   effectively   b.   efficiently

   c.   economically   d.   effortlessly

19.   Heat pumps come in four types. Which of these is not one of them?

   a.   air-to-air   b.   air-to-water

   c.   water-to-water   d.   water-to-air

   e.   air-to-metal

20.   Electrically operated snow-melting equipment is available in three types. Which of the following is not one of them?

   a.   boiler heated-fluid type

   b.   embedded piping system

   c.   embedded electrical resistance-heating elements

   d.   radiative systems infrared lamps

21.   Infrared snow-melting systems tend to be somewhat more expensive than embedded-cable or wire designs, in terms of both installed cost and _____ cost.

   a.   total   b.   inspecting

   c.   operating   d.   maintenance

22. The ASHRAE Handbook recommends cabling rated at 6 to _____ watts per linear foot for gutters and downspouts.

a. 10   b. 16

c. 26   d. 36

23. A deeper burying of the snow-melting piping embedment will require proportionally _____ heat because a greater mass of concrete is present.

a. less   b. more

c. the same   d. about the same

24. A good rule of thumb to use in the placing of tubes of a snow-melting system is to space them _____ inches center to center.

a. 10   b. 12

c. 16   d. 24

25. In snow-melting equipment, piping mains should be covered with (a, an) _____ blanket below the level of the slab.

a. insulation   b. coating

c. plastic   d. metal

26. Finned tubes and convectors, typically mounted inside an enclosure, provide heat by _____ currents passing over the fins.

a. radiation   b. convection

c. irradiation   d. convention

e. none of these

27. Comfort heating with steam or hot water from boilers is known as:

a. steam heating   b. hot-water heating

c. hydronic heating   d. humidity heating

28. Residential baseboard elements generally consist of _____ inch to 1-inch tubing in aluminum fins.

a. 3/8   b. 3/4

c. 1/4   d. 1/2

29. Cabinet convectors, as the name implies, function within metal enclosures of varying dimensions and outlet _____.

a. spacing   b. size

c. makeup   d. configuration

30. An _____ operation for combination cooling and heating is the use of cabinet fan-coil units with chilled water in summer and perimeter finned tube elements in conjunction with the cabinet units with warm water in winter.

a. efficient   b. economical

c. uneconomical   d. none of these

TRUE-FALSE EXAM

1. In small buildings a single loop may suffice, running from the boiler and back to it indirectly.

True   False

2. Finned tube radiation is widely used for total comfort heating.

True   False

3. Fin spacing on convectors usually varies in height and depth from 24 to about 60 fins per foot.

True   False

4. Comfort heating with steam or hot-water boilers is known as hydronic heating.

True   False

5. When infrared is used for total heating, the surrounding air conducts heat away from the objects, and space eventually reaches equilibrium.

True   False

6. Gas-fired infrared units don't require any more precautions to protect from fire hazards than electrically powered units.

True   False

7. Control systems are available for infrared units in three voltages, 120, 25, and millivolts.

True   False

8.   Electrical energy is ideally suited for space heating.

      True   False

9.   Certain convective heating units use metallic heating elements.

      True   False

10.   Radiant convector wall panels have glass electric heating panels.

      True   False

11.   Heat pumps use electricity more efficiently than resistance heaters.

      True   False

12.   Heat pumps come in five types.

      True   False

13.   Older hot-water heating systems used gravity to circulate the hot water.

      True   False

14.   All hot-water systems require some type of pumping to overcome friction losses of the flowing water.

      True   False

15.   ASME and ASHRAE rules should be followed when dealing with pressure vessels.

      True   False

MULTIPLE-CHOICE ANSWERS

1. B	7. D	13. D	19. E	25. A
2. A	8. C	14. D	20. A	26. B
3. B	9. A	15. D	21. C	27. C
4. D	10. B	16. F	22. B	28. D
5. C	11. D	17. D	23. B	29. D
6. C	12. D	18. B	24. B	30. B

TRUE-FALSE ANSWERS

1. F	5. T	9. T	13. T
2. T	6. F	10. T	14. T
3. T	7. T	11. T	15. T
4. T	8. T	12. F	

Chapter 3

BOILERS, BURNERS, AND BURNER SYSTEMS

BOILER CODES AND STANDARDS

Boilers must be installed and operated in accordance with applicable codes and standards. The local code authorities will refer to one or more of the following industry codes:

- American Society of Mechanical Engineers (ASME)
- American Gas Association (AGA)
- Hydronics Institute (HYDI)
- American Boiler Manufacturers' Association (ABMA)
- Insurance-companies codes, such as that of Factory Mutual (FM), particularly with respect to burner systems.
- Underwriters' Laboratories (UL) for certification of some of the control and safety devices, such as relief valves

The proposed design of the boiler system should be submitted to the owner's insurance company to ensure compliance with requirements.

Steam Pressure

Steam-pipe systems are categorized by the working pressure of the steam they supply. There are five steam systems:

- High-pressure system   above 100 psig
- Medium-pressure system   100 psig to 15 psig
- Low-pressure system   15 psig to 0 psig
- Vapor systems   subatmospheric pressures
- Vacuum systems   subatmospheric pressures

Vapor systems use the condensing process to reach subatmospheric pressures. Vacuum systems need a mechanically operated vacuum pump to reach sub-atmospheric pressures.

MULTIPLE-CHOICE EXAM

1. The optimum air-fuel ratio for natural gas in gas burners under standard conditions is approximately _____ to 1.

    a. 1.5   b. 4.5

    c. 9.5   d. 10.5

2. The most common fuel gas used in comfort heating is:

    a. butane   b. propane

    c. natural   d. coke oven

3. Fuel oil used for oil burners is a no. 6 that is black and _____.

    a. thin   b. thick

    c. both of these   d. neither of these

4. No. 1 fuel oil for oil burners is very _____ and colorless and resembles kerosene.

    a. opaque   b. jelly-like

    c. thick   d. thin

5. Atomization is the process whereby oil is _____ into fine droplets.

    a. mixed   b. broken

    c. vaporized   d. heated

6. The basic devices used for fuel control are:

    a. pumps   b. relief valves

    c. metering valves   d. pressure reducing valves

    e. all of the above   f. none of the above

7. Small boiler systems—those having capacities of 10,000 pounds per hour of steam or less—are often equipped with underfeed stokers in which the fuel is forced into the combustion area by a ram or _____.

    a. screw   b. stoker

    c. rake   d. hopper

8. Spreader stoker systems are often used in boilers designed to generate steam in quantities of
    _____ pounds per hour and higher.

    a. 25,000   b. 50,000

    c. 75,000   d. 100,000

9. Fuel-air-ratio control equipment can be classified in to three general groups: atmospheric, fixed-flow,
    and _____ flow equipment.

    a. variable-   b. low-

    c. high-   d. none of these

10. Two of the most commonly measured combustion products are oxygen and _____.

    a. carbon dioxide   b. carbon monoxide

    c. soot   d. ashes

11. The presence of _____ _____ indicates that there is incomplete combustion.

    a. carbon monoxide   b. carbon dioxide

    c. black soot   d. white ash

12. Natural gas, propane, and butane all have a flame temperature of approximately:

    a. 3000 degrees F.   b. 2700 degrees F.

    c. 3700 degrees F.   d. 4000 degrees F.

13. Regardless of fuel, the three required ingredients of combustion are:

    a. fuel, heat, and oxygen   b. fuel, air, and oxygen

    c. fuel, gas, and oxygen   d. fuel, gas, and air

14. A well-designed oil burner has the following functions and/or features:

    a. The air-fuel ratio must be held constant

    b. All the oil must come in contact with air at the right time

    c. There must be a proper flame pattern

    d. None of these

    e. All of these

15. Solid fuels such as wood and coal are being burned in boilers in a number of designs. The most pop-
    ular and best known are _____.

    a. stokers   b. mass burners

    c. boiler makers   d. coal burners

16. Two of the most commonly used stoker systems used in intermediate industrial boilers are a chain grate and a _____ _____.

   a. solid grate   b. variable grate

   c. traveling grate   d. stalled grate

17. One of the major waste products for solid-fuel boilers is:

   a. coal embers   b. solid fuels

   c. flue gases   d. ashes

18. Where are pulverized-fuel burner systems used?

   a. large buildings   b. electricity-generating plants

   c. small buildings   d. schools

19. In addition to the requirements of the International Fuel Gas Code, the installation of boilers must be done in accordance with the manufacturer's instructions and the:

   a. International Builder's Code   b. International Mechanical Code

   c. ASHRAE Standards   d. International Code Council

20. Small ceramic kilns with a maximum interior volume of _____ cubic feet that are used for hobby and noncommercial purposes must be installed in accordance with the manufacturer's installation instructions and the provisions of the International Fuel Gas Code.

   a. 10   b. 20

   c. 30   d. 40

21. According to the International Fuel Gas Code, the aggregate input rating of all un-vented appliances installed in a room or space shall not exceed _____ BTUs per hour per cubic foot of volume. If the room or space in which the equipment is installed is directly connected to another room or space by a doorway, archway, or other opening of comparable size that cannot be closed, the volume of the adjacent room or space can be included in the calculations.

   a. 40   b. 20

   c. 10   d. 30

22. The Code specifies that unvented room heaters must be equipped with an oxygen-depletion-sensitive safety shutoff system. The system should shut off the gas supply to the main and pilot burners when the oxygen in the surrounding atmosphere is depleted to the percent concentration specified by the manufacturer, but no lower than:

   a. 12 percent   b. 10 percent

   c. 18 percent   d. 36 percent

23. Suspended-type unit heaters must be supported by elements that are designed and constructed to accommodate weight and _____ loads.

    a. dynamic   b. heavy

    c. light   d. torque

24. Suspended-type unit heaters must be installed with clearances to combustible materials of not less than 18 inches at the sides, 12 inches at the bottom, and _____ inches above the top if the unit heater has an internal draft hood or 1 inch above the top of the sloping side of the vertical draft hood.

    a. 2   b. 4

    c. 8   d. 6

25. There are three types of fuel-air-ratio control equipment:

    a. atmospheric, fixed-flow, and variable-flow

    b. fixed-flow, variable-flow, and flue flow

    c. variable flow, low-flow, and fixed-flow

    d. none of the above

26. For a more responsive steam supply when using solid fuels, _____ fuel is used.

    a. pulverized   b. coal

    c. corn-shuck   d. wood

27. Oxygen _____ systems have generally been recognized as the best method for increasing the accuracy of all systems. They are used with all the different control systems.

    a. injection   b. depletion

    c. trim   d. modulation

28. In solid fuel burners, the speed at which the fuel bed travels is automatically adjusted under the control of the boiler pressure. As the fuel bed travels, the combustion air is fed to a plenum chamber beneath the stoker, flows upward through the fuel bed, and is simultaneously controlled by

_____ _____.

    a. air updrafts   b. air jets

    c. air dampers   d. damper fans

29. Floor-mounted unit heaters must be installed with clearances to combustible materials at the back and one side only of not less than _____ inches. Where the flue gases are vented horizontally, the clearance must be measured from the draft hood or vent instead of the rear wall of the unit heater. Floor-mounted unit heaters must not be installed on combustible floors unless listed for such installation.

   a. 12   b. 10

   c. 8   d. 6

30. Water-heater installation requirements relating to sizing, relief valves, drain pans, and scald protection must be in accordance with the:

   a. International Builder's Code

   b. International Mechanical Code

   c. International Fuel Gas Code

   d. International Plumbing Code

TRUE-FALSE EXAM

1. A plenum supplied as part of the air-conditioning equipment must be installed in accordance with the equipment manufacturer's instructions.

   True   False

2. One or more unvented room heaters must not be used as the sole source of comfort heating in a dwelling unit.

   True   False

3. Atmospheric burners are ignited by electric or electronic systems.

   True   False

4. A forced-draft burner uses an electric fan to deliver combustion air and the fuel/air mixture to the boiler.

   True   False

5. The amount of heat given off during complete combustion of a known amount of fuel is the heat of exhaustion.

   True   False

6. The viscosity of oil is the measure of its resistance to flow.

    True   False

7. Solid-fuel stoker-type combustion systems are seldom used in HVAC projects unless very special circumstances exist.

    True   False

8. The most common fuel gas is propane.

    True   False

9. The amount of oxygen left over after combustion is excess O_2 or excess air.

    True   False

10. Every burner generates a flame front in the boiler furnace.

    True   False

11. The combustion process in small boilers is completed as the fuel moves in a sideward direction over the stoker surfaces. This process is called tuyeres.

    True   False

12. Gasification and liquefacation designs for boilers hold promise in the future.

    True   False

13. Because few parts are needed for an atmospheric burner, it is used in most homes for heating and hot water.

    True   False

14. A metering system that uses only a pressure differential offers a major increase in accuracy over the single point control.

    True   False

15. Boilers operating as hot-water generators must be pressurized to prevent the formation of steam.

    True   False

MULTIPLE-CHOICE ANSWERS

1. B	7. D	13. D	19. E	25. A
2. A	8. C	14. D	20. A	26. B
3. B	9. A	15. D	21. C	27. C
4. D	10. B	16. F	22. B	28. D
5. C	11. D	17. D	23. B	29. D
6. C	12. D	18. B	24. B	30. B

TRUE-FALSE ANSWERS

1. F	5. T	9. T	13. T
2. T	6. F	10. T	14. T
3. T	7. T	11. T	15. T
4. T	8. T	12. F	

—NOTES—

Part B

PIPING, DUCTWORK, AND VENTILATION

HVAC Licensing Study Guide

Chapter 4
PIPING SYSTEMS

PIPE AND PIPE MATERIALS

There are seven classifications of pipe materials contained in the code:

1. Water-service pipe: This type of pipe may be made of asbestos cement, brass, cast iron, copper, plastic, galvanized iron, or steel.

2. Water-distribution pipe: This type of pipe may be made of brass, copper, plastic, or galvanized steel.

3. Above-ground drainage and vent pipe: This type of pipe may be made of brass, cast iron, type K, L, M, or DWV copper, galvanized steel, or plastic.

4. Underground drainage and vent pipe: This type of pipe can be made of brass, cast iron, type K, L, M, or DWV copper, galvanized steel, or plastic.

5. Building-sewer pipe: This type of pipe may be made of asbestos cement, bituminized fiber, cast iron, type K or L copper, concrete, plastic, or vitrified clay.

6. Building-storm-sewer pipe: This type of pipe may be made of asbestos cement, bituminized fiber, cast iron, concrete, type K, L, M, or DWV copper, or vitrified tile.

7. Subsoil-drain pipe: This type of pipe may be made of asbestos cement, bituminized fiber, cast iron, plastic, styrene rubber, or vitrified clay.

More specific uses of piping for heating purposes are included in the following material. Also make sure to review your knowledge of insulating pipes.

Insulation

Insulation is needed to prevent the penetration of heat through a wall or air hole into a cooled space. There are several insulation materials, such as wood, plastic, concrete, and brick. Each has its application. However, more effective materials are constantly being developed and made available.

Sheet Insulation

Vascocel® is an expanded, closed-cell, sponge rubber that is made in a continuous sheet 36 inches wide. It comes in a wide range of thicknesses: (3/8, 1/2, and 3/4 inches). This material is designed primarily for insulating oversize pipes, large tanks and vessels, and other similar medium- and low-temperature areas. Because of its availability on continuous rolls, this material lends itself ideally to application on large air ducts and irregular shapes.

This material is similar to its companion product, Vascocel® tubing. It may be cut and worked with ordinary hand tools such as scissors or a knife. The sheet stock is easily applied to clean, dry surfaces with an adhesive. (See Figure 4.1.) The k factor (heat-transfer coefficient) of this material is 0.23. It has some advantages over other materials: it is resistant to water penetration, water absorption, and physical abrasion.

Tubing Insulation

Insulation tape is a special synthetic rubber and cork compound designed to prevent condensation on pipes and tubing. It is usually soft and pliable. Thus, it can be molded to fit around fittings and connections. There are many uses for this type of insulation. It can be used on hot or cold pipes or tubing. It is used in residential buildings, air-conditioning units, and commercial installations. It comes in 2-inch-wide rolls that are 30 feet long. The tape is thick. If stored or used in temperatures under 90 degrees F [32.2 degrees C], its lifetime is indefinite.

Foam-insulation tape is made specifically for wrapping cold pipes to prevent pipe sweat. (See Figure 4.2.) It can be used to hold down heat loss on hot pipes below 180 degrees F [82.2 degrees C]. It can be cut in

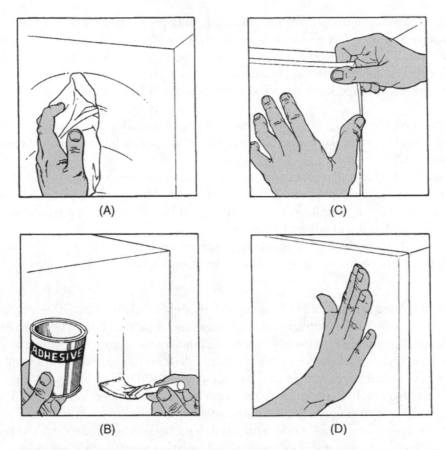

(A)

(C)

(B)

(D)

FIGURE 4.1 Installing sheet insulation. (A) Prepare the surfaces for application of the sheet insulation by wiping with a soft, dry cloth to remove any dust or foreign matter. Use a solvent to remove grease or oil. (B) Apply the adhesive in a thin, even coat to the surface to be insulated. (C) Position the sheet of insulation over the surface and then simply smooth it in place. The adhesive is a contact type. The sheet must be correctly positioned before it contacts the surface. (D) Check for adhesion of ends and edges. The surface can be painted.

FIGURE 4.2 Foam insulation tape. *(Virginia Chemical)*

pieces and easily molded around fittings and valves. It adheres to itself and clean metal surfaces. It is wrapped over pipes with about 1/4 inch overlap on each successive lap. Remember one precaution: never wrap two or more parallel runs of tubing or pipe together, leaving air voids under the tape. Fill the voids between the pipes with Permagum® before wrapping. This will prevent moisture from collecting in the air spaces. This foam-insulation tape has a unicellular composition. The k factor is 0.26 at 75 degrees F [23.9 degrees C].

Permagum® is a non hardening, water-resistant sealing compound. It is formulated to be non staining and non bleeding and to have excellent adhesion to most clean surfaces. It comes in containers in either slugs or cords. (See Figure 4.3.)

This sealer is used to seal metal to metal joints in air conditioners, freezers, and coolers. It can seal metal to wood joints and set plastic and glass windows in wood or metal frames. It can be used to seal electrical or wire entries in air-conditioning installations or in freezers. It can be worked into various spaces. It comes with a paper backing so that it will not stick to itself.

Extrusions are simple to apply. Unroll the desired length and smooth it into place. It is soft and pliable. The bulk slug material can be formed and applied by hand or with tools such as a putty knife.

Pipe Fittings

Pipe fittings are insulated for a number of reasons. Methods of insulating three different fittings are shown in Figure 4.4. In most cases it is advisable to clean all joints and waterproof them with cement. A mixture of hot crude paraffin and granulated cork can be used to fill the cracks around the fittings.

Figure 4.5 shows a piece of rock-cork insulation. It is molded from a mixture of rock wool and waterproof binder. Rock wool is made from limestone that has been melted about 3,000 degrees F [1,649 degrees C]. It is then blown into fibers by high pressure steam. Asphaltum is the binder used to hold it into a moldable form. This insulation has approximately the same insulation qualities as cork. It can be made waterproof when coated with asphalt. Some more modern materials have been developed to give the same or better insulation qualities. Vascocel® tubing can be used in the insulation of pipes. Pipe wraps are available to give good insulation and prevent dripping, heat loss, or heat gain.

Figure 4.6 shows a fitting insulated with preshrunk wool felt. This is a built-up thickness of pipe covering made of two layers of hair felt. The inside portion is covered with plastic cement before the insulation material is applied. After the application, waterproof tape and plastic cement should be added for protection against moisture infiltration. This type of insulation is used primarily on pipes located inside a building. If the pipe is located outside, another type of insulation should be used.

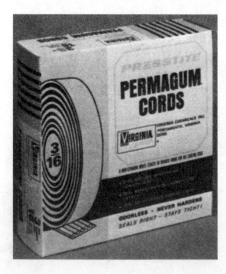

FIGURE 4.3 Slugs of insulation material and cords are workable into locations where sheet material cannot fit. *(Virginia Chemicals)*

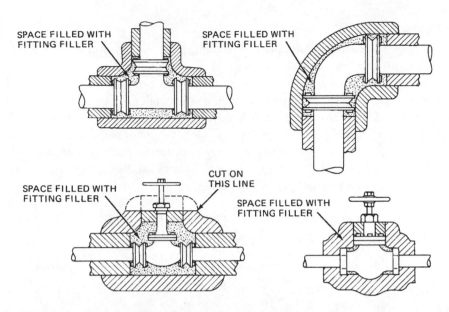

FIGURE 4.4   Pipe fittings covered with cork-type insulation. On one of the valves the top section can be removed if the packing needs replacing.

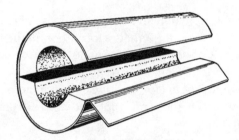

FIGURE. 4.5   Pipe insulation.

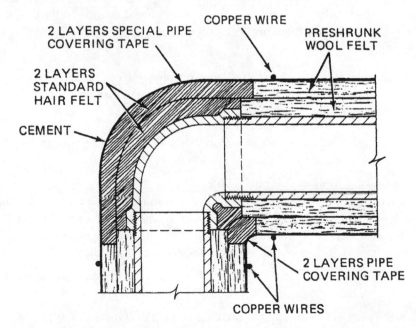

FIGURE 4.6   Insulated pipe fittings.

Refrigeration Piping

Various materials have been used for insulation purposes in the refrigeration field. It is this equipment that service people are most often called to repair or maintain. It is therefore necessary for the present-day repair-person to be acquainted with the older types of insulation that may be encountered during the workday.

The success of any refrigeration plant depends largely on the proper design of the refrigeration piping and a thorough understanding of the necessary accessories and their functions in the system. In sizing refrigerant lines, it is necessary to consider the optimum sizes with respect to economics, friction losses, and oil return. It is desirable to have line sizes as small as possible from the cost standpoint. On the other hand, suction- and discharge-line pressure drops cause a loss of compressor capacity, and excessive liquid-line pressure drops may cause flashing of the liquid refrigerant with consequent faulty expansion-valve operation.

Refrigerant piping systems, to operate successfully, should satisfy the following:

- Proper refrigerant feed to the evaporators should be ensured.

- Refrigerant lines should be of sufficient size to prevent an excessive pressure drop.

- An excessive amount of lubricating oil should be prevented from being trapped in any part of the system.

- Liquid refrigerant should be prevented from entering the compressor at all times.

Pressure-Drop Considerations

Pressure drop in liquid lines is not as critical as it is in suction and discharge lines. The important thing to remember is that the pressure drop should not be so great as to cause gas formation in the liquid line and/or insufficient liquid pressure at the liquid-feed device. A system should normally be designed so that the pressure drop due to friction in the liquid line is not greater than that corresponding to a 1-to 2-degree change in saturation temperature. Friction pressure drops in the liquid line include a drop in accessories, such as the solenoid valve, strainer-drier, and hand valves, as well as in the actual pipe and fittings from the receiver outlet to the refrigerant feed device at the evaporator.

Friction pressure drop in the suction line means a loss in system capacity because it forces the compressor to operate at a lower suction pressure to maintain the desired evaporating temperature in the coil. It is usually standard practice to size the suction line to have a pressure drop due to friction no greater than the equivalent of a 1- to 2-degree change in saturation temperature.

Liquid Refrigerant Lines

The liquid lines do not generally present any design problems. Refrigeration oil in liquid form is sufficiently miscible with commonly used refrigerants to assure adequate mixture and positive oil return. The following factors should be considered when designating liquid lines:

- The liquid lines, including the interconnected valves and accessories, must be of sufficient size to prevent excessive pressure drops.

- When interconnecting condensing units with condenser receivers or evaporative condensers, the liquid lines from each unit should be brought into a common liquid line. Each unit should join the common liquid line as far below the receivers as possible, with a minimum of 2 feet preferred. The common liquid line should rise to the ceiling of the machine room. The added head of liquid is provided to prevent, as far as possible, hot gas blowing back from the receivers.

- All liquid lines from the receivers to the common line should have equal pressure drops in order to provide, as nearly as possible, equal liquid flow and prevent the blowing of gas.

- Remove all liquid-line filters from the condensing units and install them in parallel in the common liquid line at the ceiling level.

- Hot gas blowing from the receivers can be condensed in reasonable quantities by liquid subcoolers, as specified for the regular condensing units, with a minimum lift of 60 feet at 80 degrees F condensing medium temperature.

- Interconnect all the liquid receivers of the evaporative condensers above the liquid level to equalize the gas pressure.

- The common and interconnecting liquid line should have an area equal to the sum of the areas of the individual lines. Install a hand shutoff valve in the liquid line from each receiver. Where a reduction in pipe size is necessary in order to provide sufficient gas velocity to entrain oil up the vertical risers at partial loads, greater pressure drops will be imposed at full load. These can usually be compensated for by over-sizing the horizontal and down corner lines to keep the total pressure drop within the desired limits.

Interconnection of Suction Lines

When designing suction lines, the following important considerations should be observed:

- The lines should be of sufficient capacity to prevent any considerable pressure drop at full load.

- In multiple-unit installations, all suction lines should be brought to a common manifold at the compressor.

- The pressure drop between each compressor and main suction line should be the same in order to ensure a proportionate amount of refrigerant gas to each compressor as well as a proper return of oil to each compressor.

- Equal pipe lengths, sizes, and spacing should be provided.

- All manifolds should be level.

- The inlet and outlet pipes should be staggered.

- Never connect branch lines at a cross or tee.

- A common manifold should have an area equal to the sum of the areas of the individual suction lines.

- The suction lines should be designed so as to prevent liquid from draining into the compressor during shutdown of the refrigeration system.

Discharge Lines

The hot-gas loop accomplishes two functions in that it prevents gas that may condense in the hot-gas line from draining back into the heads of the compressor during the off cycles, and it prevents oil leaving one compressor from draining down into the head of an idle machine. It is important to reduce the pressure loss in hot-gas lines because losses in these lines increase the required compressor horsepower per ton of refrigeration and decrease the compressor capacity. The pressure drop is kept to a minimum by sizing the lines generously to avoid friction losses but still making sure that refrigerant-line velocities are sufficient to entrain and carry oil at all load conditions. In addition, the following pointers should be observed:

- The compressor hot-gas discharge lines should be connected as shown in Figure 4.7.

- The maximum length of the risers to the horizontal manifold should not exceed 6 feet.

- The manifold size should be at least equal to the size of the common hot-gas line to the evaporative condenser.

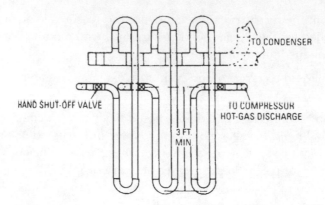

FIGURE 4.7   One way to connect hot-gas discharge lines.

- If water-cooled condensers are interconnected, the hot-gas manifolds should be at least equal to the size of the discharge of the largest compressor.

- If evaporative condensers are interconnected, a single gas line should be run to the evaporative condensers, and the same type of manifold provided at the compressors should be installed.

- Always stagger and install the piping at the condensers.

- When the condensers are above the compressors, install a loop with a minimum depth of 3 feet in the hot-gas main line.

- Install a hand shutoff valve in the hot-gas line at each compressor.

Water Valves

The water-regulating valve is the control used with water-cooled condensers. When installing water valves, the following should be observed:

- The condenser water for interconnected compressor condensers should be applied from a common water line.

- Single automatic water valves or multiple valves in parallel (Figure 4.8) should be installed in the common water line.

- Pressure-control tubing from the water valves should be connected to a common line, which in turn should be connected to one of the receivers or to the common liquid line.

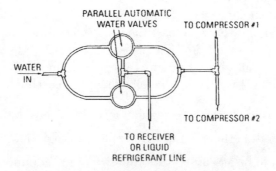

FIGURE 4.8   A method of interconnecting water valves.

Multiple-Unit Installation

Multiple compressors operating in parallel must be carefully piped to ensure proper operation. The suction piping at parallel compressors should be designed so that all compressors run at the same suction pressure and oil is returned in equal proportions to the running compressors. All suction lines should be brought into a common suction header in order to return the oil to each crankcase as uniformly as possible.

The suction header should be run above the level of the compressor suction inlets so that oil can drain into the compressors by gravity. The header should not be below the compressor suction inlets because it can become an oil trap. Branch suction lines to the compressors should be taken off from the side of the header. Care should be taken to make sure that the return mains from the evaporators are not connected into the suction header so as to form crosses with the branch suction lines to the compressors. The suction header should be run full size along its entire length. The horizontal takeoffs to the various compressors should be the same size as the suction header. No reduction should be made in the branch suction lines to the compressors until the vertical drop is reached.

Figure 4.9 shows the suction and hot-gas header arrangements for two compressors operating in parallel. Takeoffs to each compressor from the common suction header should be horizontal and from the side to ensure equal distribution of oil and prevent accumulating liquid refrigerant in an idle compressor in case of slop-over.

Piping Insulation

Insulation is required for refrigeration piping to prevent moisture condensation and prevent heat gain from the surrounding air. The desirable properties of insulation are that it should have a low coefficient of heat transmission, be easy to apply, have a high degree of permanency, and provide protection against air and moisture infiltration. Finally, it should have a reasonable installation cost.

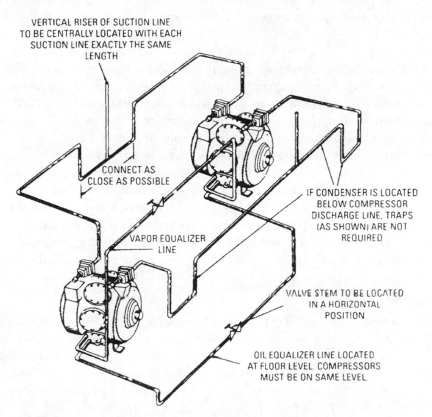

FIGURE 4.9 Connections for the suction and hot-gas headers in a multiple compressor installation.

The type and thickness of insulation used depends on the temperature difference between the surface of the pipe and the surrounding air and also on the relative humidity of the air. It should be clearly understood that although a system is designed to operate at a high suction temperature, it is quite difficult to prevent colder temperatures occurring from time to time. This may be due to a carrying over of some liquid from the evaporator or the operation of an evaporator pressure valve. Interchangers are preferable to insulation in this case.

One of the safest pipe insulations available is molded cork or rock cork of the proper thickness. Hair-felt insulation may be used, but great care must be taken to have it properly sealed. For temperatures above 40 degrees F, wool felt or a similar insulation may be used, but here again, success depends on the proper seal against air and moisture infiltration.

Liquid refrigerant lines carry much higher-temperature refrigerant than suction lines; and if this temperature is above the temperature of the space through which they pass, no insulation is usually necessary. However, if there is danger of the liquid lines going below the surrounding air temperatures and causing condensation, they should be insulated when condensation will be objectionable. If they must unavoidably pass through highly heated spaces, such as those adjacent to steam pipes, through boiler rooms, etc., then the liquid lines should also be insulated to ensure a solid column of liquid to the expansion valve.

There were four types of insulation in use before the discovery of modern insulation materials. Those you may encounter that were in general use for refrigerator piping, are cork, rock cork, wool felt, and hair felt.

Cork Insulation

Cork pipe covering is prepared by pressing dried and granulated cork in metal molds. The natural resins in the cork bind the entire mass into its new shape. In the case of the cheaper cork, an artificial binder is used. The cork may be molded to fit pipe and fittings, or it may be made into flat boards of varying sizes and thickness. Cork has a low thermal conductivity. The natural binder in the material itself makes cork highly water resistant, and its structure ensures a low capillarity. It can be made practically impervious to water by surfacing with odorless asphalt.

All fittings in the piping, as well as the pipe itself, should be thoroughly insulated to prevent heat gain in order to protect the pipe insulation from moisture infiltration and deterioration and eliminate condensation problems. Molded cork covering made especially for this purpose is available for all common types of fittings. Each covering should be the same in every respect as the pipe insulation, with the exception of the shape, and should be formed so that it joins to the pipe insulation with a break. Typical cork fitting covers are furnished in three standard thicknesses for ice water, brine, and special brine.

To secure maximum efficiency and long life from cork covering, it must be correctly applied and serviced as well as properly selected. Hence, it is essential that the manufacturer's recommendations and instructions be followed in detail. The following general information is a summary of the data.

All pipelines should be thoroughly cleaned, dried, and free from all leaks. It is also advisable to paint the piping with waterproof paint before applying the insulation, although this is not recommended by all manufacturers. All joints should be sealed with waterproof cement when applied. Fitting insulation should be applied in substantially the same manner, with the addition of a mixture of hot crude paraffin and granulated cork used to fill the space between the fittings, as shown in Figure 4.4.

Rock-Cork Insulation

Rock-cork insulation is manufactured commercially by molding a mixture of rock wool and a waterproof binder into any shape or thickness desired. The rock wool is made from limestone melted at about 3,000 degrees F and then blown into fibers by high-pressure steam. It is mixed with an asphalt binder and molded into the various commercial forms. The heat conductivity is about the same as cork, and the installed price may be less. Because of its mineral composition, it is odorless, vermin-proof, and free from decay. Like cork, it can be made completely waterproof by surfacing with odorless asphaltum. The pipe covering fabricated from rock wool and a binder is premolded in single-layer sections 36 inches long to fit all standard pipe sizes and is usually furnished with a factory-applied waterproof jacket.

When pipelines are insulated with rock-cork covering, the fittings are generally insulated with built-up rock wool impregnated with asphalt. This material is generally supplied in felted form, having a nominal thickness of about 1 inch and a width of about 18 inches. It can be readily adapted to any type of fitting and is efficient as an insulator when properly applied.

Before applying the formed rock-cork insulation it is first necessary to thoroughly clean and dry the piping and then paint it with waterproof asphalt paint. The straight lengths of piping are next covered with the insulation, which has the two longitudinal joints and one end joint of each section coated with plastic cement. The sections are butted tightly together with the longitudinal joints at the top and bottom and temporarily held in place by staples. The plastic cement should coat that part of the exterior area of each section to be covered by the waterproof lap and the lap should be pressed smoothly into it. The end joints should be sealed with a waterproof fabric embedded in a coat of the plastic cement. Each section should then be secured permanently in place with three to six loops of copper-plated annealed steel wire.

Wool-Felt Insulation

Wool felt is a relatively inexpensive type of pipe insulation and is made up of successive layers of waterproof wool felt that are indented in the manufacturing process to form air spaces. The inner layer is a waterproof asphalt-saturated felt, while the outside layer is an integral waterproof jacket. This insulating material is satisfactory when it can be kept air- and moisture-tight. If air is allowed to penetrate, condensation will take place in the wool felt, and it will quickly deteriorate. Thus, it is advisable to use it only where temperatures above 40 degrees F are encountered and when it is perfectly sealed. Under all conditions, it should carry the manufacturer's guarantee for the duty that it is to perform.

After all the piping is thoroughly cleaned and dried, the sectional covering is usually applied directly to the pipe with the outer layer slipped back and turned so that all joints are staggered. The joints should be sealed with plastic cement, and the flap of the waterproof jacket should be sealed in place with the same material. Staples and copper-clad steel wire should be provided to permanently hold the insulation in place, and then the circular joints should be covered with at least two layers of waterproof tape to which plastic cement is applied.

Pipe fittings should be insulated with at least two layers of hair felt (Figure 4.6) built up to the thickness of the pipe covering; but before the felt is placed around the fittings, the exposed ends of the pipe insulation should be coated with plastic cement.

After the felt is in place, two layers of waterproof tape and plastic cement should be applied for protection from moisture infiltration.

Insulation of this type is designed for installation in buildings where it will be normally protected against outside weather conditions. When outside pipes are to be insulated, one of the better types of pipe covering should be used. In all cases, the manufacturer's recommendations should be followed as to the application.

Hair-Felt Insulation

Hair-felt insulation is usually made from pure cattle hair that has been especially prepared and cleaned. It is a very good insulator against heat, having a low thermal conductivity. Its installed cost is somewhat lower than cork; but it is more difficult to install and seal properly, and hence its use must be considered a hazard with average workmanship. Prior to installation, the piping should be cleaned and dried and then prepared by applying a thickness of waterproof paper or tape wound spirally, over which the hair felt of approximately 1-inch thickness is spirally wound for the desired length of pipe. It is then tightly bound with jute twine, wrapped with a sealing tape to make it entirely airtight, and finally painted with waterproof paint. If more than one thickness of hair felt is desired, it should be built up in layers with tar paper between. When it is necessary to make joints around fittings, the termination of the hair felt should be tapered down with sealing tape, and the insulation applied to the fittings should overlap this taper, thus ensuring a permanently tight fit.

The important point to remember is that this type of insulation must be carefully sealed against any air or moisture infiltration, and even then difficulties may occur after it has been installed. At any point where air infiltration (or "breathing," as it is called) is permitted to occur, condensation will start and travel great distances along the pipe, even undermining the insulation that is properly sealed.

There are several other types of pipe insulation available, but they are not used extensively. These include various types of wrapped and felt insulation. Whatever insulation is used, it should be critically examined to see whether it will provide the protection and permanency required of it; otherwise, it should never be considered. Although all refrigerant piping, joints, and fittings should be covered, it is not advisable to do so until the system has been thoroughly leak-tested and operated for a time.

Pressure drop in the various parts of commercial refrigeration systems due to pipe friction and proper dimensioning to obtain the best operating results are important considerations when equipment is installed.

By careful observation of the foregoing detailed description of refrigeration piping and methods of installation, the piping problem will be greatly simplified and result in proper system operation.

MULTIPLE-CHOICE EXAM

1. It is a good practice to install suction _____ side on all oil systems to remove foreign materials that could damage the pump.

   a. sightglass
   b. filters
   c. strainers
   d. none of these

2. Duplex _____ allow the ability to inspect and clean one side of the unit without shutting down the flow of oil.

   a. pumps
   b. heaters
   c. filters
   d. strainers

3. Steam-heater lines should be left _____ to allow the steam to desuperheat prior to entering the heater.

   a. open
   b. closed
   c. insulated
   d. uninsulated

4. Which type of valve is used most commonly for controlling the quantity of fuel-gas flow to the burner?

   a. angle
   b. check
   c. butterfly
   d. standard

5. Why would it be advisable to use a strainer to protect regulators and other control equipment:

   a. the gas may have dirt or chips mixed with it
   b. there may be water or vapors in the gas line
   c. there may be solids other than dirt and chips mixed in with the gas
   d. to take out any and all impurities

6. Sometimes a local gas utility is not able to provide sufficient gas pressure to meet the needs of the boiler. What is the appropriate piece of equipment to add:

   a. gas valves
   b. gas compressor or booster
   c. line checker
   d. line backer

7. The first step in designing a gas piping system is to properly _____ components and piping.

   a. size
   b. acquire
   c. measure
   d. order

8. Which of the following is taken into consideration when a regulator for gas lines to burners is being chosen:

a. pressure rating   b. capacity

c. sharp lockup   d. regulator location

e. all of these

9. What should be done to newly installed piping systems before turning on the gas or oil supply:

a. they should be checked for dirt

b. they should be cleaned and flushed

c. they should be checked for moisture

d. none of these

10. What does corrosion at a joint in a hot-water piping system indicate:

a. good system operation

b. normal operation for an older system

c. a leak

d. none of these

11. Refrigerant pipe systems are constructed of _____ pipe for freons.

a. copper   b. iron

c. plastic   d. galvanized

12. Which type of pipe do steam and condensate pipe systems normally use:

a. black steel pipe   b. galvanized pipe

c. plastic pipe   d. flexible pipe

13. What type of piping would you find in a natural-gas line?

a. black steel and plastic pipe   b. galvanized pipe

c. iron pipe   d. flexible stainless steel

14. Soon after steam leaves the boiler, it starts to lose some of its heat to any surface with a lower temperature. It then begins to condense into _____.

a. vapor   b. bubbles

c. liquid steam   d. water

15. An entire steam-distribution system must remain free of _____ and condensate.

    a. air   b. bubbles

    c. water   d. various vapors

16. Most valve-leakage problems are covered by three categories: through-the-valve, outside-the-valve, and outside-to- _____ the valve.

    a. inside-   b. under-

    c. over-   d. around-

17. When a valve fails to shut off completely, you cannot use extra _____ to close it.

    a. hammer handles   b. leverage bars

    c. force   d. wrenches

18. Strainers should be installed in the piping system at locations that will allow removal of _____ for cleaning.

    a. screens   b. gaskets

    c. the ends   d. none of these

19. What should be installed in the piping system at various points to protect from damage during normal service activities around the equipment:

    a. valves   b. thermometers

    c. gauges   d. meters

20. What type of gauge should be permanently installed in a piping system with a pigtail, or where applicable, a snubbing device:

    a. temporary   b. permanent

    c. pressure   d. oil

21. What type of connections do you use in piping systems to prevent the transmission of vibration from equipment to piping in other parts of the building:

    a. standard   b. flexible

    c. plastic   d. stainless copper

22. If only one union is used in the piping system, it should be placed on the _____ side of the steam trap.

    a. inlet   b. discharge

    c. outlet   d. charge

23. When pressure-differential-flow switches are required, they should be installed in the piping system with a minimum of _____ pipe diameters of straight pipe on each side of the pressure-sensing point.

    a. 40   b. 10

    c. 30   d. 20

24. What are the two types of flow switches used in piping:

    a. piston and flow-through

    b. paddle and flow

    c. piston and paddle

    d. pressure and piston

25. Automatic air vents should be installed with a manual _____ valve to make it possible to install or repair the air vent.

    a. pressure   b. separator

    c. isolation   d. connector

26. Dirt legs are cleaned periodically by using a _____ valve or by removing the end cap and free-blowing until clear.

    a. check   b. relief

    c. pressure   d. blow-down

27. The piping components of _____ systems have a much longer life cycle than heat-transfer components.

    a. HVAC   b. piping

    c. hydronic   d. refrigeration

28. The maximum amount of nitrogen that can be held in solution is expressed in percentage of water _____.

    a. gallons   b. area

    c. volume   d. condensation

29. What types of pumps are used for the circulation of cooling-tower water:

    a. centrifugal   b. centripetal

    c. vacuum   d. condensate

30. Brine is pumped or circulated by using a _____ pump.

    a. centrifugal   b. centripetal

    c. vacuum   d. condensate

TRUE-FALSE EXAM

1. When pipe threads are cut too long, they allow the pipe to enter too far into the fitting or system component.

   True   False

2. Steam and condensate pipe systems are normally made of black steel pipe.

   True   False

3. Fiberglass-reinforced plastic (FRP) is normally used for condensate systems.

   True   False

4. Natural-gas piping systems are normally constructed of black steel or plastic pipe. These systems are usually maintenance-free when properly installed.

   True   False

5. Pipes should be clear of walls, ceilings, and floors to permit welding or soldering of joints or valves.

   True   False

6. A minimum of 12 to 30 inches of clearance should be provided for pipes installed near walls, ceiling, or floors.

   True   False

7. Unions are installed in an accessible location in the piping system to permit dismantling of piping or removal of equipment.

   True   False

8. Chilled-water piping systems are normally constructed of copper or black steel pipe.

   True   False

9. The following list contains the various piping systems: hot water, chilled water, and condensing water.

   True   False

10. Condensation will collect on a chilled-water system's pipe surface if insulation is defective.

True   False

11. In condensing-water-piping systems chemical treatment and cleaning are necessary maintenance to keep the pipes and components clean.

True   False

12. Flanges should be installed in an accessible location in a piping system to permit dismantling of piping or removal of equipment components.

True   False

13. Oil slicks at joints are an indication of leakage in a refrigerant pipe system and should be repaired.

True   False

14. Refrigerant piping systems should be purged with dry nitrogen when brazing to eliminate internal debris.

True   False

15. Piping should be stored on racks or dunnage, off the ground, and protected from the elements.

True   False

MULTIPLE-CHOICE ANSWERS

1.	C	7.	A	13.	A	19.	B	25.	C
2.	D	8.	E	14.	D	20.	C	26.	D
3.	D	9.	B	15.	A	21.	B	27.	D
4.	C	10.	C	16.	A	22.	B	28.	C
5.	A	11.	A	17.	C	23.	B	29.	A
6.	B	12.	A	18.	A	24.	C	30.	A

TRUE-FALSE ANSWERS

1.	T	5.	T	9.	T	13.	T
2.	T	6.	F	10.	T	14.	T
3.	T	7.	T	11.	T	15.	T
4.	T	8.	T	12.	T		

—NOTES—

Chapter 5
DUCTWORK SIZING

DUCTWORK FOR ADD-ON RESIDENTIAL EVAPORATORS

One of the more efficient ways of adding whole-house air conditioning is with an evaporator coil in the furnace. The evaporator coil becomes an important part of the whole system, as does the ductwork. It can be added to the existing furnace to make a total air-conditioning and heating package. There are two types of evaporator, down-flow and up-flow.

The down-flow evaporator is installed beneath a down-flow furnace. (See Figure 5.1.) Lennox makes down-flow models in 3-, 4-, and 5-ton sizes with an inverted "A" coil. (See Figure 5.2.) Condensate runs down the slanted side to the drain pan. This unit can be installed in a closet or utility room wherever the furnace is located. This type of unit is shipped factory-assembled and tested.

Up-flow evaporators are installed on top of the furnace. They are used in basement and closet installations. (See Figure 5.3.) The adapter base and the coil are shown as they would fit onto the top of an up-flow existing furnace. (See Figure 5.4.) The plenum must be removed and replaced once the coil has been placed on top of the furnace.

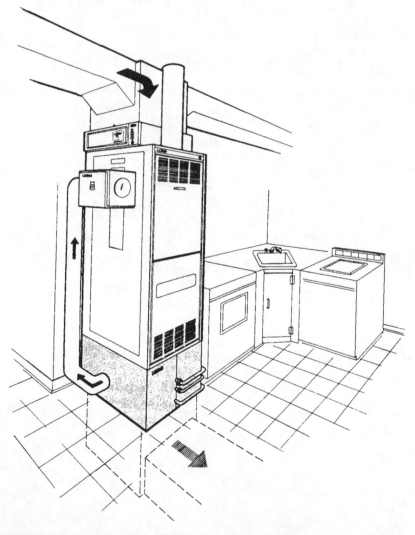

FIGURE 5.1   Utility room installation of a down-flow evaporator on an existing furnace. *(Lennox)*

FIGURE 5.2   Evaporator units for add-on cooling for up-flow and down-flow furnaces. *(Lennox)*

In most cases, use of an add-on top evaporator means that the fan motor must be changed to a higher horsepower rating. The evaporator in the plenum makes it more difficult to force air through the heating system. In some cases, the pulley size on the blower and the motor must be changed to increase the cfm (cubic feet per minute) rate moving past the evaporator.

Some motors have sealed bearings. Some blower assemblies, such as the one shown in Figure 5.5, have sealed bearings; others have sleeve bearings. In such cases, the owner should know that the motor and blower must be oiled periodically to operate efficiently.

Figure 5.6 shows how the evaporator coil sits on top of the furnace, making the up-flow type of air-conditioning unit operate properly. The blower motor is located below the heater and plenum.

The evaporator is not useful unless it is connected to a compressor and condenser. These are usually located outside the house. Figure 5.7 shows the usual outdoor compressor and condenser unit. This unit is capable of furnishing 2.5 to 5 tons of air conditioning, ranging in capacities from 27,000 to 58,000 Btu. Note that this particular unit has a U-shaped condenser coil that forms three vertical sides. The extra surface area

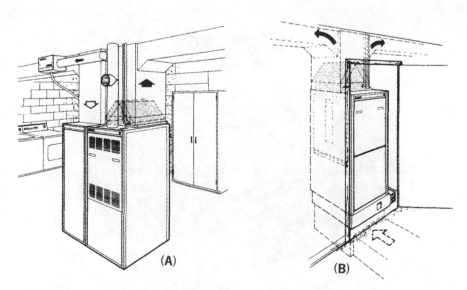

FIGURE 5.3   Installation of up-flow evaporators. (A) Basement installation with an oil furnace, return air cabinet, and automatic humidifier. (B) Closed installation of an evaporator coil with electric furnace and electronic air cleaner. *(Lennox)*

CABINET APPLICATION

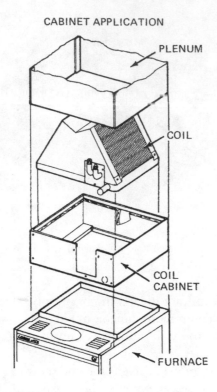

PLENUM

COIL

COIL
CABINET

FURNACE

FIGURE 5.4 Installation of an
evaporator coil on top of an existing
furnace installation. *(Lennox)*

is designed to make the unit more efficient in heat transfer. The fan, which is thermostatically operated, has
two speeds. It changes to low speed when the outside temperature is below 75 degrees F (23.9 degrees C).

Like most compressors designed for residential use, this one is hermetically sealed. The following safety
devices are built in: a suction-cooled overload protector, a pressure-relief valve, and a crankcase heater.
Controls include high- and low-pressure switches. They automatically shut off the unit if discharge pressure
becomes excessive or suction pressure falls too low.

FIGURE 5.5 Motor-driven blower unit. *(Lennox)*

FIGURE 15.6 Cutaway view of the furnace, blower, and evaporator coil on an air-conditioning and heating unit. *(Lennox)*

FIGURE 5.7 A 2.5- to 5-ton condensing unit with a one-piece, wraparound, U-shaped condenser coil. This unit has a two-speed fan and sealed-in compressor for quiet operation. *(Lennox)*

FIGURE 5.8   This unit fits in small closets or corners. It is capable of producing hot water or electric heat as well as cooling. Note the heating coils on the top of the unit. It comes in three sizes, 1.5, 2, and 2.5 tons. *(Lennox)*

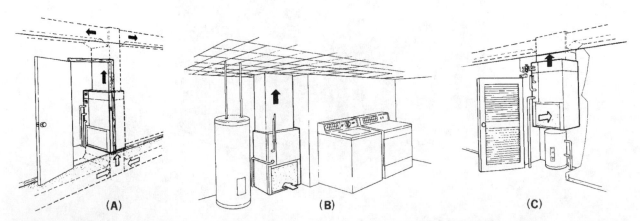

| (A) | (B) | (C) |

FIGURE 5.9   Typical installations of the blower, coil, and filter units. (A) A closet installation with electric-heat section. (B) A utility-room installation. (C) A wall-mounted installation with hot-water section. *(Lennox)*

In apartments where space is at a premium, a different type of unit is used. It differs only in size. (See Figure 5.8.) Compact units have a blower, filter and evaporator coil contained in a small package. They have electric-heat coils on top. In some cases, hot water is used to heat in the winter.

Figure 5.9 shows the various ways in which these units may be mounted. The capacity is usually 18,000 to 28,000 Btu. Note the location of the control box. This is important since most of the maintenance problems are caused by electrical rather than refrigerant malfunctions. This type of unit allows each apartment tenant to have his or her own controls.

REMOTE SYSTEMS

A remote system designed for home or commercial installation can be obtained with a complete package. It has the condensing unit, correct operating refrigerant charge, refrigerant lines, and evaporator unit.

The charge in the line makes it important to have the correct size of line for the installation. Metering control of the refrigerant in the system is accomplished by the exact sizing (bore and length) of the liquid line. The line must be used as delivered. It cannot be shortened.

Lennox has a refrigerant-flow-control (RFC) system. It is a very accurate means of metering refrigerant in the system. It must never be tampered with during installation. The whole principle of the RFC system involves matching the evaporator coil to the proper length and bore of the liquid line. This is believed by the manufacturer to be superior to the capillary-tube system. The RFC equalizes pressures almost instantly after the compressor stops. Therefore, it starts unloaded, eliminating the need for any extra controls. In addition, a precise amount of refrigerant charge is added to the system at the factory, resulting in trouble-free operation.

The condensing unit is shipped with a complete refrigerant charge. The condensing unit and evaporator are equipped with flared liquid and suction lines for quick connection. The compressor is hermetically sealed.

The unit may be built in and weatherproofed as a rigid part of the building structure. (See Figure 5.10.) The condensing unit can be free-standing on rooftops or slabs at grade level.

Figure 5.11 shows a condensing unit designed for the apartment developer and volume builder. It comes in 1-, 1.5-, and 2-ton sizes. Cooling capacities range from 17,000 through 28,000 Btu. An aluminum grille protects the condenser while offering low resistance to air discharge. The fan is mounted for

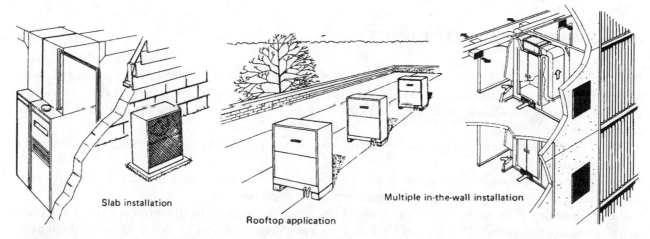

Slab installation

Rooftop application

Multiple in-the-wall installation

FIGURE 5.10 Three typical condensing applications. *(Lennox)*

FIGURE 5.11   Apartment house or residential condensing unit.
It can be installed through the wall, free standing or at grade level,
or on the roof. *(Lennox)*

less noise. It also reduces the possibility of air re-circulation back through the condenser when it is closely banked for multiple installations. When mounted at grade level, this also keeps hot-air discharge from damaging grass and shrubs.

SINGLE-PACKAGE ROOFTOP UNITS

Single rooftop units can be used for both heating and cooling for industrial and commercial installations. Figure 5.12 shows such a unit. It can provide up to 1.5 million Btu if water heat is used. It can also include optional equipment to supply heat, up to 546,000 Btu, using electricity. Such units can use oil, gas, or propane for heating fuels.

These units require large amounts of energy to operate. It is possible to conserve energy by using more sensitive controls.

Highly sensitive controls monitor supply air. They send signals to the control module. It, in turn, cycles the mechanical equipment to match the output to the load condition.

An optional device for conserving energy is available. It has a "no load" band thermostat that has a built-in differential of 6 degrees F (3.3 degrees C). This gives the system the ability to "coast" between the normal control points without consuming any primary energy within the recommended comfort-setting range.

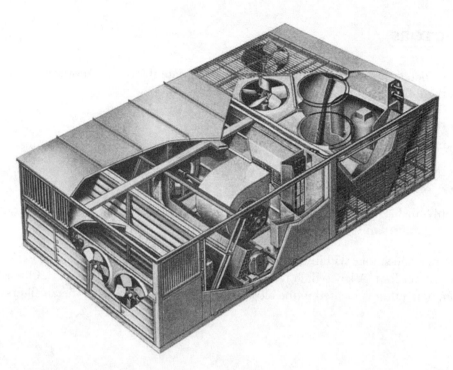

FIGURE 5.12   Single-zone rooftop system. This unit is used for industrial and commercial-market applications. Cooling capacity ranges from 8 through 60 tons. *(Lennox)*

Another feature that is prevalent is a refrigerant heat-reclaim coil. It can reduce supermarket heating costs significantly. A reheat coil can be factory-installed downstream from the evaporator coil. It will use the condenser heat to control humidity and prevent overcooling. A unit of this size is designed for a large store or supermarket. Figure 5.13 shows how the rooftop model is mounted for efficient distribution of the cold air. Since cold air is heavy, it will settle quickly to floor level. Hot air rises and stays near the ceiling in a room. Thus it is possible for this warmer air to increase the temperature of the cold air from the conditioner before it comes into contact with the room's occupants.

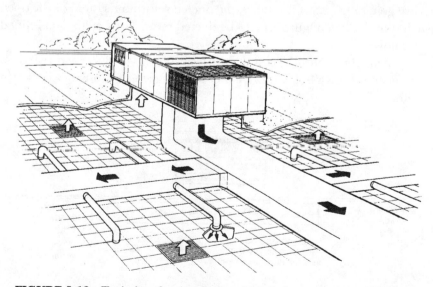

FIGURE 5.13   Typical rooftop installation of the single-zone system. *(Lennox)*

SMOKE DETECTORS

Photocell detectors detect smoke within the system. They actuate the blower motor controls and other devices to perform the following:

- Shut off the entire system

- Shut down the supply blower, close outside air, and return air dampers and runs

- Run supply-air blower, open outside-air dampers, close return-air dampers, and stop return-air blower or exhaust fans

- Run supply-air blower, open outside-air dampers, close return-air dampers, and run return-air blower or exhaust fan

Actuation occurs when smoke within the unit exceeds a density that is sufficient to obscure light by a factor of 2 or 4 percent per foot. A key switch is used for testing. Two detectors are used. One is located in the return air section. The other is located in the blower section downstream from the air filters.

FIRESTATS

Firestats are furnished as standard equipment. Manual reset types are mounted in the return air and the supply air stream. They will shut off the unit completely when either firestat detects excessive air temperatures.

On this type of unit, the blowers are turned by one motor with a shaft connecting the two fans. There are three small fan motors and blades mounted to exhaust air from the unit. There are four condenser fans. The evaporator coil is slanted. There are also two condenser coils mounted at angles and two compressors. The path for the return air is through the filters and evaporator coil back to the supply-air ductwork.

RETURN-AIR SYSTEMS

Return-air systems are generally one of two types: the ducted return-air system or the open plenum return-air system ("sandwich space"). (See Figure 5.14.) The ducted return-air system duct is lined with insulation, which greatly reduces noise.

The open-plenum system eliminates the cost of return-air returns or ducts and is extremely flexible. In a building with relocatable interior walls, it is much easier to change the location of a ceiling grille than reroute a ducted return system.

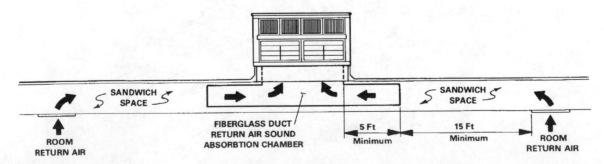

FIGURE 5.14 Return-air system for the rooftop unit. *(Lennox)*

ACOUSTICAL TREATMENT

Insulating the supply duct reduces duct loss or gain and prevents condensation. Use 1 1/2-pound density on ducts that deliver air velocities up to 1500 fpm.

Three-pound density or neoprene-coated insulation is recommended for ducts that handle air at velocities greater than 1,500 fpm. Insulation can be 1/2 or 1 inch thick on the inside of the duct.

When rooftop equipment uses the sandwich space for the return air system, a return air chamber should be connected to the air-inlet opening. Such an air chamber is shown in Figure 5.14. This reduces air-handling sound transmission through the thin ceiling panels. It should be sized not to exceed 1,500-fpm return-air velocity. The duct can be fiberglass or a fiberglass-lined metal. A ceiling return-air grille should not be installed within 15 feet of the duct inlet.

VOLUME DAMPERS

Volume dampers are important to a good system design. Lengths of supply runs vary and usually have the same cubic-foot measurements. Therefore, balancing dampers should be used in each supply branch run. The installer must furnish and install the balancing dampers. Dampers should be installed between the supply air duct and the diffuser outlet.

There are several ways in which rooftop conditioners can be installed. Figure 5.15 shows three installation methods.

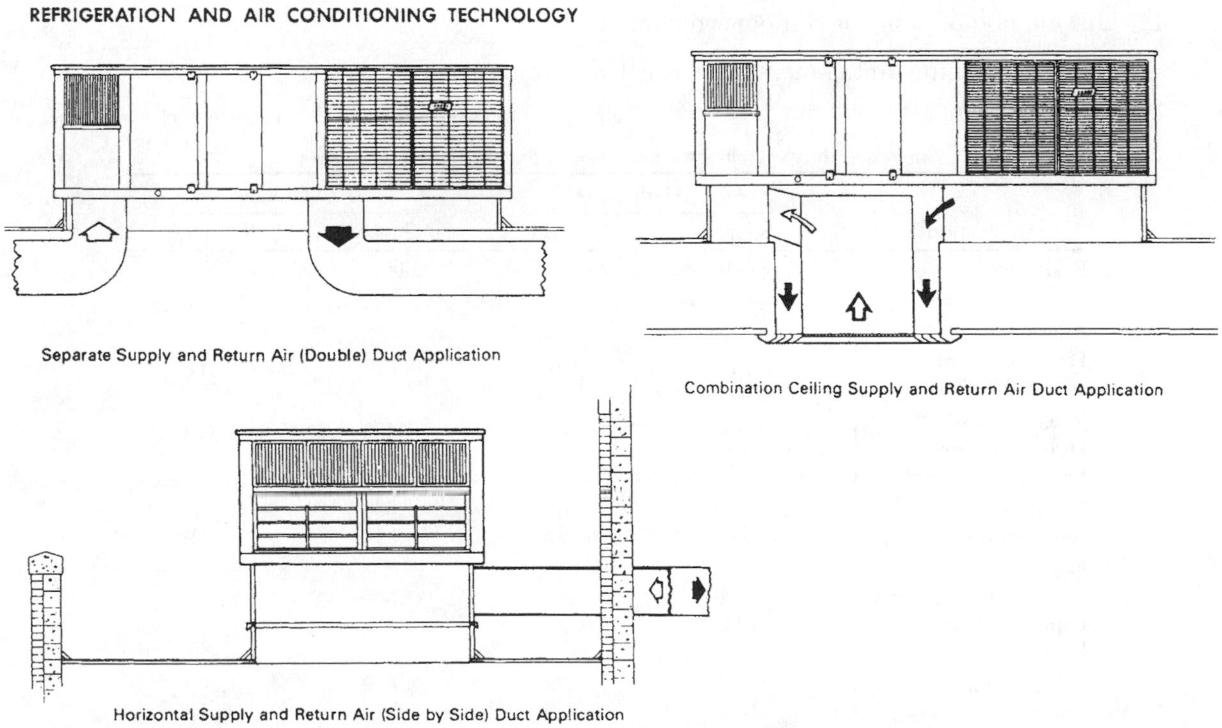

REFRIGERATION AND AIR CONDITIONING TECHNOLOGY

Separate Supply and Return Air (Double) Duct Application

Combination Ceiling Supply and Return Air Duct Application

Horizontal Supply and Return Air (Side by Side) Duct Application

FIGURE 5.15 Choice of air patterns for the rooftop unit. *(Lennox)*

MULTIPLE-CHOICE EXAM

1. Which of the following is not one of the three methods used to size duct systems?

    a. static gain   b. equal friction

    c. equal freedom of flow   d. equal velocity

2. Typical duct velocities for low-velocity systems are shown in Table 5.1. What is the air flow of the main duct return for hotel bedrooms in feet per minute

    a. 1,300   b. 2,000

    c. 1,600   d. 1,100

3. What does a duct configurator do:

    a. it figures the area of a duct

    b. it checks sizes of ducts

    c. it configures ducts

    d. it sizes and analyzes supply-duct systems

4. What is meant by equal friction in a duct system?

    a. this method of sizing supply duct systems figures constant pressure loss per unit length

    b. this method of sizing return duct systems figures constant pressure loss per unit length

    c. this method of design is computer-operated

    d. this describes the unit of measure in which air flow is gauged

TABLE 5.1   Suggested duct velocities for low-velocity duct system, ft/min (m/s).

Application	Main ducts		Branch ducts	
	Supply	Return	Supply	Return
Residences	1,000 (5.1)	800 (4.1)	600 (3)	600 (3)
Apartments Hotel bedrooms Hospital bedrooms	1,500 (7.6)	1,300 (6.6)	1,200 (6.1)	1,000 (5.1)
Private offices Director's rooms Libraries	1,800 (9.1)	1,400 (7.1)	1,400 (7.1)	1,200 (6.1)
Theaters Auditoriums	1,300 (6.6)	1,100 (5.6)	1,000 (5.1)	800 (4.1)
General offices Expensive restaurants Expensive stores Banks	2,000 (10.2)	1,500 (7.6)	1,600 (8.1)	1,200 (6.1)
Average stores Cafeterias	2,000 (10.2)	1,500 (7.6)	1,600 (8.1)	1,200 (6.1)
Industrial	2,500 (12.7)	1,800 (9.1)	2,200 (11.2)	1,600 (8.1)

5. What offsets friction losses in ducts?

    a. static pressure   b. equal friction

    c. blowers   d. fans

6. The Trane Ductulator® simplifies the manual process of sizing ductwork based on the equal-_____ methodology.

    a. friction   b. flow

    c. static pressure   d. velocity

7. What are the three primary tasks for the HVAC system of a building?

    a. to cool air, to heat air, to move air

    b. to deliver air, to return air, to make up air

    c. to make air move, to overcome friction in ductwork, to return air

    d. none of these

8. What is the suggested duct velocity for a high-velocity system used in the main duct supply of a commercial institution (See Table 5.2) in feet per minute?

    a. 1,200-1,600   b. 2,000-3,000

    c. 1,400-1,800   d. 2,500-3,800

9. Some equipment removes air from a conditioned space and has to be made up. Which of the following is not such a piece of equipment?

    a. range hood   b. spray booth

    c. clothes dryer   d. exhaust fan

    e. microwave oven

10. Proper maintenance of the HVAC system of a building is impossible if proper balance of air _____ is not ensured in all occupied building zones or conditioned spaces.

    a. flow   b. return

    c. heat   d. supply

TABLE 5.2   Suggested duct velocities for high-velocity duct system, ft/min (m/s).

	Main duct		Branch duct	
Application	Supply	Return	Supply	Return
Commercial institutions	2,500–3,800	1,400–1,800	2,000–3,000	1,200–1,600
Public buildings	(12.7–19.3)	(7.1–9.1)	(10.2–15.2)	(6.1–8.0)
Industrial	2,500–4,000	1,800–2,200	2,200–3,200	1,500–1,800
	(12.7–20.3	(9.1–11.2)	(11.2–16.3)	(7.6–9.1)

11. The HVAC system components in a building require regular _____.

a. cleaning   b. care

c. repair   d. dusting

12. There are three generally accepted methods of duct cleaning. Which of the following is not one of them?

a. contact-vacuum method   b. air-w-shing method

c. power brushing method   d. wiping with a clean rag

13. Pneumatically or _____ powered rotary brushes are used to dislodge dirt and dust particles that become airborne; they are then drawn downstream through the duct system.

a. water-   b. electrically-

c. chemically-   d. vacuum-

14. What is the factor that must be taken into consideration when planning and operating an HVAC system in elevated areas of the country?

a. sea level   b. altitude

c. humidity   d. air quality

15. Conventional vacuuming equipment may _____ extremely fine particulate matter into the building air space rather than collecting it.

a. discharge   b. suck

c. blow   d. force

TRUE-FALSE EXAM

1. The function of a duct system is to provide a means to transmit air from the air-handling equipment.

True   False

2. If the manual method is used to size the duct systems, they should be calculated by following one of the accepted procedures found in standard design handbooks.

True   False

3. The equal-friction method is the most widely used method of sizing lower-pressure, lower-velocity duct systems.

True   False

4. The static-regain design method sizes the supply-duct system to obtain uniform static pressure at all branches and outlets.

True   False

5. The Trane Duct Configurator can model both constant and variable volume systems.

True   False

6. Cleaning ducts is the job of outside contractors.

True   False

7. The HVAC system components in a building will require regular cleaning.

True   False

8. Sanitizing treatments to kill mold, mildew, bacteria, and so forth are generally administered by self-contained foggers or vaporizers or by air-driven spray units.

True   False

9. There are three generally accepted methods for duct cleaning.

True   False

10. In the power-brush method of cleaning, a vacuum-collection device is connected to the upstream end of the section being cleaned.

True   False

11. Air washing is another method of cleaning air ducts.

True   False

12. If you are hiring a contractor for duct cleaning, you should ask for proof of insurance and to see a state license.

True   False

13. A schedule for inspections of ductwork should be maintained.

True   False

14. When designing a ductwork system either manually or by computer, the effects of high altitude (over 2,500 feet) must be accounted for in the design.

True   False

15. Air flow in ductwork is measured in cubic feet/second.

True   False

MULTIPLE-CHOICE ANSWERS

1. C	4. A	7. B	10. A	13. B
2. A	5. A	8. D	11. A	14. B
3. D	6. A	9. E	12. D	15. A

TRUE-FALSE ANSWERS

1. T	5. T	9. T	13. T
2. T	6. F	10. F	14. T
3. T	7. T	11. T	15. F
4. T	8. T	12. T	

Part C

AIR CONDITIONING EQUIPMENT

Chapter 6
REFRIGERANTS

REFRIGERANTS: NEW AND OLD

Refrigerants are used in the process of refrigeration, which is a process whereby heat is removed from a substance or a space. A refrigerant is a substance that picks up latent heat when it evaporates from a liquid to a gas. This is done at a low temperature and pressure. A refrigerant expels latent heat when it condenses from a gas to a liquid at a high pressure and temperature. The refrigerant cools by absorbing heat in one place and discharging it in another area.

The desirable properties of a good refrigerant for commercial use are:

- Low boiling point

- Safe, nontoxic

- Easy to liquefy and moderate pressure and temperature

- High latent heat value

- Operation on a positive pressure

- Not affected by moisture

- Mixes well with oil

- Noncorrosive to metal

There are other qualities that all refrigerants have: molecular weight, density, compression ratio, heat value, and compression temperature. These qualities will vary with the refrigerant used. The compressor displacement and type or design will also influence the choice of refrigerant.

CLASSIFICATION

Refrigerants are classified according to their manner of absorption or extraction of heat from substances to be refrigerated:

- Class 1 refrigerants are used in the standard compression type of refrigeration systems

- Class 2 refrigerants are used as immediate cooling agents between Class 1 and Class 3 types and the substance to be refrigerated

- Class 3 refrigerants are used in the standard absorption type of refrigerating systems

Class 1 refrigerants cool by absorption or extraction of latent heat from the substances.

Class 2 refrigerants cool substances by absorbing their sensible heats. They include air, calcium-chloride brine, sodium-chloride (salt) brine, alcohol, and similar non-freezing solutions.

Class 3 refrigerants consist of solutions that contain absorbed vapors of liquefiable agents or refrigerating media. These solutions function through their ability to carry the liquefiable vapors. The vapors produce a cooling effect by the absorption of latent heat. An example is aqua ammonia, which is a solution composed of distilled water and pure ammonia.

COMMON REFRIGERANTS

Following are some of the more common refrigerants. Table 6.1 summarizes the characteristics to a selected few of the many refrigerants available for home, commercial and industrial use.

TABLE 6.1 Characteristics of typical refrigerants.

Name	Boiling Point (degrees F)	Heat of Vaporization at Boiling Point Btu/lb. 1 At.
Sulfur dioxide	14.0	172.3
Methyl chloride	-10.6	177.8
Ethyl chloride	55.6	177.0
Ammonia	-28.0	554.7
Carbon dioxide	-110.5	116.0
Freezol (isobutane)	10.0	173.5
Freon 11	74.8	78.31
Freon 12	-21.7	71.04
Freon 13	-114.6	63.85
Freon 21	48.0	104.15
Freon 22	-41.4	100.45
Freon 113	117.6	63.12
Freon 114	38.4	58.53
Freon 115	-37.7	54.20
Freon 502	-50.1	76.46

Sulfur Dioxide

Sulfur dioxide (SO_2) is a colorless gas or liquid. It is toxic, with a very pungent odor. When sulfur is burned in air, sulfur dioxide is formed. When sulfur dioxide combines with water, it produces sulfuric and sulfurous acids. These acids are very corrosive to metal. They have an adverse effect on most materials. Sulfur dioxide is not considered a safe refrigerant. As a refrigerant, sulfur dioxide operates on a vacuum to give the temperatures required. Moisture in the air will be drawn into the system when a leak occurs. This means the metal parts will eventually corrode, causing the compressor to seize.

Sulfur dioxide (SO_2) boils at 14 degrees F (-10 degrees C) and has a heat of vaporization at boiling point (1 atmosphere) of 172.3 Btu/pound. It has a latent heat value of 166 Btu per pound.

To produce the same amount of refrigeration, sulfur dioxide requires about one-third more vapor than Freon® and methyl chloride. This means that the condensing unit has to operate at a higher speed or that the compressor cylinders must be larger. Since sulfur dioxide does not mix well with oil, the suction line must be on a steady slant to the machine. Otherwise the oil will trap out, constricting the suction line. This refrigerant is not feasible for use in some locations.

Methyl Chloride

Methyl chloride (CHP) has a boiling point of -10.6 degrees F (-23.3 degrees C). Its heat of vaporization at boiling point (at 1 atmosphere) is 177.8 Btu/pound. It is a good refrigerant. However, because it will burn under some conditions, some cities will not allow it to be used. It is easy to liquefy and has a comparatively high latent heat value. It does not corrode metal when in its dry state.

However in the presence of moisture it damages the compressor. A sticky black sludge is formed when excess moisture combines with the chemical. Methyl chloride mixes well with oil. It will operate on a positive pressure as low as -10 degrees F (-23 degrees C). The amount of vapor needed to cause discomfort in a person is in proportion to the following numbers:

Carbon dioxide	100
Methyl chloride	70
Ammonia	2
Sulfur dioxide	1

That means methyl chloride is 35 times safer than ammonia and 70 times safer than sulfur dioxide.

Methyl chloride is hard to detect with the nose or eyes. It does not produce irritating effects. Therefore, some manufacturers add a 1 percent amount of acrolein, a colorless liquid with a pungent odor, as a warning agent. It is produced by destructive distillation of fats.

Ammonia

Ammonia (NH_3) is used most frequently in large industrial plants. Freezers for packing houses usually employ ammonia as a refrigerant. It is a gas with a very noticeable odor. Even a small leak can be detected with the nose. Its boiling point at normal atmospheric pressure is -28 degrees F (-33 degrees C). Its freezing point is -107.86 degrees F (-77.7 degrees C). It is very soluble in water. Large refrigeration capacity is possible with small machines. It has high latent heat (555 Btu at 18 degrees F (-7.7 degrees C). It can be used with steel fittings. Water-cooled units are commonly used to cool down the refrigerant. High pressures are used in the lines (125 to 200 pounds per square inch). Anyone inside the refrigeration unit when it springs a leak is rapidly overcome by the fumes. Fresh air is necessary to reduce the toxic effects of ammonia fumes. Ammonia is combustible when combined with certain amounts of air (about one volume of ammonia to two volumes of air). It is even more combustible when combined with oxygen. It is very toxic. Heavy steel fittings are required since pressures of 125 to 200 pounds per square inch are common. The units must be water-cooled.

Carbon Dioxide

Carbon dioxide (CO_2) is a colorless gas at ordinary temperatures. It has a slight odor and an acid taste. Carbon dioxide is non-explosive and non-flammable. It has a boiling point of 5 degrees F (-15 degrees C). A pressure of over 300 pounds per square inch is required to keep it from evaporation. To liquefy the gas, a condenser temperature of 80 degrees F (26.6 degrees C) and a pressure of approximately 1000 pounds per square inch are needed. Its critical temperature is 87.8 degrees F (31 degrees C). It is harmless to breathe except in extremely large concentrations. The lack of oxygen can cause suffocation under certain conditions of carbon-dioxide concentration.

Carbon dioxide is used aboard ships and in industrial installations. It is not used in household applications. The main advantage of using carbon dioxide for a refrigerant is that a small compressor can be used. The compressor is very small since a high pressure is required for the refrigerant. Carbon dioxide is, however, very inefficient compared to other refrigerants.

Calcium Chloride

Calcium chloride ($CaCl_2$) is used only in commercial refrigeration plants. It is used as a simple carrying medium for refrigeration.

Brine systems are used in large installations where there is danger of leakage. They are also used where the temperature fluctuates in the space to be refrigerated. Brine is cooled down by the direct expansion of the refrigerant. It is then pumped through the material or space to be cooled. Here it absorbs sensible heat.

Most modern plants operate with the brine at low temperature. This permits the use of less brine, less piping or smaller-diameter pipe, and smaller pumps. It also lowers pumping costs. Instead of cooling a large volume of brine to a given temperature, the same number of refrigeration units are used to cool a smaller volume of brine to a lower temperature. This results in greater economy. The use of extremely low-freezing brine, such as calcium chloride, is desirable in the case of the shell-type cooler. Salt brine with a minimum possible freezing point of -6 degrees F (-20.9 degrees C) may solidify under excess vacuum on the cold side of the refrigerating unit. This can cause considerable damage and loss of operating time. There are some cases in which the cooler has been ruined.

Ethyl Chloride

Ethyl chloride (C_2H_5Cl) is not commonly used in domestic refrigeration units. It is similar to methyl chloride in many ways. It has a boiling point of 55.6 degrees F (13.1 degrees C) at atmospheric pressure. Critical temperature is 360.5 degrees F (182.5 degrees C) at a pressure of 784 pounds absolute. It is a colorless liquid or gas with a pungent ethereal odor and a sweetish taste. It is neutral toward all metals. This means that iron, copper, and even tin and lead can be used in the construction of the refrigeration unit. It does, however, soften all rubber compounds and gasket material. Thus, it is best to use only lead for gaskets.

Freon® Refrigerants

Freon® refrigerants have been one of the major factors responsible for the tremendous growth of the home refrigeration and air conditioning industries. The safe properties of these products have permitted their use under conditions where flammable or more toxic refrigerants would be hazardous to use. There is a Freon refrigerant for every application—from home and industrial air conditioning to special low-temperature requirements.

The unusual combination of properties found in the Freon compounds is the basis for their wide application and usefulness. Table 6.2 presents a summary of the specific properties of some of the fluorinated products. Figure 6.1 gives the absolute pressure and gauge pressure of Freon refrigerants at various temperatures.

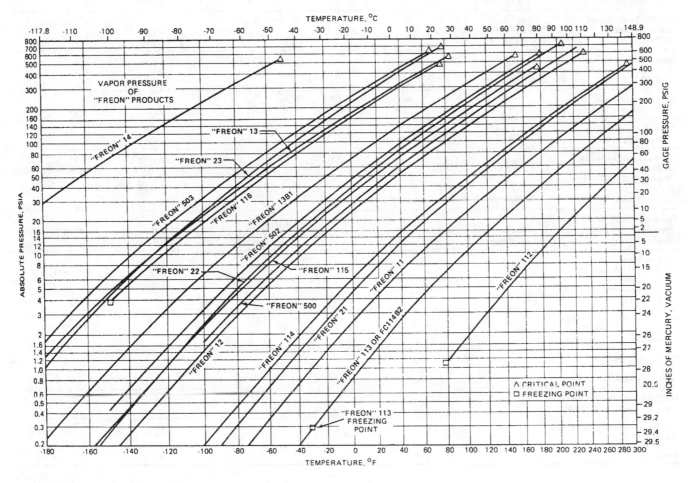

FIGURE 6.1 The absolute and gauge pressures of Freon refrigerants.

TABLE 6.2 Physical properties of Freon* products.

		Freon 11	Freon 12	Freon 13	Freon 13B1	Freon 14
Chemical formula		CCl_3F	CCl_2F_2	$CClF_3$	$CBrF_3$	CF_4
Molecular weight		137.37	120.92	104.46	148.92	88.00
Boiling point at 1 atm	°C	23.82	-29.79	-81.4	-57.75	-127.96
	°F	74.87	-21.62	-114.6	-71.95	-198.32
Freezing point	°C	-111	-158	-181[1]	-168	-184[2]
	°F	-168	-252	-294	-270	-299
Critical temperature	°C	198.0	112.0	28.9	67.0	-45.67
	°F	388.4	233.6	83.9	152.6	-50.2
Critical pressure	atm	43.5	40.6	38.2	39.1	36.96
	lbs/sq in abs	639.5	596.9	561	575	543.2
Critical volume	cc/mol	247	217	181	200	141
	cu ft/lb	0.0289	0.0287	0.0277	0.0215	0.0256
Critical density	g/cc	0.554	0.588	0.578	0.745	0.626
	lbs/cu ft	34.6	34.8	36.1	46.5	39.06
Density, liquid at 25°C (77°F)	g/cc	1.476	1.311	1.298 @ -30°C (-22°F)	1.538	1.317 @ -80°C (-112°F)
	lbs/cu ft	92.14	81.84	81.05	96.01	82.21
Density, sat'd vapor at boiling point	g/l	5.86	6.33	7.01	8.71	7.62
	lbs/cu ft	0.367	0.395	0.438	0.544	0.476
Specific heat, liquid (Heat capacity) at 25°C (77°F)	cal/(g)(°C) or Btu/(lb)(°F)	0.208	0.232	0.247 @ -30°C (-22°F)	0.208	0.294 @ -80°C (-112°F)
Specific heat, vapor, at const pressure (1 atm) at 25°C (77°F)	cal/(g)(°C) or Btu/(lb)(°F)	0.142 @ 38°C (100°F)	0.145	0.158	0.112	0.169
Specific heat ratio at 25°C and 1 atm	C_p/C_v	1.137 @ 38°C (100°F)	1.137	1.145	1.144	1.159
Heat of vaporization at boiling point	cal/g	43.10	39.47	35.47	28.38	32.49
	Btu/lb	77.51	71.04	63.85	51.08	58.48
Thermal conductivity at 25°C (77°F) Btu/(hr) (ft) (°F) liquid vapor (1 atm)		0.0506 0.00451	0.0405 0.00557	0.0378 0.00501 @ -30°C (-22°F)	0.0234 0.00534	0.0361 0.00463 @ -80°C (-112°F)
Viscosity[7] at 25°C (77°F) liquid vapor (1 atm)	centipoise centipoise	0.415 0.0107	0.214 0.0123	0.170 0.0119 @ (-30°C) (-22°F)	0.157 0.0154	0.23 0.0116 @ (-80°C) (-112°F)
Surface tension at 25°C (77°F) dynes/cm		18	9	14 @ -73°C -100°F	4	4 @ -73°C (-100°F)
Refractive index of liquid at 25°C (77°F)		1.374	1.287	1.199 @ -73°C (-100°F)	1.238	1.151 @ -73°C (-100°F)
Relative dielectric strength[8] at 1 atm and 25°C (77°F) (nitrogen = 1)		3.71	2.46	1.65	1.83	1.06
Dielectric constant liquid vapor (1 atm)[9a]		2.28 @ 29°C 1.0036 @ 24°C[9b]	2.13 @ 29°C (84°F) 1.0032	1.0024 @ 29°C (84°F)		1.0012 @ 24.5°C (76°F)
Solubility of "Freon" in water at 1 atm and 25°C (77°F)	wt %	0.11	0.028	0.009	0.03	0.0015
Solubility of water in "Freon" at 25°C (77°F)	wt %	0.011	0.009		0.0095 @ 21°C (70°F)	
Toxicity		Group 5a[12]	Group 6[12]	Probably Group 6[13]	Group 6[12]	Probably Group 6[13]

(continued)

TABLE 6.2 Physical properties of Freon* products. *(Continued)*

		Freon 21	Freon 22	Freon 23	Freon 112	Freon 113	Freon 114
Chemical formula		$CHCl_2F$	$CHClF_2$	CHF_3	$CCl_2F–CCl_2F$	$CCl_2F–CClF_2$	$CClF_2–CClF_2$
Molecular weight		102.93	86.47	70.01	203.84	187.38	170.93
Boiling point at 1 atm	°C	8.92	-40.75	-82.03	92.8	47.57	3.77
	°F	48.06	-41.36	-115.66	199.0	117.63	38.78
Freezing point	°C	-135	-160	-155.2	26	-35	-94
	°F	-211	-256	-247.4	79	-31	-137
Critical temperature	°C	178.5	96.0	25.9	278	214.1	145.7
	°F	353.3	204.8	78.6	532	417.4	294.3
Critical pressure	atm	51.0	49.12	47.7	34[3]	33.7	32.2
	lbs/sq in abs	750	721.9	701.4	500	495	473.2
Critical volume	cc/mol	197	165	133	370[3]	325	293
	cu ft/lb	0.0307	0.0305	0.0305	0.029	0.0278	0.0275
Critical density	g/cc	0.522	0.525	0.525	0.55[3]	0.576	0.582
	lbs/cu ft	32.6	32.76	32.78	34	36.0	36.32
Density, liquid	g/cc	1.366	1.194	0.670	1.634 @ 30°C (86°F)[b]	1.565	1.456
at 25°C (77°F)	lbs/cu ft	85.28	74.53	41.82	102.1	97.69	90.91
Density, sat'd vapor	g/l	4.57	4.72	4.66	7.02[5]	7.38	7.83
at Boiling Point	lbs/cu ft	0.285	0.295	0.291	0.438	0.461	0.489
Specific heat, liquid (heat capacity) at 25°C (77°F)	cal/(g)(°C) or Btu/(lb)(°F)	0.256	0.300	0.345 @ -30°C -22°F		0.218	0.243
Specific heat, vapor, at const pressure (1 atm) at 25°C (77°F)	cal/(g)(°C) or Btu/(lb)(°F)	0.140	0.157	0.176		0.161 @ 60°C (140°F)	0.170
Specific-heat ratio at 25°C and 1 atm	C_p/C_v	1.175	1.184	@ 1.191 0 pressure		1.080 @ 60°C (140°F)	1.084
Heat of vaporization	cal/g	57.86	55.81	57.23	37 (est)	35.07	32.51
at boiling point	Btu/b	104.15	100.45	103.02	67	63.12	58.53
Thermal conductivity[1] at 25°C (77°F) Btu/(hr)(ft)(°F)						0.0434	
liquid		0.0592	0.0507	0.0569 @ -30°C (-22°F)		0.0044	0.0372
vapor (1 atm)		0.00506	0.00609	0.0060	0.040	(0.5 atm)	0.0060
Viscosity[1] at 25°C (77°F)							
liquid	centipoise	0.313	0.198	0.167 @ -30°C (-22°F)	1.21[6]	0.68	0.36
vapor (1 atm)	centipose	0.0114	0.0127	0.0118		0.010 (0.1 atm)	0.0112
Surface tension at 25°C (77°F) dynes/cm		18	8	15 @ -73°C (-100°F)	23 @ 30°C (86°F)	17.3	12
Refractive index of liquid at 25°C (77°F)		1.354	1.256	1.215 @ -73°C (-100°F)	1.413	1.354	1.288
Relative-dielectric strength[8] at 1 atm and 25°C (77°F) (nitrogen = 1)		1.85	1.27	1.04	5 (est)	3.9 (0.44 atm)	3.34
Dielectric constant liquid Vapor (1 atm)[9a]		5.34 @ 28°C 1.0070 @ 30°C	6.11 @ 24°C 1.0071 @ 25.4°C	1.0073 @ 25°C[9b]	2.54 @ 25°C (77°F)	2.41 @ 25°C (77°F)	2.26 @ 25°C 1.0043 @ 26.8°C
Solubility of "Freon" in water wt % at 1 atm and 25°C (77°F)		0.95	0.30	0.10	0.012 (Sat'n Pres)	0.017 (Sat'n Pres)	0.013
Solubility of water in "Freon" wt % at 25°C (77°F)		0.13	0.13			0.011	0.009
Toxicity		much less than Group 4, somewhat more than Group 5[12]	Group 5a[12]	probably Group 6[13]	probably less than Group 4, more than Group 5[13]	much less than Group 4, somewhat more than Group 5[12]	Group 6[12]

(continued)

TABLE 6.2 Physical properties of Freon* products. *(Continued)*

FC 114B2	Freon 115	Freon 116	Freon 500	Freon 502	Freon 503
$CBrF_2$–$CBrF_2$	$CClF_2$–CF_3	CF_3–CF_3	a	b	c
259.85	154.47	138.01	99.31	111.64	87.28
47.26	-39.1	-78.2	-33.5	-45.42	-87.9
117.06	-38.4	-108.8	-28.3	-49.76	-126.2
-110.5	-106[10]	-100.6	-159		
-166.8	-159	-149.1	-254		
214.5	80.0	19.7[4]	105.5	82.2	19.5
418.1	175.9	67.5	221.9	179.9	67.1
34.4	30.8	29.4[4]	43.67	40.2	43.0
506.1	453	432	641.9	591.0	632.2
329	259	225	200.0	199	155
0.0203	0.0269	0.0262	0.03226	0.02857	0.0284
0.790	0.596	0.612	0.4966	0.561	0.564
49.32	37.2	38.21	31.0	35.0	35.21
2.163	1.291	1.587 } [4] @ -73°C (-100°F)	1.156	1.217	1.233 } @ -30°C (-22°F)[7]
135.0	80.60	99.08 }	72.16	75.95	76.95 }
	8.37	9.01[4]	5.278	6.22	6.02
	0.522	0.562	0.3295	0.388	0.374
0.166	0.285	0.232 @ -73°C (-100°F)[4]	0.258	0.293	0.287 @ -30°C (-22°F)
	0.164	0.182[11] @ 0 pressure	0.175	0.164	0.16
	1.091	1.085 (est) @ 0 pressure	1.143	1.132	1.21 @ -34°C (-30°F)
25 (est)	30.11	27.97	48.04	41.21	42.86
45 (est)	54.20	50.35	86.47	74.18	77.15
	0.0302	0.045 } @ -73°C (-100°F)		0.0373	0.0430 @ -30°C (-22°F)[7]
0.027	0.00724	0.0098 }	0.0432	0.00670	
0.72	0.193	0.30	0.192	0.180	0.144 @ -30°C (-22°F)
	0.0125	0.0148	0.0120	0.0126	
18	5	16 @ -73°C (-100°F)	8.4	5.9	6.1 @ -30°C (-22°F)
1.367	1.214	1.206 @ -73°C (-100°F)	1.273	1.234	1.209 @ -30°C (-22°F)
4.02 (0.44 atm)	2.54	2.02		1.3	
2.34 @ 25°C (77°F)	1.0035 @ 27.4°C	1.0021 @ 23°C (73°F)		6.11 @ 25°C 1.0035 (0.5 atm)	
	0.006				0.042
			0.056 Group 5a	0.056	
Group 5a[12]	Group 6[12]	probably Group 6[13]		Group 5a[12]	probably Group 6[13]

*FREON is Du pont's registered trademark for its fluorocarbon products
a. CCl_2F_2/CH_3CHF_2 (73.8/26.2% by wt.)
b. $CHClF_2$/$CClF_2CF_3$ (48.8/51.2% by wt.)
c. CHF_3/$CClF_3$ (40/60% by wt.)

MOLECULAR WEIGHTS

Compounds containing fluorine in place of hydrogen have higher molecular weights and often have unusually low boiling points. For example, methane (CH_4), with a molecular weight of 16, has a boiling point of -258.5 degrees F (-161.4 degrees C) and is nonflammable. Freon® 14 (CF_4) has a molecular weight of 88 and a boiling point of -198.4 degrees F (-128 degrees C) and is nonflammable. The effect is even more pronounced when chlorine is also present. Methylene chloride (CH_2Cl_2) has a molecular weight of 85 and boils at 105.2 degrees F (40.7 degrees C), while Freon 12 (CCl_2F_2, molecular weight 121) boils at -21.6 degrees F (-29.8 degrees C). It can be seen that Freon compounds are high-density materials with low boiling points, low viscosity, and low surface tension. Freon products have boiling points covering a wide range of temperatures. (See Table 6.3.)

The high molecular weight of the Freon compounds also contributes to low vapor, specific heat values, and fairly low latent heats of vaporization. Tables of thermodynamic properties including enthalpy, entropy, pressure, density, and volume for the liquid and vapor are available from manufacturers.

Freon compounds are poor conductors of electricity. In general, they have good dielectric properties.

TABLE 6.3 Fluorinated products and their molecular weight and boiling point.

Freon Products			Boiling Point	
Product	Formula	Molecular Weight	°F	°C
Freon 14	CF_4	88.0	-198.3	-128.0
Freon 503	$CHF_3/CClF_3$	87.3	-127.6	-88.7
Freon 23	CHF_3	70.0	-115.7	-82.0
Freon 13	$CClF_3$	104.5	-114.6	-81.4
Freon 116	$CF_3–CF_3$	138.0	-108.8	-78.2
Freon 13B1	$CBrF_3$	148.9	-72.0	-57.8
Freon 502	$CHClF_2/CClF_2–CF_3$	111.6	-49.8	-45.4
Freon 22	$CHClF_2$	86.5	-41.4	-40.8
Freon 115	$CClF_2–CF_3$	154.5	-37.7	-38.7
Freon 500	CCl_2F_2/CH_3CHF_2	99.3	-28.3	-33.5
Freon 12	CCl_2F_2	120.9	-21.6	-29.8
Freon 114	$CClF_2–CClF_2$	170.9	38.8	3.8
Freon 21	$CHCl_2F$	102.9	48.1	8.9
Freon 11	CCl_3F	137.4	74.9	23.8
Freon 113	$CCl_2F–CClF_2$	187.4	117.6	47.6
Freon 112	$CCl_2F–CCl_2F$	203.9	199.0	92.8
Other Fluorinated Compounds				
FC 114B2	$CBrF_2–CBrF_2$	259.9	117.1	47.3
1,1-Difluoroethane*	CH_3, CHF_2	66.1	13.0	25.0
1,1,1-Chlorodifluoroethane†	$CH_3–CClF_2$	100.5	14.5	-9.7
Vinyl fluoride	$CH_2–CHF$	46.0	-97.5	-72.0
Vinylidene fluoride	$CH_2–CF_2$	64.0	-122.3	-85.7
Hexafluoroacetone	CF_3COCF_3	166.0	-18.4	-28.0
Hexafluoroisopropanol	$(CF_3)_2CHOH$	168.1	136.8	58.2

* Propellant or refrigerant 152a
† Propellant or refrigerant 142b
Copyright 1969 by E. I. du Pont de Nemours and Company, Wilmington, Delaware 19898

FLAMMABILITY

None of the Freon compounds are flammable or explosive. However, mixtures with flammable liquids or gases may be flammable and should be handled with caution. Partially halogenated compounds may also be flammable and must be individually examined.

TOXICITY

One of the most important qualities of the Freon fluorocarbon compounds is their low toxicity under normal conditions of handling and usage. However, the possibility of serious injury or death exists under unusual or uncontrolled exposures or in deliberate abuse by inhalation of concentrated vapors. The potential hazards of fluorocarbons are summarized in Table 6.4.

SKIN EFFECTS

Liquid fluorocarbons with boiling points below 32 degrees F (0 degrees C) may freeze the skin, causing frostbite on contact. Suitable protective gloves and clothing give insulation protection. Eye protection should be used. In the event of frostbite, warm the affected area quickly to body temperature. Eyes should be flushed

TABLE 6.4 Potential hazards of fluorocarbons.

Condition	Potential Hazard	Safeguard
Vapors may decompose in flames or in contact with hot surfaces.	Inhalation of toxic decomposition products.	Good ventilation. Toxic decomposition products serve as warning agents.
Vapors are four to five times heavier than air. High concentrations may tend to accumulate in low places.	Inhalation of concentrated vapors can be fatal.	Avoid misuse. Forced-air ventilation at the level of vapor concentration. Individual breathing devices with air supply. Lifelines when entering tanks or other confined areas.
Deliberate inhalation to produce intoxication.	Can be fatal.	Do not administer epinephrine or other similar drugs.
Some fluorocarbon liquids tend to remove natural oils from the skin.	Irritation of dry, sensitive skin.	Gloves and protective clothing.
Lower boiling liquids may be splashed on skin.	Freezing.	Gloves and protective clothing.
Liquids may be splashed into eyes.	Lower boiling liquids may cause freezing. Higher boiling liquids may cause temporary irritation and if other chemicals are dissolved, may cause serious damage.	Wear eye protection. Get medical attention. Flush eyes for several minutes with running water.
Contact with highly reactive metals.	Violent explosion may occur.	Test the proposed system and take appropriate safety precautions.

copiously with water. Hands may be held under armpits or immersed in warm water. Get medical attention immediately. Fluorocarbons with boiling points at or above ambient temperature tend to dissolve protective fat from the skin. This leads to skin dryness and irritation, particularly after prolonged or repeated contact. Such contact should be avoided by using rubber or plastic gloves. Eye protection and face shields should be used if splashing is possible. If irritation occurs following contact, seek medical attention.

ORAL TOXICITY

Fluorocarbons are low in oral toxicity as judged by single-dose administration or repeated dosing over long periods. However, direct contact of liquid fluorocarbons with lung tissue can result in chemical pneumonitis, pulmonary edema, and hemorrhage. Fluorocarbons 11 and 113, like many petroleum distillates, are fat solvents and can produce such an effect. If products containing these fluorocarbons were accidentally or purposely ingested, induction of vomiting would be contraindicated.

CENTRAL-NERVOUS-SYSTEM (CNS) EFFECTS

Inhalation of concentrated fluorocarbon vapors can lead to CNS effects comparable to the effects of general anesthesia. The first symptom is a feeling of intoxication. This is followed by a loss of coordination and unconsciousness. Under severe conditions, death can result. If these symptoms are felt, the exposed individual should immediately go or be moved to fresh air. Medical attention should be sought promptly. Individuals exposed to fluorocarbons should NOT be treated with adrenaline (epinephrine).

CARDIAC SENSITIZATION

Fluorocarbons can, in sufficient vapor concentration, produce cardiac sensitization. This is a sensitization of the heart to adrenaline brought about by exposure to high concentrations of organic vapors. Under severe exposure, cardiac arrhythmias may result from sensitization of the heart to the body's own levels of adrenaline. This is particularly so under conditions of emotional or physical stress, fright, panic, and so forth. Such cardiac arrhythmias may result in ventricular fibrillation and death. Exposed individuals should immediately go or be removed to fresh air, where the hazard of cardiac effects will rapidly decrease. Prompt medical attention and observation should be provided following accidental exposures. A worker adversely affected by fluorocarbon vapors should NOT be treated with adrenaline (epinephrine) or similar heart stimulants, since these would increase the risk of cardiac arrhythmias.

THERMAL DECOMPOSITION

Fluorocarbons decompose when exposed directly to high temperatures. Flames and electrical-resistance heaters, for example, will chemically decompose fluorocarbon vapors. Products of this decomposition in air include halogens and halogen acids (hydrochloric, hydrofluoric, and hydrobromic), as well as other irritating compounds. Although much more toxic than the parent fluorocarbon, these decomposition products tend to irritate the nose, eyes, and upper-respiratory system. This provides a warning of their presence. The practical hazard is relatively slight. It is difficult for a person to remain voluntarily in the presence of decomposition products at concentrations where physiological damage occurs.

When such irritating decomposition products are detected, the area should be evacuated and ventilated. The source of the problem should be corrected.

REACTION OF FREON TO REFRIGERATION MATERIALS

Metals

Most of the commonly used construction metals, such as steel, cast iron, brass, copper, tin, lead, and aluminum, can be used satisfactorily with Freon compounds under normal conditions of use. At high temperatures some of the metals may act as catalysts for the breakdown of the compound. The tendency of metals to promote thermal decomposition of Freon compounds is in the following general order, with those metals that least promote thermal decomposition listed first:

Inconel®

Stainless steel

Nickel

1340 steel

Aluminum

Copper

Bronze

Brass

Silver

The above order is only approximate. Exceptions may be found with individual Freon compounds or for special conditions of use.

Magnesium alloys and aluminum containing more than 2 percent magnesium are not recommended for use in systems containing Freon compounds if water may be present. Zinc is not recommended for use with Freon 113. Experience with zinc and other Freon compounds has been limited, and no unusual reactivity has been observed. However, it is more chemically reactive than other common construction metals, so it would seem wise to avoid its use with the Freon compounds unless adequate testing is carried out.

Some metals may be questionable for use in applications requiring contact with Freon compounds for long periods of time or unusual conditions of exposure. These metals, however, can be cleaned safely with Freon solvents. Cleaning applications are usually for short exposures at moderate temperatures.

Most halocarbons may react violently with highly reactive materials, such as sodium, potassium, and barium in their free metallic form. Materials become more reactive when finely ground or powdered. In this state, magnesium and aluminum may react with fluorocarbons, especially at higher temperatures. Highly reactive materials should not be brought into contact with fluorocarbons until a careful study is made and appropriate safety precautions are taken.

Plastics

A brief summary of the effect of Freon compounds on various plastic materials follows. However, compatibility should be tested for specific applications. Differences in polymer structure and molecular weight, plasticizers, temperature, and pressure may alter the resistance of the plastic toward the Freon compound:

- Teflo-TFE-fluorocarbon: no swelling observed when submerged in Freon liquids, but some diffusion found with Freon 12 and Freon 22

- Polychlorotrifloororoethylene: slight swelling, but generally suitable for use with Freon compounds

- Polyvinyl alcohol: not affected by the Freon compounds but very sensitive to water; used especially in tubing with an outer protective coating

- Vinyl: resistance to the Freon compounds depends on vinyl type and plasticizer, and considerable variation is found; samples should be tested before use

- Orlon-acrylic fiber: generally suitable for use with the Freon compounds

- Nylon: generally suitable for use with Freon compounds but may tend to become brittle at high temperatures in the presence of air or water; tests at 250 degrees F (121 degrees C) with Freon 12 and Freon 22 showed the presence of water or alcohol to be undesirable, so adequate testing should be carried out

- Polyethylene: may be suitable for some applications at room temperatures; however, it should be thoroughly tested since greatly different results have been found with different samples

- Lucite®-acrylic resin (methacrylate polymers): dissolved by Freon 22 but generally suitable for use with Freon 12 and Freon 114 for short exposure; with long exposure it tends to crack, craze, and become cloudy; use with Freon 113 and Freon 11 may be questionable

- Cast Lucite acrylic resin: much more resistant to the effect of solvents than extruded resin; can probably be used with most of the Freon compounds

- Polystyrene: Considerable variation found in individual samples but generally not suited for use with Freon compounds; some applications might be acceptable with Freon 114

- Phenolic resins: usually not affected by Freon compounds but composition of resins of this type may be quite different; samples should be tested before use

- Epoxy resins: resistant to most solvents and entirely suitable for use with the Freon compounds

- Cellulose acetate or nitrate: suitable for use with Freon compounds

- Delrin-acetal resin: suitable for use with Freon compounds under most conditions

- Elastomers: considerable variation is found in the effect of Freon compounds, depending on the particular compound and elastomer type, but in nearly all cases a satisfactory combination can be found; in some instances the presence of other materials, such as oils, may give unexpected results, so preliminary testing of the system involved is recommended

REFRIGERANT PROPERTIES

Refrigerants can be characterized by a number of properties. These properties are pressure, temperature, volume, density, enthalpy, flammability, ability to mix with oil, moisture reaction, odor, toxicity, leakage tendency, and leakage detection.

Freon refrigerants R-11, R-12, and R-22, plus ammonia and water, will be discussed in terms of the above-mentioned categories. Freon R-11, R-12, and R-22 are common Freon refrigerants. The number assigned to ammonia is R-717, while water has the number R-718.

Pressure

The pressure of a refrigeration system is important. It determines how sturdy the equipment must be to hold the refrigerant. The refrigerant must be compressed and sent to various parts of the system under pressure. The main concern is keeping the pressure as low as possible. The ideal low-side, or evaporating, pressure should be as near atmospheric pressure (14.7 pounds per square inch) as possible. This keeps the price of the equipment down. It also puts positive pressure on the system at all points. With a small pressure it is possible to prevent air and moisture from entering the system. In the case of a vacuum or a low pressure, it is possible for a leak to suck in air and moisture. Note the five refrigerants and their pressures in Table 6.5.

Freon R-11 is used in very large systems because it requires more refrigerant than others—even though it has the best pressure characteristics of the group. Several factors must be considered before a suitable refrigerant is found. There is no ideal refrigerant for all applications.

Temperature

Temperature is important in selecting a refrigerant for a particular job. The boiling temperature is that point at which a liquid is vaporized upon the addition of heat. This, of course, depends upon the refrigerant and the absolute pressure at the surface of the liquid and vapor. Note that in Table 6.6, R-22 has the lowest boiling temperature. Water (R-718) has the highest boiling temperature. Atmospheric pressure is 14.7 pounds per square inch.

Once again, there is no ideal atmospheric boiling temperature for a refrigerant. However, temperature–pressure relationships are important in choosing a refrigerant for a particular job.

Volume

Specific volume is defined as the definite weight of a material. Usually expressed in terms of cubic feet per pound, the volume is the reciprocal of the density. The specific volume of a refrigerant is the number of cubic feet of gas that is formed when one pound of the refrigerant is vaporized. This is an important factor to be considered when choosing the size of refrigeration-system components. Compare the specific volumes (at 5 degrees F) of the five refrigerants we have chosen. Freon R-12 and R-22 (the most often used refrigerants) have the lowest specific volumes as vapors. (Refer to Table 6.7.)

Density

Density is defined as the mass or weight per unit of volume. In the case of a refrigerant, it is given in pounds per cubic foot (pounds/cubic foot). Note in Table 6.8 that the density of R-11 is the greatest. The density of R-717 (ammonia) is the least.

TABLE 6.5 Operating pressures.

Refrigerant	Evaporating Pressure (PSIG) at 5°F	Condensing Pressure (PSIG) at 86°F
R-11	24.0 in. Hg	3.6
R-12	11.8	93.2
R-22	28.3	159.8
R-717	19.6	154.5
R-718	29.7	28.6

TABLE 6.6 Refrigerants in order of boiling point.

ASHRAE number	Type of Refrigerant	Class of Refrigerant	Boiling Point °F (°C)
123	Single component	HCFC	82.2 (27.9)
11	Single component	CFC	74.9 (23.8)
245fa	Single component	HFC	59.5 (15.3)
236fa	Single component	HFC	29.5 (-1.4)
134a	Single component	HFC	-15.1 (-26.2)
12	Single component	CFC	-21.6 (-29.8)
401A	Zeotrope	HCFC	-27.7 (-32.2)
500	Azeotrope	CFC	-28.3 (-33.5)
409A	Zeotrope	HCFC	-29.6 (-34.2)
22	Single component	HCFC	-41.5 (-40.8)
407C	Zeotrope	HFC	-46.4 (-43.6)
502	Azcotrope	CFC	-49.8 (-45.4)
408A	Zeotrope	HCFC	-49.8 (-45.4)
404A	Zeotrope	HFC	-51.0 (-46.1)
507	Azetrope	HFC	-52.1 (-46.7)
402A	Zeotrope	HCFC	-54.8 (-48.2)
410A	Zeotrope	HFC	-62.9 (-52.7)
13	Single component	CFC	-114.6 (-81.4)
23	Single component	HFC	-115.7 (-82.1)
508B	Azeotrope	HFC	-125.3 (-87.4)
503	Azeotrope	CFC	-126.1 (-87.8)

Enthalpy

Enthalpy is the total heat in a refrigerant. The sensible heat plus the latent heat makes up the total heat. Latent heat is the amount of heat required to change the refrigerant from a liquid to a gas. The latent heat of vaporization is a measure of the heat per pound that the refrigerant can absorb from an area to be cooled. It is therefore a measure of the cooling potential of the refrigerant circulated through a refrigeration system. (See Table 6.9.) Latent heat is expressed in Btu per pound.

Flammability

Of the five refrigerants mentioned, the only one that is flammable is ammonia. None of the Freon compounds are flammable or explosive. However, mixtures with flammable liquids or gases may be flammable and should be handled with caution. Partially halogenated compounds may also be flammable and must be individually examined. If the refrigerant is used around fire, its flammability should be carefully considered. Some city codes specify which refrigerants cannot be used within city limits.

Capability of Mixing with Oil

Some refrigerants mix well with oil. Others, such as ammonia and water, do not. The ability to mix with oil has advantages and disadvantages. If the refrigerant mixes easily, parts of the system can be lubricated eas-

TABLE 6.7 Specific volumes at 5°F.

Refrigerant	Liquid Volume (cubic feet/lb)	Vapor Volume (cubic feet/lb)
R-11	0.010	12.27
R-12	0.011	1.49
R-22	0.012	1.25
R-717	0.024	8.15
R-718 (water)	0.016	12,444.40

TABLE 6.8   Liquid density at 86°F.

Refrigerant	Liquid Density (lb/ft^3)
R-11	91.4
R-12	80.7
R-22	73.4
R-717	37.2
R-718	62.4

ily by the refrigerant-oil mixture. The refrigerant will bring the oil back to the compressor and moving parts for lubrication.

There is a disadvantage to the mixing of refrigerant and oil. If it is easily mixed, the refrigerant can mix with the oil during the off cycle and then carry off the oil once the unit begins to operate again. This means that the oil needed for lubrication is drawn off with the refrigerant. This can cause damage to the compressor and moving parts. With this condition, there is foaming in the compressor crankcase and loss of lubrication. In some cases the compressor is burned out. Procedures for cleaning a burned-out motor will be given later.

Moisture and Refrigerants

Moisture should be kept out of refrigeration systems, since it can corrode parts of the system. Whenever low temperatures are produced, the water or moisture can freeze. If freezing of the metering device occurs, then refrigerant flow is restricted or cut off. The system will have a low efficiency or none at all. The degree of efficiency will depend upon the amount of icing or the part affected by the frozen moisture.

All refrigerants will absorb water to some degree. Those that absorb very little water permit free water to collect and freeze at low-temperature points. Those that absorb a high amount of moisture will form corrosive acids and corrode the system. Some systems will allow water to be absorbed and frozen. This causes corrosion.

Hydrolysis is the reaction of a material, such as Freon 12 or methyl chloride, with water. Acid materials are formed. The hydrolysis rate for the Freon compounds as a group is low compared with other halogenated compounds. Within the Freon group, however, there is considerable variation. Temperature, pressure, and the presence of other materials also greatly affect the rate. Typical hydrolysis rates for the Freon compounds and other halogenated compounds are given in Table 6.10.

With water alone at atmospheric pressure, the rate is too low to be determined by the analytical method used. When catalyzed by the presence of steel, the hydrolysis rates are detectable but still quite low. At saturation pressures and a higher temperature, the rates are further increased.

Under neutral or acidic conditions, the presence of hydrogen in the molecule has little effect on the hydrolytic stability. However, under alkaline conditions compounds containing hydrogen, such as Freon 22 and Freon 21, tend to be hydrolyzed more rapidly.

TABLE 6.9   Enthalpy (Btu/lb. at 5°F [-15°C]).

Refrigerant	Liquid Enthalpy	+	Latent Heat of Vaporization	=	Vapor Enthalpy
R-11	8.88	+	84.00	=	92.88
R-12	9.32	+	60.47	=	78.79
R-22	11.97	+	93.59	=	105.56
R-717	48.30	+	565.00	=	613.30
R-718 (at 40°F)	8.05	+	1,071.30	=	1,079.35

TABLE 6.10 Hydrolysis rate in water grams/litre of water/year.

Compound	1 atm Pressure 86°F		Saturation Pressure 122°F
	Water Alone	With Steel	With Steel
CH_3Cl	*	*	110
CH_2Cl_2	*	*	55
Freon 113	<0.005	ca. 50[†]	40
Freon 11	<0.005	ca. 10[†]	28
Freon 12	<0.005	0.8	10
Freon 21	<0.01	5.2	9
Freon 114	<0.005	1.4	3
Freon 22	<0.01	0.1	*
Freon 502	<0.01[††]	<0.01[††]	

[*] Not measured
[†] Observed rates vary
[††] Estimated

Odor

The five refrigerants are characterized by odor or the absence of odor. Freon R-11, R-12, and R-22 have a slight odor. Ammonia (R-717) has a very acrid odor and can be detected even in small amounts. Water (R-718), of course, has no odor.

A slight odor is needed in a refrigerant so that leakage can be detected. A strong odor may make it impossible to service equipment. Special gas masks may be needed. Some refrigerated materials may be ruined if the odor is too strong. About the only time that an odor is preferred in a refrigerant is when a toxic material is used. A refrigerant that may be very inflammable should also have an odor so that leakage can be detected easily to prevent a fire or explosion.

Toxicity

Toxicity is the characteristic of a material that makes it intoxicating or poisonous. Some refrigerants can be very toxic to humans; others may not be toxic at all. The halogen refrigerants (R11, R-12, and R-22) are harmless in their normal condition or state. However, they form a highly toxic gas when an open flame is used around them.

Water, of course, is not toxic. Ammonia can be toxic if present in sufficient quantities. Make sure the manufacturer's recommended procedures for handling are followed when working with refrigerants.

Tendency to Leak

The size of the molecule makes a difference in the tendency of a refrigerant to leak. The greater the molecular weight, the larger the hole must be for the refrigerant to escape. A check of the molecular weight of a refrigerant will indicate the problem it may present to a sealed refrigeration system. Table 6.11 shows that R-11 has the least tendency to leak, whereas ammonia is more likely to leak.

TABLE 6.11 Molecular Weight of Selected Refrigerants

Refrigerant	Molecular Weight
R-11	137.4
R-12	120.9
R-22	86.5
R-717 (ammonia)	17.0
R-718 (water)	18.0

DETECTING LEAKS

There are several tests used to check for leaks in a closed refrigeration system. Most of them are simple. Following are some useful procedures:

- Hold the joint or suspected leakage point under water and watch for bubbles
- Coat the area suspected of leakage with a strong solution of soap; if a leak is present, soap bubbles will be produced

Sulfur Dioxide

To detect sulfur dioxide leaks, an ammonia swab may he used. The swab is made by soaking a sponge or cloth, tied onto a stick or piece of wire, in aqua ammonia. Household ammonia may also be used. A dense white smoke forms when the ammonia comes in contact with the sulfur dioxide. The usual soap-bubble or oil test may be used when no ammonia is available.

If ammonia is used, check for leakage in the following ways:

- Burn a sulfur stick in the area of the leak: if there is a leak, a dense white smoke will be produced; the stronger the leak, the denser the white smoke
- Hold a wet litmus paper close to the suspected leak area: if there is a leak, the ammonia will cause the litmus paper to change color.

Freon

Refrigerants that are halogenated hydrocarbons (Freon compounds) can be checked for leakage with a halide leak test. This involves holding a torch or flame close to the leak area. If there is a refrigerant leak, the flame will turn green. The room should be well ventilated when the torch test is performed.

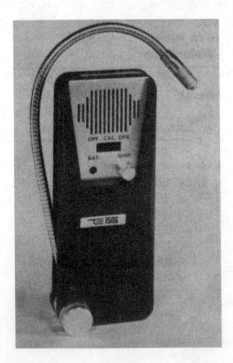

FIGURE 6.2 A hand-held electronic leak detector. *(Thermal Engineering)*

FIGURE 6.3   A halide gas leak detector. *(Turner)*

There is also an electronic detector for such refrigerant leaks. The detector gives off a series of rapid clicks if the refrigerant is present. The higher the concentration of the refrigerant, the more rapid the clicks. (See Figure 6.2.)

Carbon Dioxide

Leaks can be detected with a soap solution if there is internal pressure on the part to be tested. When carbon dioxide is present in the condenser water, the water will turn yellow with the addition of bromothymol blue.

Ammonia

Leaks are detected (in small amounts of ammonia) when a lit sulfur candle is used. The candle will give off a very thick, white smoke when it contacts the ammonia leak. The use of phenolphthalein paper is also considered a good test. The smallest trace of ammonia will cause the moistened paper strip to turn pink. A large amount of ammonia will cause the paper to turn a vivid scarlet.

Methyl Chloride

Leaks are detected by a leak-detecting halide torch. (See Figure 6.3.) Some torches use alcohol for fuel and produce a colorless flange. When a methyl-chloride leak is detected, the flame turns green. A brilliant blue flame is produced when large or stronger concentrations are present. The room should be well ventilated when the torch test is conducted. The combustion of the refrigerant and the flame produces harmful chemicals. If a safe atmosphere is not present, the soap-bubble or oil test should be used to check for leaks.

As mentioned, methyl chloride is hard to detect with the nose or eyes. It does not produce irritating effects. Therefore, some manufacturers add a 1 percent amount of acrolein as a warning agent. Acrolein is a colorless liquid (C_3H_4O) with a pungent odor.

OZONE-DEPLETING REFRIGERANTS

In 1987 the Montreal Protocol, an international environmental agreement, established requirements that began the worldwide phaseout of ozone-depleting chlorofluorocarbons. These requirements were later modified, leading to the phaseout, in 1996, of CFC production in all developed nations. In 1992 an amendment to the Montreal Protocol established a schedule for the phaseout of HCFCs (hydrochlorofluorocarbons).

HCFCs are substantially less damaging to the ozone layer than CFCs. However, they still contain ozone-destroying chlorine. The Montreal Protocol, as amended, is carried out in the U.S. through Title VI of the Clean Air Act. This act is implemented by the Environmental Protection Agency.

An HCFC known as R-22 has been the refrigerant of choice for residential heat-pump and air-conditioning systems for more than four decades. Unfortunately for the environment, release of R-22 resulting from system leaks contributes to ozone depletion. In addition, the manufacture of R-22 results in a byproduct that contributes significantly to global warming.

As the manufacture of R-22 is phased out over the coming years as part of the agreement to end production of HCFCs, manufacturers of residential air-conditioning systems are beginning to offer equipment that uses ozone-friendly refrigerants. Many homeowners may be misinformed about how much longer R-22 will be available to service their central A/C systems and heat pumps. The EPA document assists consumers in deciding what to consider when purchasing a new A/C system or heat pump or repairing an existing system.

Under the terms of the Montreal Protocol, the U. S. agreed to meet certain obligations by specific dates. These will affect the residential heat-pump and air -conditioning industry.

In accordance with the terms of the protocol, the amount of all HCFCs that can be produced nationwide was to be reduced by 35 percent by January 1, 2004. In order to achieve this goal, the U.S. ceased production of HCFC-141b, the most ozone-damaging of this class of chemicals, on January 1, 2003. This production ban should greatly reduce nationwide use of HCFCs as a group and make it likely that the 2004 deadline will have a minimal effect on R-22 supplies.

After January 1, 2010, chemical manufacturers may still produce R-22 to service existing equipment but not for use in new equipment. As a result, heating, ventilation and air-conditioning (HVAC) manufacturers will only be able to use preexisting supplies of R-22 in the production of new air conditioners and heat pumps. These existing supplies will include R-22 recovered from existing equipment and recycled by licensed reclaimers.

Use of existing refrigerant, including refrigerant that has been recovered and recycled, will be allowed beyond January 1, 2020 to service existing systems. However, chemical manufacturers will no longer be able to produce R-22 to service existing air conditioners and heat pumps.

Implications for Consumers

What does the R-22 phase out mean for consumers? The following paragraphs are an attempt to answer this question.

The Clean Air Act does not allow any refrigerant to be vented into the atmosphere during installation, service, or retirement of equipment. Therefore, R-22 must be:

- recovered and recycled (for reuse in the same system)
- reclaimed (reprocessed to the same purity levels as new R-22)
- destroyed

After 2020 the servicing of R-22-based systems will rely on recycled refrigerants. It is expected that reclamation and recycling will ensure that existing supplies of R-22 will last longer and be available to service a greater number of systems. As noted above, chemical manufacturers will be able to produce R-22 for use in new A/C equipment until 2010, and they can continue production of R-22 until 2020 for use in servicing that equipment. Given this schedule, the transition away from R-22 to the use of ozone-friendly refrigerants should be smooth. For the next 20 years or more, R-22 should continue to be available for all systems that require R-22 for servicing.

While consumers should be aware that prices of R-22 may increase as supplies dwindle over the next 20 or 30 years, the EPA believes that consumers are not likely to be subjected to major price increases within a short time period. Although there is no guarantee that service costs of R-22 will not increase, the lengthy phaseout period means that market conditions should not be greatly affected by the volatility and resulting refrigerant price hikes that have characterized the phaseout of R-12, the refrigerant used in automotive air-conditioning systems, which has been replaced by R-134a.

Alternatives for residential air conditioning will be needed as R-22 is gradually phased out. Non-ozone-depleting alternative refrigerants are being introduced. Under the Clean Air Act, EPA reviews alternatives to ozone-depleting substances like R-22 in order to evaluate their effects on human health and the environment. The EPA has reviewed several of these alternatives and has compiled a list of acceptable substitutes. One of these substitutes is R-410A, a blend of hydrofluorocarbon (HFC) substances that does not contribute to depletion of the ozone layer but, like R-22, does contribute to global warming. R-410A is manufactured and sold under various trade names, including Genetron AZ 20, SUVA 410A®, and Puron. Additional refrigerants on the list of acceptable substitutes include R-134a and R-407C. These two refrigerants are not yet available for residential applications in the U.S. but are commonly found in residential A/C systems and heat pumps in Europe. EPA will continue to review new non-ozone-depleting refrigerants as they are developed.

Existing units using R-22 can continue to be serviced with R-22. There is no EPA requirement to change or convert R-22 units for use with a non-ozone-depleting substitute refrigerant. In addition, the new substitute refrigerants cannot be used without making some changes to system components. As a result, service technicians who repair leaks will continue to charge R-22 into the system as part of that repair.

The transition away from ozone-depleting R-22 to systems that rely on replacement refrigerants like R-410A has required redesign of heat-pump and air-conditioning systems. New systems incorporate compressors and other components specially designed for use with specific replacement refrigerants. With these significant product and production process changes, testing and training must also change. Consumers should be aware that dealers of systems that use substitute refrigerants should be schooled in installation and service techniques required for use of that substitute refrigerant.

Servicing Your System

Along with prohibiting the production of ozone-depleting refrigerants, the Clean Air Act also mandates the use of common sense in handling refrigerants. By containing and using refrigerants responsibly—that is, by recovering, recycling, and reclaiming and by reducing leaks—their ozone-depletion and global-warming consequences are minimized. The Clean Air Act outlines specific refrigerant containment and management practices for HVAC manufacturers, distributors, dealers, and technicians. Properly installed home-comfort systems rarely develop refrigerant leaks, and with proper servicing a system using R-22, R-410A, or another refrigerant will minimize its impact on the environment. While EPA does not mandate repairing or replacing small systems because of leaks, system leaks can not only harm the environment but also result in increased maintenance costs.

One important thing a homeowner can do for the environment, regardless of the refrigerant used, is to select a reputable dealer that employs service technicians who are EPA-certified to handle refrigerants. Technicians often call this certification "Section 608 certification," referring to the part of the Clean Air Act that requires minimizing releases of ozone-depleting chemicals from HVAC equipment.

Purchasing New Systems

Another important thing a homeowner can do for the environment is to purchase a highly energy-efficient system. Energy-efficient systems result in cost savings for the homeowner. Today's best air conditioners use much less energy to produce the same amount of cooling as air conditioners made in the mid-1970s. Even if your air conditioner is only 10 years old, you may save significantly on your cooling energy costs by replacing it with a newer, more efficient model. Products with EPA's "Energy Star" label can save homeowners 10 to 40 percent on their heating and cooling bills every year. These products are made by most major manufacturers and have the same features as standard products but also incorporate energy saving technology. Both R-22 and R-410A systems may have the Energy Star® label. Equipment that displays the Energy Star® label must have a minimum seasonal energy efficiency ratio (SEER). The higher the SEER specification, the more efficient the equipment.

Energy efficiency, along with performance, reliability, and cost, should be considered in making a decision. And don't forget that when purchasing a new system, you can also speed the transition away from ozone-depleting R-22 by choosing a system that uses ozone-friendly refrigerants.

Recycling Refrigerants

Several regulations have been issued under Section 608 of the Clean Air Act to govern the recycling of refrigerants in stationary systems and to end the practice of venting refrigerants to the air. These regulations also govern the handling of halon fire-extinguishing agents. A Web site and both the regulations themselves and fact sheets are available from the EPA Stratospheric Ozone Hotline at 1-800-296-1996. The handling and recycling of refrigerants used in motor-vehicle air-conditioning systems is governed under section 609 of the Clean Air Act.

In 2005 EPA finalized a rule amending the definition of refrigerant to make certain that it only includes substitutes that consist of a class I or class II ozone-depleting substance (ODS). This rule also amended the venting prohibition to make certain that it remains illegal to knowingly vent non exempt substitutes that do not consist of a class I or class II ODS, such as R-134a and R-410A.

In the same year EPA published a final rule extending the required leak-repair practices and the associated reporting and record-keeping requirements to owners and/or operators of comfort-cooling, commercial-refrigeration, or industrial-process refrigeration appliances containing more than 50 pounds of a substitute refrigerant if the substitute contains a class I or class II ozone-depleting substance (ODS). In addition, EPA defined leak rate in terms of the percentage of the appliance's full charge that would be lost over a consecutive 12-month period if the current rate of loss were to continue over that period. EPA now requires calculation of the leak rate whenever a refrigerant is added to an appliance.

In 2004 EPA finalized a rule sustaining the Clean Air Act prohibition against venting hydrofluorocarbon (HFC) and perfluorocarbon (PFC) refrigerants. This rule makes the knowing venting of HFC and PFC refrigerants during the maintenance, service, repair, and disposal of air-conditioning and refrigeration equipment (i.e., appliances) illegal under Section 608 of the Clean Air Act. The ruling also restricts the sale of HFC refrigerants that consist of an ozone-depleting substance (ODS) to EPA-certified technicians. However, HFC refrigerants and HFC refrigerant blends that do not consist of an ODS are not covered under "The Refrigerant Sales Restriction," a brochure that documents the environmental and financial reasons to replace CFC chillers with new, energy-efficient equipment. A partnership of governments, manufacturers, NGOs (nongovernmental organizations), and others have endorsed the brochure to eliminate uncertainty and underscore the wisdom of replacing CFC chillers.

TABLE 6.12 Trigger Leak Rates

Appliance Type	Trigger Leak Rate
Commercial refrigeration	35%
Industrial process refrigeration	35%
Comfort cooling	15%
All other appliances	15%

Leak Repair

The leak-repair requirements, promulgated under Section 608 of the Clean Air Act Amendments of 1990, are that when an owner or operator of an appliance that normally contains a refrigerant charge of more than 50 pounds discovers that the refrigerant is leaking at a rate that would exceed the applicable trigger rate during a 12-month period, the owner or operator must take corrective action.

For all appliances that have a refrigerant charge of more than 50 pounds, the trigger leak rates for a 12-month period shown in Table 6.12 are applicable (see Table 6.12).

In general, owners or operators must either repair leaks within 30 days from the date the leak was discovered or develop a dated retrofit/retirement plan within 30 days and complete actions under that plan within one year from the plan's date. However, for industrial-process refrigeration equipment and some federally owned chillers, additional time may be available.

Industrial-process refrigeration is defined as complex customized appliances used in the chemical, pharmaceutical, petrochemical, and manufacturing industries. These appliances are directly linked to the industrial process. This sector also includes industrial ice machines, appliances used directly in the generation of electricity, and ice rinks. If at least 50 percent of an appliance's capacity is used in an industrial-process refrigeration application, the appliance is considered industrial-process refrigeration equipment and the trigger rate is 35 percent.

Industrial-process refrigeration equipment and federally owned chillers must conduct initial and follow-up verification tests at the conclusion of any repair efforts. These tests are essential to ensure that the repairs have been successful. In cases where an industrial-process shutdown is required, a repair period of 120 days is substituted for the normal 30-day repair period. Any appliance that requires additional time may be subject to record-keeping/reporting requirements.

Additional time is permitted for conducting leak repairs if the necessary repair parts are unavailable or if other applicable federal, state, or local regulations make a repair within 30 or 120 days impossible. If own-

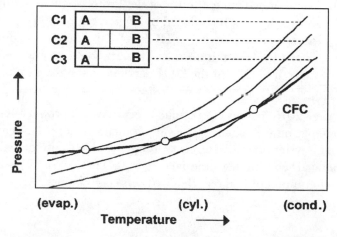

FIGURE 6.4 New refrigerants variable: composition. *(National Refrigerants)*

ers or operators choose to retrofit or retire appliances, a retrofit or retirement plan must be developed within 30 days of detecting a leak rate that exceeds the trigger rates. A copy of the plan must be kept on site. The original plan must be made available to EPA upon request. Activities under the plan must be completed within 12 months from the date of the plan. If a request is made within 6 months from the expiration of the initial 30-day period, additional time beyond the 12-month period is available for owners or operators of industrial-process refrigeration equipment and federally owned chillers in the following cases: EPA will permit additional time to the extent reasonably necessary if a delay is caused by the requirements of other applicable federal, state, or local regulations or if a suitable replacement refrigerant, in accordance with the regulations promulgated under Section 612, is not available; and EPA will permit one additional 12-month period if an appliance is custom-built and the supplier of the appliance or a critical component has quoted a delivery time of more than 30 weeks from when the order was placed (assuming the order was placed in a timely manner). In some cases EPA may provide additional time beyond this extra year where a request is made by the end of the ninth month of the extra year.

The owners or operators of industrial-process refrigeration equipment or federally owned chillers may be relieved from the retrofit or repair requirements if:

- second efforts to repair the same leaks that were subject to the first repair efforts are successful
- within 180 days of the failed follow-up verification test, the owners or operators determine the leak rate is below 35 percent; in this case, the owners or operators must notify EPA as to how this determination was made and submit the information within 30 days of the failed verification test

For all appliances subject to the leak-repair requirements, the timelines may be suspended if the appliance has undergone system mothballing. System mothballing means the intentional shutting down of a refrigeration appliance undertaken for an extended period of time where the refrigerant has been evacuated from the appliance or the affected isolated section of the appliance to at least atmospheric pressure. However, the timelines pick up again as soon as the system is brought back online.

PHASE-OUT SCHEDULE FOR HCFCS, INCLUDING R-22

Under the terms of the Montreal Protocol, the United States agreed to meet certain obligations by specific dates. That will affect the residential heat pump and air-conditioning industry.

January 1, 2004 In accordance with the terms of the Protocol, the amount of all the HCFCs that can be produced nationwide must be reduced by 35 percent by 2004. In order to achieve this goal, the United States has ceased production of HCFC-141 b, the most ozone damaging of this class of chemicals, on January 1, 2003. this production ban should greatly reduce nationwide me of HCFCs as a group and make it likely that the 2004 deadline will have a minimal effect on R-22 supplies.

January 1, 2010 After 2010, chemical manufacturers may still produce R-22. But this is to service existing equipment and not for use in new equipment. As a result, heating, ventilation and air-conditioning (HVAC) system manufacturers will only be able to use preexisting supplies of R-22 in the production of new air conditioners and heat pumps. These existing supplies will include R-22 recovered from existing equipment and recycled by licensed reclaimers.

January 1, 2020   Use of existing refrigerant, including refrigerant that has been recovered and re-cycled, will be allowed beyond 2020 to service existing systems. However, chemical manufacturers will no longer be able to produce R-22 to service existing air conditioners and heat pumps.

What does the R-22 phase out mean for consumers? The following paragraphs are an attempt to answer this question.

AVAILABILITY OF R-22

The Clean Air Act does not allow any refrigerant to be vented into the atmosphere during installation, service, or retirement of equipment. Therefore, R-22 must be:

- Recovered and recycled (for reuse in the same system)
- Reclaimed (reprocessed to the same purity levels as new R-22)
- Destroyed

After 2020, the servicing of R-22-based systems will rely on recycled refrigerants. It is expected that reclamation and recycling will ensure that existing supplies of R-22 will last longer and be available to service a greater number of systems. As noted earlier, chemical manufacturers will be able to produce R-22 for use in new air-conditioning equipment until 2010, and they can continue production of R-22 until 20,20 for use in servicing that equipment. Given this schedule, the transition away from R-22 to the use of ozone-friendly refrigerants should be smooth. For the next 20 years or more, R-22 should continue to be available for all systems that require R-22 for servicing.

MULTIPLE-CHOICE EXAM

1. Chemicals used for refrigerants can be broken down into two groups:
    a. organic and inorganic
    b. liquid and solid
    c. water and ice
    d. none of these

2. Hydrocarbons include propane, ethane, and _____.
    a. methane
    b. isobutane
    c. butane
    d. ethane

3. R-12 refrigerant boils at _____ degrees F.
    a. 212
    b. −21.6
    c. −28.3
    d. 32

4. A refrigerant with a low boiling point will have a _____ vapor pressure and vice versa.
    a. high
    b. low
    c. cold
    d. medium

5. HCFC-22 is almost identical to _____.
    a. HFC-22
    b. HFC-12
    c. CFC-12
    d. CFC-13

6. Future refrigerants will be limited to _____ with no chlorine.
    a. LCPs
    b. CFCs
    c. HCFCs
    d. HFCs

7. ASHRAE 34 is the standard for which the International Mechanical Code classifies _____.
    a. refrigerants
    b. water
    c. liquids
    d. solids

8. Recovered refrigerants must be filtered and _____ before being used to recharge the unit from which removed.
    a. dried
    b. reconfigured
    c. decontaminated
    d. purified

9. Azeotropes do not exhibit the temperature glides of zeotropic mixtures. New HFC azeotropes are now being accepted in _____ use.

   a. commercial
   b. industrial
   c. agricultural
   d. residential

10. Temperature glide is the difference between the dew point and the bubble points of the refrigerant at any given _____.

    a. time
    b. pressure
    c. suction
    d. point

11. Which of these flammability classifications of refrigerants is based on ASHRAE Standard 34-2001?

    a. Class 1 nonflammable
    b. Class 2 moderately flammable
    c. Class 3 highly flammable
    d. none of these
    e. all of these

12. The reliability of a refrigeration system is largely dependent on the hardware design, installation, and _____.

    a. application
    b. utilization
    c. lubrication system
    d. refrigerant used

13. What refrigerant can be either a colorless gas or liquid?

    a. sulfur trioxide
    b. sulfur oxide
    c. sulfur dioxide
    d. sulfur

14. Which refrigerant is used most frequently in large industrial plants?

    a. ammonia
    b. sulfur dioxide
    c. calcium chloride
    d. carbon dioxide

15. Which refrigerant is not commonly used in domestic refrigeration units?

    a. ethyl chloride
    b. methyl chloride
    c. ammonia
    d. water

16. Inhalation of concentrated fluorocarbon vapors can lead to CNS or central _____ system effects such as those encountered during general anesthesia.

    a. nonsense
    b. narcosis
    c. nervous
    d. noncondensable

17. The oil separator collects _____ oil that escapes with the compressor-discharge gas.

    a. engine   b. lubricating

    c. compressor   d. condenser

18. What are three basic properties of lubricating oils?

    a. viscosity, lubricity, and chemical stability

    b. viscosity, lubricity, and chemicals

    c. viscosity, density, and slickness

    d. viscosity, open cells, and water

19. The blended refrigerant R-134a is a long-term, HFC alternative with similar properties to: _____.

    a. R-12   b. R-22

    c. R-20   d. R-21

20. Where do you commonly find R-134a refrigerant used today?

    a. in ice-cream freezers   b. in truck refrigeration systems

    c. in commercial installations   d. in automobile A/C systems

21. What is another refrigerant, other than 134a, that can be substituted for R-12?

    a. R-401A   b. R-22

    c. R-414   d. R-717

22. When working with refrigerants the use of modular, solid-state electronics, and built-in diagnostic testing reduces the amount of _____ analysis that must be performed in order to isolate malfunctions in the latest equipment.

    a. assistant   b. work

    c. refrigerant   d. trouble

23. What does toxicity mean?

    a. nonpoisonous   b. safe to drink

    c. poisonous   d. safe to breathe

24. Where will technicians get their Freon after 2020?

    a. EPA   b. recyclers

    c. manufacturers   d. local wholesalers

25. What does the Montreal Protocol have to do with the future of refrigerants?

    a. little or nothing   b. changes in composition to eliminate chloride

    c. stops production altogether   d. none of the above

26. What kind of toxic gas is produced when Freon is heated?

    a. propane   b. boron

    c. chlorine   d. phosgene

27. Which refrigerant is a blend of R-22, R-125, and R-290?

    a. R-402A   b. R-500

    c. R-400   d. R-22

28. Refrigerant cylinders should be stored in a dry, sheltered, and well- _____ area.

    a. ventilated   b. known

    c. built   d. saturated

29. All common refrigerants have satisfactorily high critical temperatures except carbon dioxide and _____.

    a. ethane   b. butane

    c. propane   d. methane

30. The refrigerant effect per pound of refrigerant under standard ton conditions determines the amount of refrigerant to be evaporated per _____.

    a. minute   b. hour

    c. second   d. day

TRUE-FALSE EXAM

1.   Higher-pressure refrigerants can also be effectively utilized in centrifugal chillers.
     True   False

2.   Vapor-compression systems can use a wide variety of refrigerants.
     True   False

3.   Virtually all vapor-compression refrigeration systems require lubricants to permit reliable compressor operation.
     True   False

4.   As a refrigerant, water has the disadvantage of low freezing and boiling points.
     True   False

5.   An azeotropic refrigerant is only a true azeotrope at one temperature for a given composition.
     True   False

6.   Generally speaking, refrigerants with a low boiling point have more favorable thermodynamic properties.
     True   False

7.   Flammability of a refrigerant is classified by ASHRAE Standard 34-2001.
     True   False

8.   For automotive air conditioning many of the controls and safety switches are related to the high-side pressure.
     True   False

9.   Miscibility defines the temperature region where refrigerant and oil will mix or separate.
     True   False

10. Solubility determines whether the refrigerant will over-thicken the oil.
     True   False

11. One of the requirements of an ideal refrigerant is that it must be nontoxic.
     True   False

12. Vaporized refrigerants such as ammonia and sulfur dioxide bring about irritation and congestion of the lungs and bronchial organs and produce violent coughing and vomiting.
True   False

13. In refrigeration it is often an easier and cheaper retrofit job to match evaporator pressures to R-12 and split the glide.
True   False

14. You should store full and empty refrigerant cylinders away from one another in order to eliminate confusion.
True   False

15. It is OK to drop refrigerant cylinders in the process of recharging a system.
True   False

MULTIPLE-CHOICE ANSWERS

1.	A	7.	A	13.	C	19.	A	25.	B
2.	B	8.	A	14.	A	20.	D	26.	D
3.	B	9.	A	15.	A	21.	A	27.	A
4.	A	10.	D	16.	C	22.	D	28.	A
5.	B	11.	E	17.	B	23.	C	29.	A
6.	D	12.	A	18.	A	24.	B	30.	A

TRUE-FALSE ANSWERS

1.	T	5.	T	9.	T	13.	T
2.	T	6.	F	10.	F	14.	T
3.	T	7.	T	11.	T	15.	F
4.	F	8.	T	12.	T		

Chapter 7

COOLING AND DISTRIBUTION SYSTEMS

The condenser is a heat-transfer device. It is used to remove heat from hot refrigerant vapor. Using some method of cooling, the condenser changes the vapor to a liquid. There are three basic methods of cooling the condenser's hot gases. The method used to cool the refrigerant and return it to the liquid state serves to categorize the two types of condensers: air-cooled and water-cooled. Cooling towers are also used to cool the refrigerant.

Most commercial or residential home air-conditioning units are air-cooled. Water is also used to cool the refrigerant. This is usually done if there is an adequate supply of fairly clean water. Industrial applications rely upon water to cool the condenser gases. The evaporative process is also used to return the condenser gases to the liquid state. Cooling towers use the evaporative process.

CONDENSERS

Air-Cooled Condensers

Figure 7.1 illustrates the refrigeration process within an air-cooled condenser. Figure 7.2 shows some of the various types of compressors and condensers mounted as a unit. These units may be located outside the cooled space. Such a location makes it possible to exhaust the heated air from the cooled space. Note that the condenser has a large-bladed fan that pushes air through the condenser fins. The fins are attached to coils of copper or aluminum tubing. The tubing houses the liquid and the gaseous vapors. When the blown air contacts the fins, it cools them. The heat from the compressed gas in the tubing is thus transferred to the cooler fin.

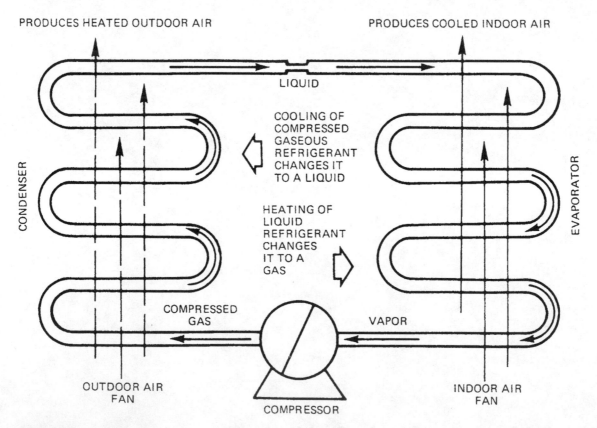

FIGURE 7.1 Refrigeration cycle.

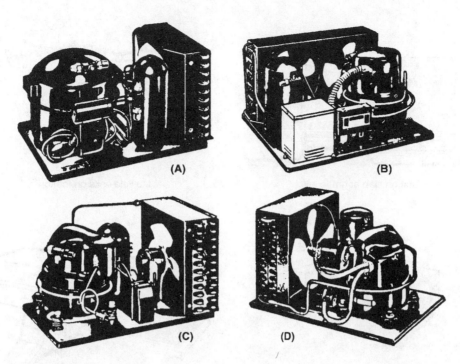

FIGURE 7.2   A condenser, fan, and compressor, self-contained in one unit. *(Tecumseh)*

Heat given up by the refrigerant vapor to the condensing medium includes both the heat absorbed in the evaporator and the heat of compression. Thus, the condenser always has a load that is the sum of these two heats. This means that the compressor must handle more heat than that generated by the evaporator. The quantity of heat (in Btu) given off by the condenser is rated in heat per minute per ton of evaporator capacity. These condensers are rated at various suction and condensing temperatures.

The larger the condenser area exposed to the moving airstream, the lower the temperature of the refrigerant when it leaves the condenser. The temperature of the air leaving the vicinity of the condenser will vary with the load inside the area being cooled. If the evaporator picks up the additional heat and transfers it to the condenser, the condenser must transmit this heat to the air passing over the surface of the fins. The temperature rise in the condensing medium passing through the condenser is directly proportional to the condenser load. It is inversely proportional to the quantity and specific heat of the condensing medium.

To exhaust the heat without causing the area being cooled to heat up again, it is common practice to locate the condenser outside of the area being conditioned. For example, for an air-conditioned building, the condenser is located on the rooftop or on an outside slab at grade level. (See Figure 7.3.)

Some condensers are cooled by natural airflow. This is the case for domestic refrigerators. Such natural convection condensers use either plate surface or finned tubing. (See Figure 7.4.)

Air-cooled condensers that use fans are classified as chassis-mounted and remote. The chassis-mounted type is shown in Figure 7.2. The compressor, fan, and condenser are mounted as one unit. The remote type is shown in Figure 7.3. Remote air-cooled condensers can be obtained in sizes that range from 1 to 100 tons. The chassis-mounted type is usually limited to 1 ton or less.

Water-Cooled Condensers

Water is used to cool condensers. One method is to cool them with water from the city water supply and then exhaust the water into the sewer. This method can be expensive and in some instances is not allowed by law. When there is a sewer problem, a limited sewer-treatment-plant capacity, or drought, it is impractical to use this cooling method.

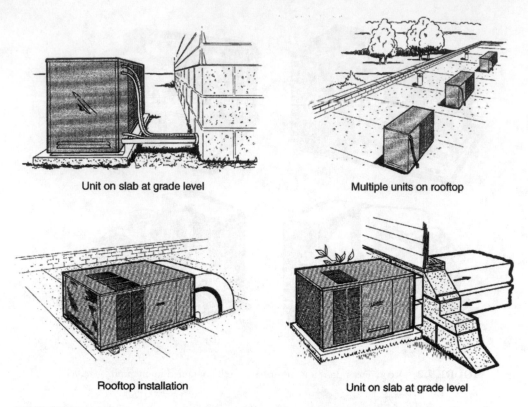

Unit on slab at grade level

Multiple units on rooftop

Rooftop installation

Unit on slab at grade level

FIGURE 7.3 Condensers mounted on rooftops and at grade level. *(Lennox)*

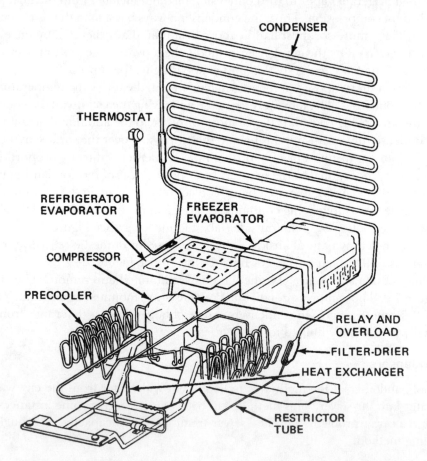

CONDENSER

THERMOSTAT

REFRIGERATOR
EVAPORATOR

FREEZER
EVAPORATOR

COMPRESSOR

PRECOOLER

RELAY AND
OVERLOAD

FILTER-DRIER

HEAT EXCHANGER

RESTRICTOR
TUBE

FIGURE 7.4 Flat, coil-type condenser, with natural air circulation, used in refrigeration in the home. *(Sears)*

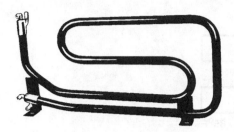

FIGURE 7.5 Co-axial, water-cooled condenser, used with refrigeration and air-conditioning units where space is limited.

The use of recirculation to cool the water for reuse is more practical. However, in recirculation, the power required to pump the water to the cooling location is part of the expense of operating the unit.

There are three types of water-cooled condensers:

•   Double-tube

•   Shell and coil

•   Shell and tube

The double-tube type consists of two tubes, one inside the other (see Figure 7.5). Water is piped through the inner tube. Refrigerant is piped through the tube that encloses the inner tube. The refrigerant flows in the opposite direction from the water (see Figure 7.6).

This type of coaxial water-cooled condenser is designed for use with refrigeration and air-conditioning condensing units where space is limited. These condensers can be mounted vertically, horizontally, or at any angle.

This type of condenser can also be used with cooling towers. It performs at peak heat of rejection with a water-pressure drop of not more than five pounds per square inch, utilizing flow rates of three gallons per minute per ton.

The typical counterflow path shows the refrigerant going in at 105 degrees F (41 degrees C) and the water at 85 degrees F (30 degrees C) and leaving at 95 degrees F (35 degrees C) (see Figure 7.7).

The counterswirl design, shown in Figure 7.6, gives heat-transfer performance of superior quality. The tube construction provides for excellent mechanical stability. The water-flow path is turbulent. This provides a scrubbing action that maintains cleaner surfaces. The construction method shown also has very high system pressure resistance.

The water-cooled condenser shown in Figure 7.5 can be obtained in a number of combinations. Some of these combinations are listed in Table 7.1. Copper tubing is suggested for use with fresh water and with cooling towers. The use of cupronickel is suggested when salt water is used for cooling purposes.

Convolutions to the water tube result in a spinning, swirling water flow that inhibits the accumulation of deposits on the inside of the tube. This contributes to the antifouling characteristics in this type of condenser. Figure 7.8 shows the various types of construction.

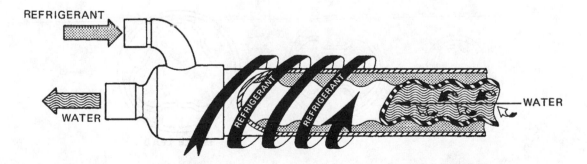

FIGURE 7.6 A typical counterflow path inside a coaxial water-cooled condenser. *(Packless)*

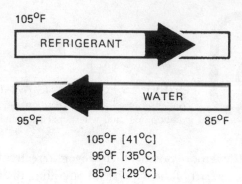

105°F [41°C]
95°F [35°C]
85°F [29°C]

FIGURE 7.7 Water and refrigerant temperatures in a counterflow, water-cooled condenser. *(Packless)*

TABLE 7.1 Some possible metal combinations in water-cooled condensers.

Shell Metal	Tubing Metal
Steel	Copper
Copper	Copper
Steel	Cupronickel
Copper	Cupronickel
Steel	Stainless steel
Stainless steel	Stainless steel

This type of condenser may be added as a booster to standard air-cooled units. Figure 7.9 shows some of the configurations of this type of condenser—the spiral, helix, and trombone. Note the input for the water and for the refrigerant. Using a cooling tower to furnish water to contact the outside tube can further cool the condensers. A water tower can also be used to cool the water sent through the inside tube for cooling purposes. This type of condenser is usable where refrigeration or air-conditioning requirements are 1/3 to 3 tons.

Placing a bare or finned tube inside a steel shell makes the shell-and-coil condenser (see Figure 7.10). Water circulates through the coils. Refrigerant vapor is injected into the shell. The hot vapor contacts the cooler tubes and condenses. The condensed vapor drains from the coils and drops to the bottom of the tank or shell. From there it is recirculated through the refrigerated area by way of the evaporator. In most cases, placing chemicals into the water cleans the unit. The chemicals have a tendency to remove the deposits that build up on the tubing walls.

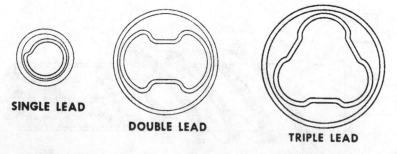

FIGURE 7.8 Different types of tubing fabrication, located inside the coaxial-type water-cooled condenser. *(Packless)*

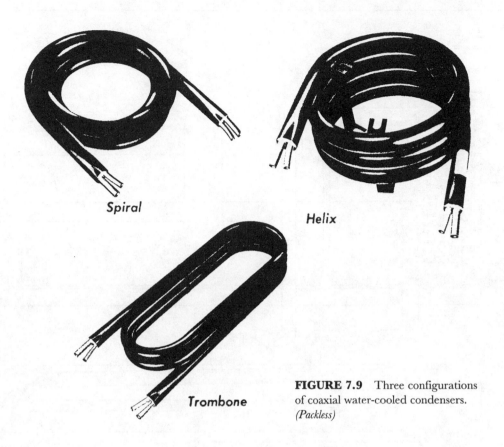

Spiral

Helix

Trombone

FIGURE 7.9 Three configurations of coaxial water-cooled condensers. *(Packless)*

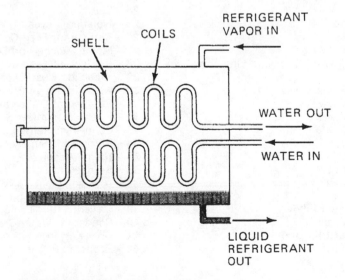

FIGURE 7.10 The shell-and-coil condenser.

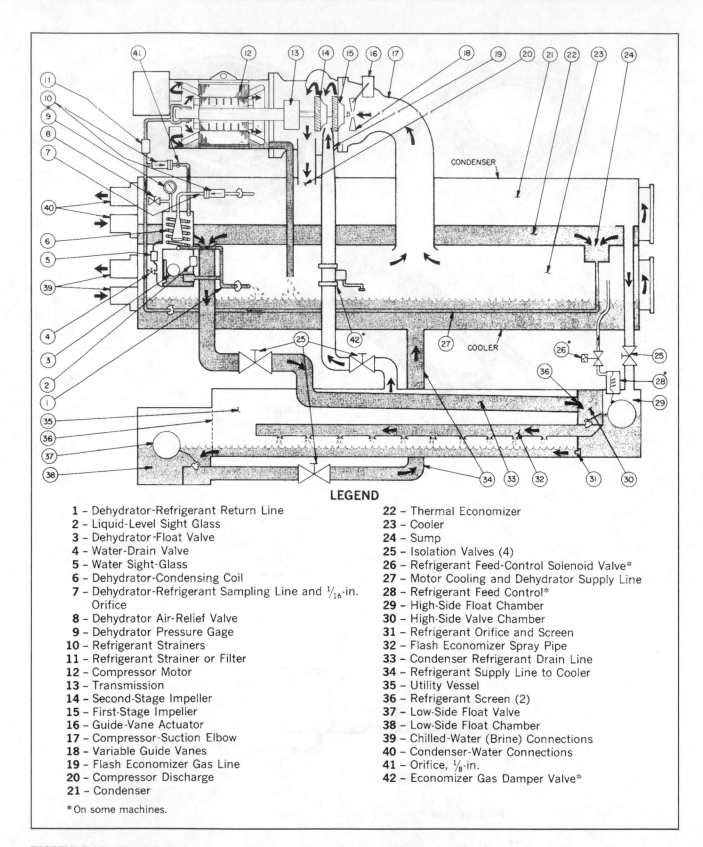

LEGEND

1 – Dehydrator-Refrigerant Return Line
2 – Liquid-Level Sight Glass
3 – Dehydrator-Float Valve
4 – Water-Drain Valve
5 – Water Sight-Glass
6 – Dehydrator-Condensing Coil
7 – Dehydrator-Refrigerant Sampling Line and $\frac{1}{16}$-in. Orifice
8 – Dehydrator Air-Relief Valve
9 – Dehydrator Pressure Gage
10 – Refrigerant Strainers
11 – Refrigerant Strainer or Filter
12 – Compressor Motor
13 – Transmission
14 – Second-Stage Impeller
15 – First-Stage Impeller
16 – Guide-Vane Actuator
17 – Compressor-Suction Elbow
18 – Variable Guide Vanes
19 – Flash Economizer Gas Line
20 – Compressor Discharge
21 – Condenser

22 – Thermal Economizer
23 – Cooler
24 – Sump
25 – Isolation Valves (4)
26 – Refrigerant Feed-Control Solenoid Valve*
27 – Motor Cooling and Dehydrator Supply Line
28 – Refrigerant Feed Control*
29 – High-Side Float Chamber
30 – High-Side Valve Chamber
31 – Refrigerant Orifice and Screen
32 – Flash Economizer Spray Pipe
33 – Condenser Refrigerant Drain Line
34 – Refrigerant Supply Line to Cooler
35 – Utility Vessel
36 – Refrigerant Screen (2)
37 – Low-Side Float Valve
38 – Low-Side Float Chamber
39 – Chilled-Water (Brine) Connections
40 – Condenser-Water Connections
41 – Orifice, $\frac{1}{8}$-in.
42 – Economizer Gas Damper Valve*

*On some machines.

FIGURE 7.11   The chiller, compressor, condenser, and cooler are combined in one unit. *(Carrier)*

CHILLERS

A chiller is part of a condenser. Chillers are used to cool water or brine solutions. The cooled (chilled) water or brine is then fed through pipes to evaporators. This cools the area in which the evaporators are located. This type of cooling, using chilled water or brine, can be used in large air-conditioning units. It can also be used for industrial processes where cooling is required for a particular operation.

Figure 7.11 illustrates such an operation. Note how the compressor sits atop the condenser. Chillers are the answer to requirements of 200 to 1,600 tons of refrigeration. They are used for process cooling, comfort air conditioning, and nuclear-power-plant cooling. In some cases they are used to provide ice for skating rinks. The arrows in Figure 7.11 indicate the refrigerant flow and the water or brine flow through the large pipes. Figure 7.12 shows the machine in a cutaway view. The following explanation of the various cycles will provide a better understanding of the operation of this type of equipment.

Refrigeration Cycle

The machine compressor continuously draws large quantities of refrigerant vapor from the cooler at a rate determined by the size of the guide vane opening. This compressor suction reduces the pressure within the cooler, allowing the liquid refrigerant to boil vigorously at a fairly low temperature, typically 30 to 35 degrees F (-1 to 2 degrees C).

Liquid refrigerant obtains the energy needed for the change to vapor by removing heat from the water in the cooler tubes. The cold water can then be used in the air-conditioning process.

After removing heat from the water, the refrigerant vapor enters the first stage of the compressor. There it is compressed and flows into the second stage of the compressor, where it is mixed with flash-economizer gas and further compressed.

Compression raises the refrigerant temperature above that of the water flowing through the condenser tubes. When the warm (typically 100 to 105 degrees F [38 to 41 degrees C]) refrigerant vapor contacts the condenser tubes, the relatively cool condensing water (typically 85 to 95 degrees F [29 to 35 degrees C]) removes some of the heat and the vapor condenses into a liquid.

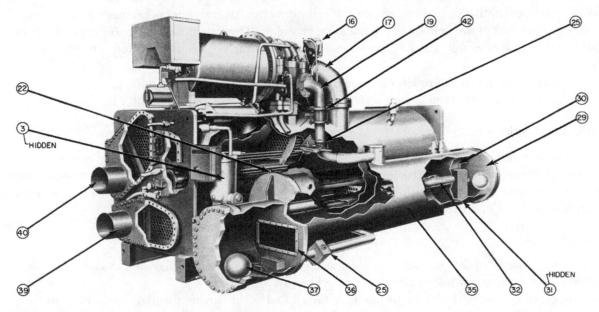

FIGURE 7.12 Cutaway view of a chiller.

Further heat removal occurs in the group of condenser tubes that forms the thermal economizer. Here, the condensed liquid refrigerant is subcooled by contact with the coolest condenser tubes. These are the tubes that contain the entering water.

The subcooled liquid refrigerant drains into a high-side valve chamber. This chamber maintains the proper fluid level in the thermal economizer and meters the refrigerant liquid into a flash economizer chamber. Pressure in this chamber is intermediate between condenser and cooler pressures. At this lower pressure some of the liquid refrigerant flashes to gas, cooling the remaining liquid. The flash gas, having absorbed heat, is returned directly to the compressor's second stage. Here, it is mixed with gas already compressed by the first-stage impeller. Since the flash gas must pass through only half the compression cycle to reach condenser pressure, there is a savings in power.

The cooled liquid refrigerant in the economizer is metered through the low-side valve chamber into the cooler. Because pressure in the cooler is lower than economizer pressure, some of the liquid flashes and cools the remainder to evaporator (cooler) temperature. The cycle is now complete.

Motor Cooling Cycle

Refrigerant liquid from a sump in the condenser (No. 24 in Figure 7.11) is subcooled by passage through a line in the cooler (No. 27 in Figure 7.11). The refrigerant then flows externally through a strainer and variable orifice (No. 11 in Figure 7.11) and enters the compressor motor end. Here, it sprays and cools the compressor rotor and stator. It then collects in the base of the motor casing. Here, it drains into the cooler. Differential pressure between the condenser and cooler maintains the refrigerant flow.

Dehydrator Cycle

The dehydrator removes water and noncondensable gases. It indicates any water leakage into the refrigerant. (See No. 6 in Figure 7.11.)

This system includes a refrigerant condensing coil and chamber, water-drain valve, purging valve, pressure gauge, refrigerant float valve, and refrigerant piping.

A dehydrator sampling line continuously picks up refrigerant vapor and contaminants, if any, from the condenser. Vapor is condensed into a liquid by the dehydrator-condensing coil. Water, if present, separates and floats on the refrigerant liquid. The water level can be observed through a sight glass.

Water may be withdrawn manually at the water drain valve. Air and other non-condensable gases collect in the upper portion of the dehydrator-condensing chamber. The dehydrator gauge indicates the presence of air or other gases through a rise in pressure. These gases may be manually vented through the purging valve.

A float valve maintains the refrigerant liquid level and pressure difference necessary for the refrigerant condensing action. Purified refrigerant is returned to the cooler from the dehydrator float chamber.

Lubrication Cycle

The oil pump and oil reservoir are contained within the uni-shell. Oil is pumped through an oil filter-cooler that removes heat and foreign particles. A portion of the oil is then fed to the compressor motor-end bearings and seal. The remaining oil lubricates the compressor transmission, compressor thrust and journal bearings, and seal. Oil is then returned to the reservoir to complete the cycle.

Controls

The cooling capacity of the machine is automatically adjusted to match the cooling load by changes in the position of the compressor inlet guide vanes. (See Figure 7.13.)

A temperature-sensing device in the circuit of the chilled water leaving the machine cooler continuously transmits signals to a solid-state module in the machine control center. The module in turn transmits the amplified and modulated temperature signals to an automatic guide vane actuator.

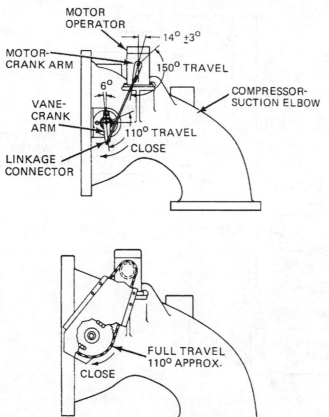

FIGURE 7.13 Vane motor-crank angles. These are shown as No. 16 and No. 17 in Figure 7.11.

A drop in the temperature of the chilled water leaving the circuit causes the guide vanes to move toward the closed position. This reduces the rate of refrigerant evaporation and vapor flow into the compressor. Machine capacity decreases. A rise in chilled water temperature opens the vanes. More refrigerant vapor moves through the compressor, and the capacity increases.

The modulation of the temperature signals in the control center allows precise control of guide-vane response regardless of the system load.

Solid-State Capacity Control

In addition to amplifying and modulating the signals from chilled water sensor to vane actuator, the solid-state module in the control center provides a means for preventing the compressor from exceeding full load amperes. It also provides a means for limiting motor current down to 40 percent of full load amperes to reduce electrical demand rates.

A throttle-adjustment screw eliminates guide-vane hunting. A manual capacity-control knob allows the operator to open, close, or hold the guide-vane position when desired.

COOLING TOWERS

Cooling towers are used to conserve or recover water. In one design the hot water from the condenser is pumped to the tower. There, it is sprayed into the tower basin. The temperature of the water decreases as

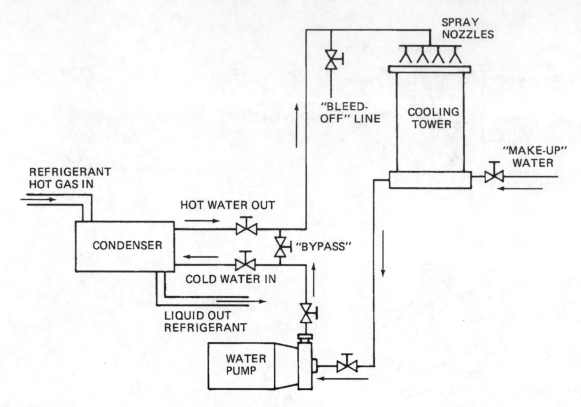

FIGURE 7.14   Recirculating water system using a tower.

it gives up heat to the air circulating through the tower. Some of the towers are rather large, since they work with condensers yielding 1,600 tons of cooling capacity. (See Figure 7.14.) Most of the cooling that takes place in the tower results from the evaporation of part of the water as it falls through the tower. The lower the wet bulb temperature of the incoming air, the more efficient the air is in decreasing the temperature of the water being fed into the tower. The following factors influence the efficiency of the cooling tower:

- Mean difference between vapor pressure of the air and pressure in the tower water

- Length of exposure time and amount of water surface exposed to air

- Velocity of air through the tower

- Direction of air flow relative to the exposed water surface (parallel, transverse, or counter)

Theoretically, the lowest temperature to which the water can be cooled is the temperature of the air (wet bulb) entering the tower. However, in practical terms it is impossible to reach the temperature of the air. In most instances, the temperature of the water leaving the tower will be no lower than 7 to 10 degrees F (4 to 6 degrees C) above the air temperature.

The range of the tower is the temperature of the water going into the tower and the temperature of the water coming out of the tower. This range should be matched to the operation of the condenser for maximum efficiency.

Cooling-Systems Terms

The following terms apply to cooling tower systems.

Cooling range is the number of degrees Fahrenheit through which the water is cooled in the tower. It is the difference between the temperature of the hot water entering the tower and the temperature of the cold water leaving the tower.

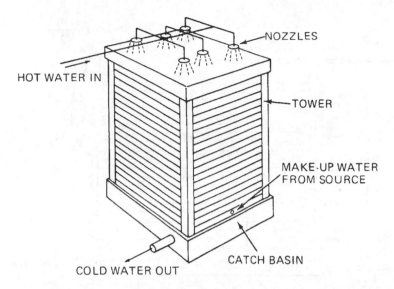

FIGURE 7.15 Natural-draft cooling tower.

Approach is the difference in degrees Fahrenheit between the temperature of the cold water leaving the cooling tower and the wet-bulb temperature of the surrounding air.

Heat load is the amount of heat "thrown away" by the cooling tower in Btu per hour (or per minute). It is equal to the pounds of water circulated multiplied by the cooling range.

Cooling-tower pump head is the pressure required to lift the returning hot water from a point level with the base of the tower to the top of the tower and force it through the distribution system.

Drift is the small amount of water lost in the form of fine droplets retained by the circulating air. It is independent of and in addition to evaporation loss.

Bleed-off is the continuous or intermittent wasting of a small fraction of circulating water to prevent the buildup and concentration of scale-forming chemicals in the water.

Makeup is the water required to replace the water that is lost by evaporation, drift, and bleed-off.

Design of Cooling Towers

Classified by the air-circulation method used, there are two types of cooling towers: natural-draft or mechanical-draft. Figure 7.15 shows the operation of the natural-draft cooling tower, and Figure 7.16 shows the operation of the mechanical-draft cooling tower. The forced-draft cooling tower shown in Figure 7.17 is just one example of the mechanical-draft designs available today.

Cooling-tower ratings are given in tons. This is based on heat-transfer capacity of 250 Btu per minute per ton. The normal wind velocity taken into consideration for tower design is 3 miles per hour. The wet-bulb temperature is usually 80 degrees F (27 degrees C) for design purposes. The usual flow of water over the tower is 4 gallons per minute for each ton of cooling desired. Several charts are available showing current design technology. Manufacturers supply the specifications for their towers. However, there are some important points to remember when use of a tower is being considered:

- In tons of cooling, the tower should be rated at the same capacity as the condenser

- The wet-bulb temperature must be known

- The temperature of the water leaving the tower and entering the condenser should be known

Towers present some maintenance problems. These stem primarily from the water used in the cooling system. Chemicals are employed to control the growth of bacteria and other substances. Scale in the pipes and on parts of the tower must also be controlled. Chemicals are used for each of these controls.

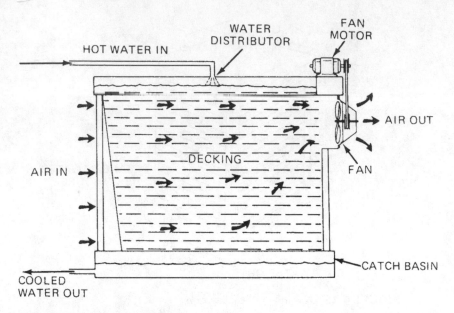

FIGURE 7.16 Small induced-draft cooling tower.

EVAPORATIVE CONDENSERS

The evaporative condenser is a condenser and a cooling tower combined. Figure 7.18 illustrates how the nozzles spray water over the cooling coil to cool the fluid or gas in the pipes. This is a very good water-conservation tower. In the future this system will probably become more popular. The closed-circuit cooler should see increased use because of dwindling water supplies and more expensive treatment problems. The function of this cooler is to process the fluid in the pipes. This is a sealed, contamination-free system. Instead of allowing the water to drop onto slats or other deflectors, this unit sprays the water directly onto the cooling coil.

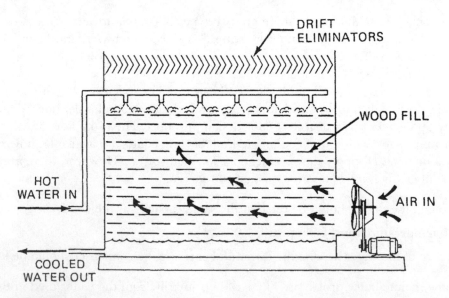

FIGURE 7.17 Forced-draft cooling tower.

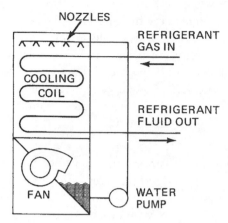

FIGURE 7.18 Evaporator cooler has no fill deck. The water cools the process fluid directly. *(Marley)*

NEW DEVELOPMENTS

All-metal towers with housing, fans, fill, piping, and structural members made of galvanized or stainless steel are now being built. Some local building codes are becoming more restrictive with respect to fire safety. Low maintenance is another factor in the use of all-metal towers.

Engineers are beginning to specify towers less subject to deterioration due to environmental conditions. Thus, all-steel or all-metal towers are called for. Already, galvanized-steel towers have made inroads into the air-conditioning and refrigeration market. Stainless-steel towers are being specified in New York City, northern New Jersey, and Los Angeles. This is due primarily to a polluted atmosphere, which can lead to early deterioration of non-metallic towers and, in some cases, metals.

Figure 7.19 shows a no-fans design for a cooling tower. Large quantities of air are drawn into the tower by cooling water as it is injected through spray nozzles at one end of a venturi plenum. No fans are needed. Effective mixing of air and water in the plenum permits evaporative-heat transfer to take place without the fill required in conventional towers.

The cooled water falls into the sump and is pumped through a cooling-water circuit to return for another cycle. The name applied to this design is Baltimore Aircoil. In 1981 towers rated at 10 to 640 tons with 30 to

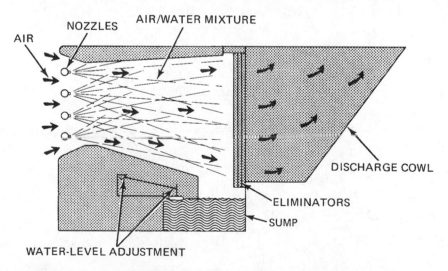

FIGURE 7.19 Cooling tower with natural draft properties. There are no moving parts in the cooling tower. *(Marley)*

1,920 gallons per minute were standard. Using prestrainers in the high-pressure flow has minimized the nozzle-clogging problem. There are no moving parts in the tower. This results in very low maintenance costs.

Air-cooled condensers are reaching 1,000 tons in capacity. Air coolers and air condensers are quite attractive for use in refineries and natural-gas compressor stations. They are also used for cooling in industry, as well as for commercial air-conditioning purposes. Figure 7.19 shows how the air-cooled condensers are used in a circuit system that is completely closed. These are very popular where there is little or no water supply.

Temperature Conversion

A cooling tower is a device for cooling a stream of water. Evaporating a portion of the circulating stream does this. Such cooled water may be used for many purposes, but the main concern here is its utilization as a heat sink for a refrigeration-system condenser. A number of types of cooling towers are used for industrial and commercial purposes. They are usually regarded as a necessity for large buildings or manufacturing processes. Some of these types have already been mentioned, but the following will bring you more details on the workings of cooling towers and their differences.

Cooling-water concerns must be addressed for the health of those who operate and maintain the systems. There is the potential for harboring and growth of pathogens in the water basin or related surface. This may occur during the summer and also during idle periods. When the temperature falls to the 70 to 120 degrees F range, there are periods when the unit will not be operational and will sit idle. Dust from the air will settle in the water and create an organic medium for the culture of bacteria and pathogens. Algae will grow in the water—some need sunlight, others grow without. Some bacteria feed on iron. The potential for pathogenic culture is there, and cooling-tower design should include some kind of filtration and/or chemical sterilization of the water.

Types of Towers

The atmospheric type of tower does not use a mechanical device, such as a fan, to create air flow through the tower. There are two main types of atmospheric towers—large and small. The large hyperbolic towers are equipped with "fill" since their primary applications are with electric power plants. The steam-driven alternator has very high-temperature steam to reduce to water or liquid state.

Atmospheric towers are relatively inexpensive. They are usually applied in very small sizes. They tend to be energy-intensive because of the high spray pressures required. The atmospheric towers are far more affected by adverse wind conditions than are other types. Their use with systems requiring accurate, dependable cold-water temperatures is not recommended (see Figure 7.20).

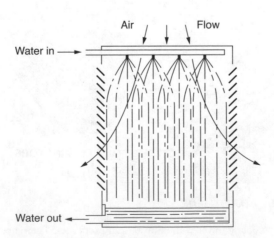

FIGURE 7.20 Atmospheric tower.

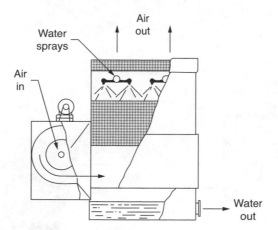

FIGURE 7.21 Forced-draft counterflow tower.

Mechanical draft towers, such as the one shown in Figure 7.21, are categorized as either forced-draft or induced-draft towers. In the forced-draft type the fan is located in the ambient airstream entering the tower. The air is also brought through or induced to enter the tower by a fan above, as in Figures 7.21 and 7.22. In the latter type air is drawn through the tower by an induced draft.

Forced-draft towers have high air-entrance velocities and low edit velocities. They are extremely susceptible to recirculation and are therefore considered to have less performance stability than induced-draft towers. There is concern in northern climates as the forced-draft fans located in the cold entering ambient airstream can become subject to severe icing. The resultant imbalance occurs when the moving air, laden with either natural or recirculated moisture, becomes ice.

Usually forced-draft towers are equipped with centrifugal blower-type fans. These fans require approximately twice the operating horsepower of propeller-type fans. They have the advantage of being able to operate against the high static pressures generated with ductwork. So equipped, they can be installed either indoors or within a specifically designed enclosure that provides sufficient separation between the air intake and discharge locations to minimize recirculation. (See Figure 7.23.)

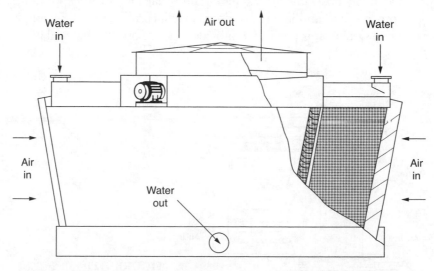

FIGURE 7.22 Induced-draft crossflow tower.

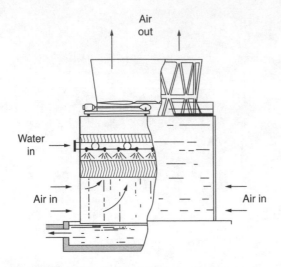

FIGURE 7.23   Induced-draft counterflow tower.

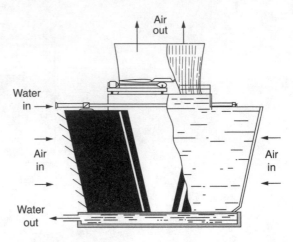

FIGURE 7.24   Double-flow crossflow tower.

Crossflow Towers

Crossflow towers, as seen in Figure 7.24, have a fill configuration through which the air flows horizontally. That means that it is across the downward fall of the water. The water being cooled is delivered to hot water inlet basins. The basins are located above the fill areas. The water is distributed to the fill by gravity through metering orifices in the basins' floor. This removes the need for a pressure-spray distribution system. and it places the resultant gravity system in full view for maintenance.

A cooling tower is a specialized heat exchanger (see Figure 7.25). The two fluids, air and water, are brought into direct contact with each other to effect the transfer of heat. In the spray-filled tower, as in Figure 7.25, this is accomplished by spraying a flowing mass of water into a rain-like pattern. Then an upward-moving mass flow of cool air is induced by the action of the fan.

Fluid Cooler

The fluid cooler is one of the most efficient systems for industrial and HVAC applications (see Figure 7.26). By keeping the cooling process fluid and in a clean, closed loop it combines the function of a cooling tower and a heat exchanger into one system. It is possible to provide superior operational and maintenance benefits.

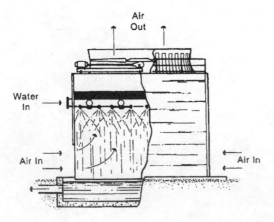

FIGURE 7.25   Spray-filled counterflow tower.

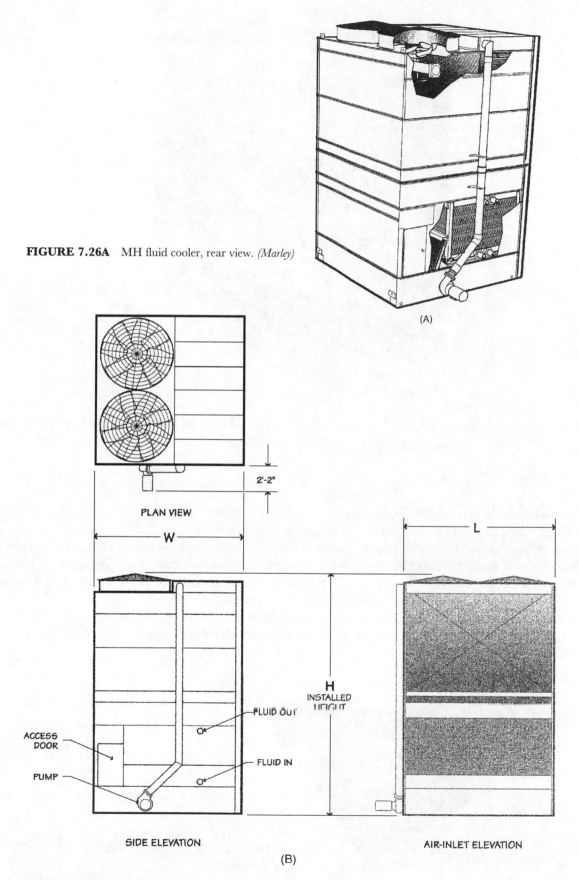

FIGURE 7.26A MH fluid cooler, rear view. *(Marley)*

(A)

PLAN VIEW

2'-2"

W

L

H
INSTALLED
HEIGHT

FLUID OUT

ACCESS
DOOR

FLUID IN

PUMP

SIDE ELEVATION

AIR-INLET ELEVATION

(B)

FIGURE 7.26B MH fluid cooler, various views. *(Marley)*

(C)

FIGURE 7.26C   MH fluid cooler, front view. *(Marley)*

The fluid-cooler coil is suitable for cooling water, oils, and other fluids. It is compatible with most oils and other fluids when the carbon-steel coil is in a closed, pressurized system. Each coil is constructed of continuous steel tubing, formed into a serpentine shape and welded into an assembly (see Figure 7.27). The complete asssembly is then hot-dipped in liquid tin to galvanize it after fabrication. The galvanized-steel coil has proven itself through the years. Paints and electrostatically applied coatings cannot seem to approach galvanization for inceasing coil longevity. The coils can also be made of stainless steel.

The fluid cooler uses a mechanically induced-draft, crossflow technology. The fill media is located above the coil. The process fluid is pumped internally through the coil. Recirculating water is cooled as it passes over

FIGURE 7.27   MH fluid-cooler coil.
(Marley)

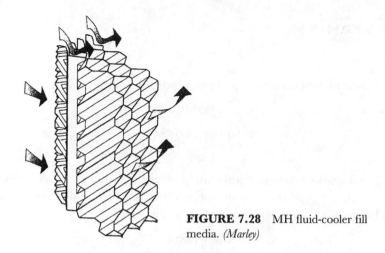

FIGURE 7.28 MH fluid-cooler fill media. *(Marley)*

the fill media (see Figure 7.28). The process fluid is thermally equalized and redistributed over the outside of the coil. A small portion of recirculating water is evaporatd by the drawn air that is passing through the coil and fill media. This cools the process fluid. The coil section rejects heat through evaporative cooling. This process uses the fresh airstream and precooled recirculating spray water. Recirculated water falls from the coil into a collection basin. From the basin it is then pumped back up to be distributed over the fill media.

For industrial and HVAC applications this is an ideal type of system. The process fluid is kept in a clean, closed loop. It combines the function of a cooling tower and a heat exchanger into one system. This improves efficiency and has many maintenance benefits. The unit shown here has a capacity ranging from 100 tons to 650 tons in a compact enclosure. It is suitable for cooling a wide range of fluids from water and glycols to quench oils and plating solutions.

MULTIPLE-CHOICE EXAM

1.   Water systems are used in air-conditioning applications for heat removal and _____.
     a. dehumidification   b. humidification
     c. cooling   d. none of these

2.   A chilled-water system works in conjunction with air-handling units or process equipment to remove the heat generated within a conditioned space or process. The terminal unit's cooling coils collect the heat and move it by _____ and convection to the water.
     a. radiation   b. absorption
     c. adsorption   d. conduction

3.   The two basic distribution systems for chilled water are:
     a. the modulated valve control and the two-pipe return control
     b. the two-pipe return and the two-pipe direct-return arrangement
     c. the one-pipe return and the one-pipe direct-return arrangement
     d. the one-pipe return and the two-pipe direct-return arrangement

4.   What is the common temperature range in degrees F for chilled-water designs?
     a. 42 to 46   b. 34 to 54
     c. 24 to 64   d. 64 to 72

5.   If the chilled-water system is subject to freezing temperatures, a mixture of water and _____ may be used.
     a. alcohol   b. salt
     c. glycol   d. none of these

6.   Corrosion inhibitors are often added to chilled-water systems to reduce corrosion and _____.
     a. mildew   b. mold
     c. moss   d. scale

7.   When antifreeze is added to the water, it generally reduces the specific heat and conductivity and increases the _____ of the solution.
     a. flow   b. efficiency
     c. viscosity   d. none of the above

8. What is added to the chilled-water system to make it a storage system?

    a. a thermally insulated storage tank

    b. an uninsulated tank

    c. a cooling tower

    d. a downstream pool arrangement

9. Chilled-water storage systems generally become economically feasible when more than _____ tons of refrigeration are needed.

    a. 200   b. 300

    c. 400   d. 500

10. The purpose of the condenser in a refrigeration system is:

    a. to cool hot refrigerant vapor

    b. to dehumidify refrigerant vapor

    c. to dry refrigerant

    d. to condense the cooled refrigerant

11. The recommended metal for tubing in systems using salt water for cooling purposes is:

    a. copper   b. cupronickel

    c. nickel   d. zinc

12. Cooling towers are necessary:

    a. to cool the water circulating in a water-cooled system

    b. to cool and dehumidify air in the tower

    c. to humidify the cooling water

    d. to dry and cool the circulating air

13. Cooling towers are rated:

    a. in tons   b. in pounds

    c. in meters   d. in gallons

14. The word venturi means:

    a. a device used to produce a crossflow

    b. a device used to produce a forced draft

    c. a device that induces a draft

    d. a constriction that lowers the pressure of a fluid and increases the velocity

15. The use of recirculation to cool water for reuse is _____ practical than air.

    a. more

    b. less

    c. much more

    d. much less

16. Most commercial or residential home air-conditioning units are _____ cooled.

    a. liquid-

    b. water-

    c. air-

    d. none of the above

17. Which of the following is NOT a type of tube condenser used for standard air-cooled units as boosters?

    a. spiral

    b. helix

    c. trombone

    d. trumpet

18. A chiller is part of a(n)_____.

    a. evaporator

    b. condenser

    c. compressor

    d. none of the above

19. The dehydrator removes _____ and noncondensable gases.

    a. water

    b. humidity

    c. air

    d. none of the above

20. The term bleed off means:

    a. a continuous or intermittent wasting of a small fraction of circulating water to prevent the buildup and concentration of scale-forming chemicals in the cooling tower system

    b. draining off part of the tower water

    c. removing contaminated water

    d. removing refrigerant

21. In designing a cooling tower, the tower should be rated at the same ton-capacity as the _____.

    a. condenser

    b. evaporator

    c. compressor

    d. all of these

    e. none of these

22. Atmospheric towers are relatively _____.

    a. expensive

    b. inexpensive

    c. sturdy

    d. low maintenance

23. The _____ cooler is one of the most efficient systems for industrial and HVAC applications.

a. water
b. refrigerant

c. fluid
d. air

24. The cooling tower is a specialized heat _____.

a. exchanger
b. cooler

c. generator
d. eliminator

25. Mechanical-draft towers are categorized as either forced-draft or _____- draft towers.

a. unforced
b. induced

c. natural
d. horizontal

26. Usually forced-draft tower are equipped with centrifugal _____ -type fans.

a. suction
b. wound

c. star-wound
d. blower

27. The fluid cooler uses a _____ induced draft.

a. naturally
b. pump-

c. mechanically
d. closed-loop-

28. The _____ has jurisdiction over water supply and protection in water-cooling towers.

a. International Plumbing Code

b. International Mechanical Code

c. International Fuel Gas Code

d. International Building Code

29. The _____ has jurisdiction over heat-exchange equipment that contains a refrigerant which is flammable, combustible, or hazardous.

a. International Fire Code

b. International Building Code

c. International Fuel Gas Code

d. International Mechanical Code

30. Plume discharges from cooling towers must be located to prevent the discharge-vapor plumes from entering occupied spaces. Plume discharges shall not be less than _____ feet above or 20 feet away from any ventilation inlet to a building.

a. 10
b. 20

c. 5
d. 30

TRUE-FALSE EXAM

1.   The condenser is a heat-transfer device.
     True   False

2.   Heat given up by the refrigerant vapor to the condensing medium includes both the heat absorbed in the evaporator and the heat of compression.
     True   False

3.   The larger the condenser area exposed to the moving airstream, the higher the temperature of the refrigerant when it leaves the condenser.
     True   False

4.   To exhaust the heat without causing the area being cooled to heat up again, it is common practice to locate the condenser outside the area being conditioned.
     True   False

5.   Air-cooled condensers that use fans are classified as chassis-mounted and remote.
     True   False

6.   Water is used to cool condensers.
     True   False

7.   There are four of water-cooled condensers.
     True   False

8.   The shell-and-coil condenser is made by placing a bare or finned tube inside a steel shell.
     True   False

9.   Chillers are used to cool water or brine.
     True   False

10. In chillers the cooled liquid refrigerant in the economizer is metered through the high-side valve chamber into the cooler.
     True   False

11. In chillers the oil pump and oil reservoir are contained within the unit shell.
     True   False

12. Cooling towers are used to conserve or recover water.
     True   False

13. Drift is the small amount of water lost in the form of fine droplets retained by the circulating air.
    True   False

14. Bleed-off is the continuous or intermittent wasting of a small fraction of circulating water to prevent the buildup and concentration of scale-forming chemicals in the water.
    True   False

15. The fluid cooler is one of the most efficient systems for industrial HVAC applications.
    True   False

MULTIPLE-CHOICE ANSWERS

1.	A	7.	C	13.	A	19.	A	25.	B
2.	D	8.	A	14.	D	20.	A	26.	D
3.	B	9.	C	15.	A	21.	A	27.	C
4.	A	10.	A	16.	C	22.	B	28.	A
5.	C	11.	B	17.	D	23.	C	29.	A
6.	D	12.	A	18.	B	24.	A	30.	C

TRUE-FALSE ANSWERS

1.	T	5.	T	9.	T	13.	T
2.	T	6.	T	10.	F	14.	T
3.	F	7.	F	11.	T	15.	T
4.	T	8.	T	12.	T		

Part D

REFRIGERATION

HVAC Licensing Study Guide

Chapter 8

REFRIGERATION EQUIPMENT AND PROCESSES

WORKING WITH WATER-COOLING PROBLEMS

Three-fourths of the earth's surface is covered with water. The earth is blanketed with water vapor, which is an indispensable part of the atmosphere. Heat from the sun shining on oceans, rivers, and lakes evaporates some water into the atmosphere. Warm, moisture-laden air rises and cools. The cooling vapor condenses to form clouds. Wind currents carry clouds over land masses where the precipitation may occur in the form of rain, snow, or sleet. Because of the sun and upper-air currents, this process is repeated again and again. Pure water has no taste and no odor. Pure water, however, is actually a rarity.

All water found in oceans, rivers, lakes, streams, and wells contains various amounts of minerals picked up from the earth. Even rainwater is not completely pure. As rain falls to earth, it washes from the air various gases and solids such as oxygen, carbon dioxide, industrial gases, dust, and even bacteria. Some of this water sinks into the earth and collects in wells or forms underground streams. The remainder runs over the ground and finds its way back into various surface-water supplies.

Water is often referred to as the universal solvent. Water runs over and through the earth and mixes with many minerals. Some of these mineral solids are dissolved or disintegrated by water.

Pure water and sanitary water are the same as far as municipalities are concerned. "Pure" in this case means that the water is free from excessive quantities of germs and will not cause disease. Mineral salts or other substances in water do not have to be removed by water-treatment plants unless they affect sanitary conditions. Mineral salts are objectionable in water used for many other purposes. These uses include generating power, heating buildings, processing materials, and manufacturing. Water fit for human consumption is not necessarily acceptable for use in boilers or cooling equipment.

Water is used in many types of cooling systems. Heat removal is the main use of water in air conditioning or refrigeration equipment. Typical uses include once-through condensers, open recirculating cooling systems employing cooling towers, evaporative condensers, chilled-water systems, and air washers. In evaporative condensers, once through systems and cooling towers, water removes heat from a refrigerant and then is either wasted or cooled by partial evaporation in air. Knowledge of impurities in water used in any of these systems aids in predicting possible problems and methods of preventing them.

Cooling towers are usually remotely located; it becomes necessary to regularly inspect and clean the tower according to the manufacturer's recommendation. The few hours each month spent on inspecting the cooling tower and maintaining it will pay dividends. The life of a tower varies according to:

- Construction materials

- Location within the system

- Location of the city or country

Generally, the premium materials of construction are:
- Wood
- Concrete
- Stainless steel
- Fiberglass

These units are expected to last from 20 to 30 years if properly cared for. The less expensive units, made of galvanized steel, will operate for 8 to 20 years. Of course, tower life will vary due to the extremes of weather, number of hours used each year, and type of water treatment. It is sufficient to say that in order to get the most use from the tower, cooling tower manufacturers want to make that tower last as long as possible.

FOULING, SCALING, AND CORROSION

Fouling reduces water flow and heat transfer. It can be caused by the collection of loose debris over pump-suction screens in sumps, growth of algae in sunlit areas, and slime in shade or dark sections of water systems. Material can clog pipes or other parts of a system after it has broken loose and been carried into the system by the water stream. ***Scaling*** also reduces water flow and heat transfer. The depositing of dissolved minerals on equipment surfaces causes scaling. This is particularly so in hot areas, where heat transfer is most important. ***Corrosion*** is caused by impurities in the water. In addition to reducing water flow and heat transfer it also damages equipment. Eventually, corrosion will reduce operational efficiency. It may lead to expensive repairs or even equipment replacement.

Impurities have at least five confirmed sources. One is the earth's atmosphere. Water falling through the air, whether it be natural precipitation or water showering through a cooling tower, picks up dust as well as oxygen and carbon dioxide. Similarly, synthetic atmospheric gases and dust affect the purity of water. Heavily industrialized areas are susceptible to such impurities being introduced into their water systems.

Decaying plant life is a source of water impurity. Decaying plants produce carbon dioxide. Other products of vegetable decay cause bad odor and taste. The byproducts of plant decay provide a nutrient for slime growth.

These three sources of impurities contaminate water with material that makes it possible for it to pick up more impurities from a fourth source, minerals. Minerals found in the soil beneath the earth's surface are probably the major source of impurities in water. Many minerals are present in subsurface soil. They are more soluble in the presence of the impurities from the first three sources, mentioned above.

Industrial and municipal wastes are a fifth major source of water impurities. Municipal waste affects bacterial count. Therefore, it is of interest to health officials but is not of primary concern from a scale or corrosion standpoint. Industrial waste, however, can add greatly to the corrosive nature of water. It can indirectly cause a higher than normal mineral content.

The correction or generation of finely divided material that has the appearance of mud or silt causes fouling. This sludge is normally composed of dirt and trash from the air. Silt is introduced with makeup water. Leaves and dust are blown in by wind and washed from the air by rain. This debris settles in sumps or other parts of cooling systems. Plant growth also causes fouling. Bacteria or algae in water will result in the formation of large masses of algae and slime. These may clog system water pipes and filters. Paper, bottles, and other trash also cause fouling.

Scaling

Prevention

There are three ways to prevent scaling. The first is to eliminate or reduce hardness minerals from the feed water. Control of factors that cause hardness salts to become less soluble is important. Hardness minerals are defined as water-soluble compounds of calcium and magnesium. Most calcium and magnesium compounds are much less soluble than are corresponding sodium compounds. By replacing the calcium and magnesium portion of these minerals with sodium, solubility of the sulfates and carbonates is improved to such a degree that scaling no longer is a problem. This is the function of a water softener.

The second method of preventing scale is by controlling water conditions that affect the solubility of scale-forming minerals. The five factors that affect the rate of scale formation are:

*   Temperature
*   TDS (total dissolved solids)
*   Hardness
*   Alkalinity
*   pH

These factors can to some extent be regulated by proper design and operation of water-cooled equipment. Proper temperature levels are maintained by ensuring a good water flow rate and adequate cooling in the tower. Water flow in recirculating systems should be approximately 3 gallons per minute per ton. Lower flow levels allow the water to remain in contact with hot surfaces of the condenser for a longer time and pick up more heat. Temperature drop across the tower should be 8 to 10 degrees F (4.5 to 5.5 degrees C) for a compression refrigeration system and 18 to 20 degrees F (10 to 11 degrees C) in most absorption systems. This cooling effect, due to evaporation, is dependent on tower characteristics and uncontrollable atmospheric conditions.

Airflow through the tower and the degree of water breakup are two factors that determine the amount of evaporation that will occur. Since heat energy is required for evaporation, the amount of water that is changed into vapor and lost from the system determines the amount of heat. That is, the number of Btu to be dissipated is the heat factor. One pound of water, at cooling tower temperatures, requires 1050 Btu to be converted from liquid to vapor. Therefore, the greater the weight of water evaporated from the system, the greater the cooling effect or temperature drop across the tower.

Total dissolved solids, hardness, and alkalinity are affected by three interrelated factors: evaporation, makeup, and bleed or blow-down rates. Water, when it evaporates, leaves the system in a pure state, leaving behind all dissolved matter. Water volume of evaporative cooling systems is held at a relatively constant figure through the use of float valves.

Fresh makeup water brings with it dissolved material. This is added to that already left behind by the evaporated water. Theoretically, assuming that all the water leaves the system by evaporation and the system volume stays constant, the concentration of dissolved material will continue to increase indefinitely. For this reason, a bleed or blow-down is used.

There is a limit to the amount of any material that can be dissolved in water. When this limit is reached, the introduction of additional material will cause either sludge or scale to form. Controlling the rate at which dissolved material is removed controls the degree to which this material is concentrated in circulating water.

Identification

Scale removal depends on the chemical reaction between scale and the cleaning chemical. Scale identification is important. Of the four scales most commonly found, only carbonate is highly reactive with cleaning chemicals generally regarded as safe for use in cooling equipment. The other scales require a pretreatment that renders them more reactive. This pretreatment depends on the type of scale to be removed. Attempting to remove a problem scale without proper pretreatment can waste time and money.

Scale identification can be accomplished in one of three ways:

- Experience

- Field tests

- Laboratory analysis

With the exception of iron scale, which is orange, it is very difficult, if not impossible, to identify scale by appearance. Experience is gained by cleaning systems in a given area over an extended period of time. In this way, the pretreatment procedure and the amount of scale remover required to remove the type of scale most often found in this area become common knowledge. Unless radical changes in feed-water quality occur, the type of scale encountered remains fairly constant. Experience is further developed through the use of the two other methods. Figure 8.1 shows a water-analysis kit.

Field tests, which are quite simple to perform, determine the reactivity of scale with the cleaning solution. Adding 1 tablespoon of liquid scale remover or 1 teaspoon of solid scale remover to 1/2 pint of water prepares a small sample of cleaning solution. A small piece of scale is then dropped into the cleaning solution.

The reaction rate usually will determine the type of scale. The reaction between scale remover and carbonate scale results in vigorous bubbling. The scale eventually dissolves or disintegrates. However, if the scale sample is of hard or flinty composition and little or no bubbling in the acid solution is observed, heat

FIGURE 8.1   Field kit for testing pH, phosphate, chromate, total hardness, calcium hardness, alkalinity, and chloride. *(Virginia Chemicals)*

should be applied. Sulfate scale will dissolve at 140 degrees F (60 degrees C). The small-scale sample should be consumed in about an hour. If the scale sample contains a high percentage of silica, little or no reaction will be observed. Iron scale is easily identified by appearance. Testing with a clean solution usually is not required.

Since this identification procedure is quite elementary and combinations of all types of scale are often encountered, it is obvious that more precise methods may be required. Such methods are most easily carried out in a laboratory. Many chemical manufacturers provide this service. Scale samples that cannot be identified in the field may be mailed to these laboratories. Here a complete breakdown and analysis of the problem scale will be performed. Detailed cleaning recommendations will be given to the sender.

Most scales are predominantly carbonate, but they may also contain varying amounts of sulfate, iron, or silica. Thus, the quantities of scale remover required for cleaning should be calculated specifically for the type of scale present. The presence of sulfate, iron, or silica also affects other cleaning procedures.

CORROSION

There are four basic causes of corrosion:

- Corrosive acids
- Oxygen
- Galvanic action
- Biological organisms

Corrosive Acids

Aggressive or strong acids, such as sulfurous, sulfuric, hydrochloric, and nitric, are found in most industrial areas. These acids are formed when certain industrial-waste gases are washed out of the atmosphere by water showering through a cooling tower. The presence of any of these acids will cause a drop in circulating-water pH. Water and carbon dioxide are found everywhere. When carbon dioxide is dissolved in water, carbonic acid is formed. This acid is less aggressive than the acids already mentioned. Because it is always present, however, serious damage to equipment can result.

Oxygen

Corrosion by oxygen is another problem. Water that is sprayed into the air picks up oxygen. This oxygen then is carried into the system. Oxygen reacts with any iron in equipment. It forms iron oxide, which is a porous material. Flaking or blistering of oxidized metal allows corrosion of the freshly exposed metal. Blistering also restricts water flow and reduces heat transfer. Reaction rates between oxygen and iron increase rapidly as temperatures increase. Thus, the most severe corrosion takes place in hot areas of equipment with iron parts.

Oxygen also affects copper and zinc. Zinc is the outer coating of galvanized material. Here, damage is much less severe because oxidation of zinc and copper forms an inert metal oxide. This sets up a protective film between the metal and the attacking oxygen.

Galvanic Action

Galvanic corrosion is the third cause of corrosion. Galvanic corrosion is basically a reaction between two different metals in electrical contact. This reaction is both electrical and chemical in nature. The following three conditions are necessary to produce galvanic action:

- Two dissimilar metals possessing different electrochemical properties must be present

- An electrolyte, a solution through which an electrical current can flow, must be present

- An electron path to connect these two metals is also required

Many different metals are used to fabricate air-conditioning and refrigeration systems. Copper and iron are two dissimilar metals. Add a solution containing ions, and an electrolyte is produced. Unless the two metals are placed in contact, no galvanic action will take place. A coupling is made when two dissimilar metals, such as iron and copper, are brought into contact with each other. This sets up an electrical path or a path for electron movement and allows electrons to pass from the copper to the iron. As current leaves the iron and reenters the solution to return to the copper, corrosion of the iron takes place. Copper-iron connections are common in cooling systems.

Greater separation of metals in the galvanic series results in their increased tendency to corrode. For example, if platinum is joined with magnesium with a proper electrolyte, then platinum would be protected and magnesium would corrode. Since they are so far apart on the scale, the corrosion would be rapid (see Table 8.1). If iron and copper are joined, we can tell by their relative positions in the series that iron would corrode, but to a lesser degree than the magnesium mentioned in the previous example. Nevertheless, corrosion would be extensive enough to be very damaging. However, if copper and silver were joined together, the copper would corrode. Consequently, the degree of corrosion is determined by the relative positions of the two metals in the galvanic series.

Improperly grounded electrical equipment or poor insulation can also initiate or accelerate galvanic action. Stray electrical currents cause a similar type of corrosion, usually referred to as electrolytic corrosion. This generally results in the formation of deep pits in metal surfaces.

TABLE 8.1 Galvanic series.

Anodic (Corroded End)	
Magnesium	Tin
Magnesium alloy	Brass
Zinc	Copper
Aluminum	Bronze
Mild steels	Copper-nickel alloys
Alloy steels	Nickel
Wrought iron	Silver
Cast iron	Gold
Soft solders	Platinum
Lead	Cathodic (Protected End)

Biological Organisms

Another cause of corrosion is biological organisms. These are algae, slime, and fungi. Slimes thrive in complete absence of light. Some slimes cling to pipes and will actually digest iron. This localized attack results in the formation of small pits, which, over a period of time, will expand to form holes.

Other slimes live on mineral impurities, especially sulfates, in water. When doing so, they give off hydrogen-sulfide gas. The gas forms weak hydrosulfuric acid. (Do not confuse this with strong sulfuric acid.) This acid slowly but steadily deteriorates pipes and other metal parts of the system. Slime and algae release oxygen into the water. Small oxygen bubbles form and cling to pipes. This oxygen may act in the same manner as a dissimilar metal and cause corrosion by galvanic action. This type of corrosion is commonly referred to as oxygen-cell corrosion.

Algae are a very primitive form of plant life. They are found almost everywhere in the world. The giant Pacific kelp are algae. Pond scums and the green matter that grows in cooling towers are also algae. Live algae range in color from yellow, red, and green to brown and gray. Like bacterial slime, they need a wet or moist environment and prefer a temperature between 40 and 80 degrees F (4 and 27 degrees C). Given these conditions, they will find mineral nourishment for growth in virtually any water supply.

Bacteria cause slime. Slime bacteria can grow and reproduce at temperatures from well below freezing (32 degrees F [0 degrees C]) to the temperature of boiling water (212 degrees F [100 degrees C]). However, they prefer temperatures between 40 and 80 degrees F (4 and 27 degrees C). They usually grow in dark places. Some types of slime also grow when exposed to light in cooling towers. The exposure of the dark-growing organisms to daylight will not necessarily stop their growth. The only condition essential to slime propagation is a wet or moist environment.

Fungi are a third biological form of corrosion. Fungi attack and destroy the cellulose fibers of wood. They cause what is known as brown rot or white rot. If fungal decay proceeds unchecked, serious structural damage will occur in a tower.

It is essential that a cooling system be kept free of biological growths as well as scale. Fortunately, several effective chemicals are available for controlling algae and slime. Modern algaecides and slimeicides fall into three basic groups: chlorinated phenols (penta-chloro-phenates), quaternary ammonium compounds, and various organo-metallic compounds.

A broad range of slime and algae control agents is required to meet the various conditions that exist in water-cooled equipment. Product selection is dependent on the following:

- The biological organism present

- The extent of the infestation

- The resistance of the existing growths to chemical treatment

- The type and specific location of the equipment to be treated

There is considerable difference of opinion in the trade as to how often algaecides should be added and whether "slug" or continuous feeding is the better method.

In treating heavy biological growths, remember that when these organisms die, they break loose and circulate through the system. Large masses can easily block screens, strainers, and condenser tubes. Some provision should be made for preventing them from blocking internal parts of the system. The best way to do this is to remove the thick, heavy growths before adding treatment. The day after treatment is completed, thoroughly drain and flush the system and clean all strainers.

One of the most critical areas of concern about cleanliness is bacteria breeding grounds. The most difficult issue to deal with is stagnant water. A system's piping should be free of "dead legs," and tower flow should be maintained. When dirt accumulates in the collection basin of a tower, it provides the right combination of supplies for the creation of *Legionella* bacteria:

- moisture

- oxygen

- warm water

- food supply

These bacteria can be found in water supplies as well as around rivers and/or streams. They are contained in water droplets and can become airborne. Humans are susceptible to them by breathing in the contaminated air. No chemicals can positively eliminate all bacteria from the water supply in a cooling tower. However, evidence exists to suggest that good maintenance along with comprehensive treatment can dramatically minimize the risk.

SYSTEM CLEANING TREATMENTS

Air conditioning or refrigeration is basically the controlled removal of heat from a specific area. The refrigerant that carries heat from the cooled space must be cooled before it can be reused. Cooling and condensation of refrigerant require the use of a cooling medium that, in many systems, is water.

There are two types of water-cooled systems. The first type uses once-through operation. The water picks up heat and is then discarded or wasted. In effect, this is 100 percent, or total, bleed. Little if any mineral concentration occurs. The scale that forms is due to the breakdown of bicarbonates by heat. These form carbonates, which are less soluble at high temperatures than at low. Such scale can be prevented through use of a treatment chemical.

The other type of water-cooled system is the type in which heat is removed from water by partial evaporation. The water is then recirculated. Water volume lost by evaporation is replaced. This type of system is more economical from the standpoint of water use. However, the concentration of dissolved minerals leads to conditions that, if not controlled and chemically treated, may result in heavy scale formation.

Evaporative Systems

One method of operating evaporative recirculating systems involves 100 percent evaporation of the water with no bleed. This, of course, causes excessive mineral concentration. Without a bleed on the system, water conditions will soon exceed the capability of any treatment chemical. A second method of operation employs a high bleed rate without chemical treatment. Scale will form and water is wasted. The third method is the reuse of water, with a bleed to control concentration of scale-forming minerals. Thus, by the addition of minimum amounts of chemical treatment, good water economy can be realized. This last approach is the

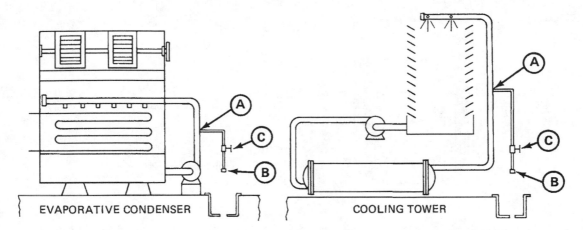

FIGURE 8.2 Connections for bleed lines for evaporative condensers and cooling towers. *(Virginia Chemicals)*

most logical and least expensive. Figure 8.2 shows how connections for bleed lines are made on evaporative condensers and cooling towers.

Scale is formed as a direct result of mineral insolubility. This, in turn, is a direct function of temperature, hardness, alkalinity, pH, and total dissolved solids. Generally speaking, as these factors increase, solubility or stability of scale-forming minerals decreases. Unlike most minerals, scale-forming salts are less soluble at high temperatures. For this reason, scale forms most rapidly on heat-exchanger surfaces.

To clean cooling towers and evaporative condensers, first determine the amount of water in the system. This is done by determining the amount of water in the sump. Measure the length, width, and water depth in feet (see Figure 8.3).

Use the following formula:

$$\text{length} \times \text{width} \times \text{water depth} \times 7.5 = \text{gallons of water in the sump}$$

Example: A sump is 5 feet long and 4 feet wide, with a water depth of 6 inches.

Complete the equation:

$$5 \times 4 \times 0.5 \times 7.5 = 75 \text{ gallons of water in the sump}$$

Next, determine the amount of water in the tank. Measure the diameter of the tank and the depth of the water in feet (see Figure 8.4).

Use the following formula:

$$\text{diameter}^2 \times \text{water depth} \times 6 = \text{gallons of water in tank}$$

Example: A tank has a diameter of 3 feet and the water is 3 feet deep.

Complete the equation:

$$3^2 \times 3 \times 6, \text{ or } 9 \times 3 \times 6 = 162 \text{ gallons of water in the tank}$$

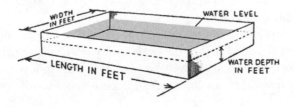

FIGURE 8.3 Method of calculating the amount of water in a rectangular tank or sump.

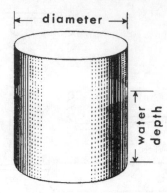

FIGURE 8.4 Calculating the
amount of water in a round tank
(Virginia Chemicals)

The two formulas above will give you the water volume in either the tank or sump. Each is figured separately since they are both part of the system's circulating water supply. There is also water in the connecting lines. These lines must be measured for total footage. Once you find the pipe footage connecting the system, you can figure its volume of water. Simply take 10 percent of the water volume in the sump for each 50 feet of pipe run. This is added to the water in the sump and the water in the tank to find the total system water volume.

For example, a system has 75 gallons of water in the sump and 162 gallons of water in the tank. The system has 160 feet of pipe:

75 gallons + 162 gallons − 237 gallons

160 feet ÷ 50 feet = 3.2

75 gallons in sump ÷ 10 = 7.5 gallons for every 50 feet in the pipes

7.5 gallons × 3.2 = 24 gallons in the total pipe system

237 gallons (in tank and sump) + 24 gallons (in pipes) = 261 gallons in the total system

This is the amount of water that must be treated to keep the system operating properly.

Now that you have determined the volume of water in the system, you can calculate the amount of chemicals needed. To clean the system:

1. Drain the sump. Flush out, or remove manually, all loose sludge and dirt. This is important because they waste the chemicals.

2. Close the bleed line and refill the sump with fresh water to the lowest level at which the circulating pump will operate (see Figure 8.5).

3. Calculate the total gallons of water in the system. Next, while the water is circulating, add starting amounts of either chemical slowly, as follows. (These amounts are for hot water systems.)
    For solid scale remover, use 5 pounds per 10 gallons of water. For regular liquid scale remover, use 1 gallon per 15 gallons of water. For concentrated liquid scale remover, use 1 gallon per 20 gallons of water (refer to Figure 8.5). The scale removers can be introduced at the water-tower distribution plate (A), the sump (B), or the water tank (C). Convenience is the keyword here. The preferred addition point is directly into the pump suction area.

4. When using liquid scale remover, add 1 ampoule of antifoam reagent per gallon of chemical. This will usually prevent excessive foaming if added before the scale remover. Extra antifoam is available in 1-pint bottles. When using the solid scale remover, stir the crystals in a plastic pail or drum until completely dissolved. Then pour slowly as a liquid. Loose crystals, if

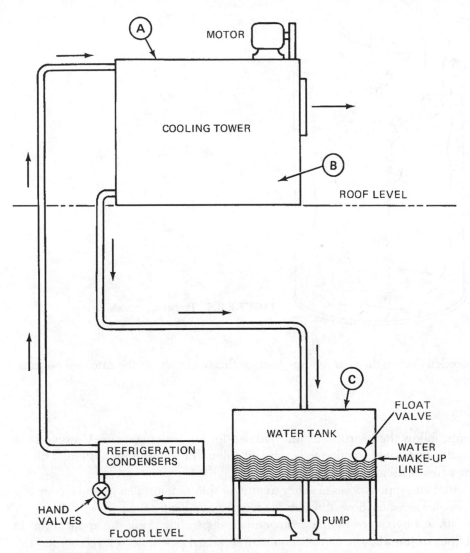

FIGURE 8.5   Forced- and natural-draft towers. (A) Water tower distribution plate, (B) sump, (C) water tank. *(Virginia Chemicals)*

allowed to fall to the bottom of the sump, will not dissolve without much stirring. If not dissolved, they might damage the bottom of the sump.

5.   Figure 8.6 shows how to prepare the crystals in the drum. Use a 55-gallon drum. Install a drain or spigot about 6 to 8 inches from the bottom. Set the drum in an upright position. Fill the drum with fresh water within 6 inches from the top, preferably warm water at about 80 degrees F (27 degrees C). Since the fine particles of the water-treatment crystals are quite irritating to the nose and eyes, immerse each plastic bag in the water. Cut the bag below the surface of the water. (See Figure 8.6.) Stir the crystals until dissolved. About 6 to 8 pounds of crystals will dissolve in each gallon of water.

6.   Drain or pump this strong solution into the system. Repeat this procedure until the required weight of the chemical has been added in concentrated solution. Then add fresh water to fill the system. The treatment should be repeated once each year for best results.

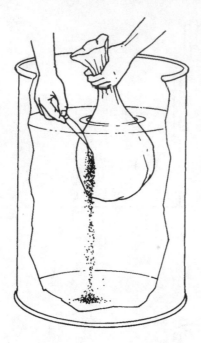

FIGURE 8.6 Preparing crystals in a drum or tank. *(Virginia Chemicals)*

If makeup water is needed during the year, be sure to treat this water also at the rate of 6 pounds per 100 gallons of water added.

Chilled Water Systems

For chilled water systems, follow the instructions just outlined for hot water systems. However, use only 3 pounds of circulating water-treatment solution for each 100 gallons of water in the system. This treatment solution should be compatible with antifreeze solutions.

For easy feeding of initial and repeat doses of water treatment solution, install a crystal feeder in a bypass line. The crystals will dissolve as water flows through the feeder. (See Figure 8.7.)

Install the feeder in either a bypass or in-line arrangement, depending upon the application. Place the feeder on a solid, level floor or foundation. Connection with standard pipe unions is recommended.

Connecting pipe threads should be carefully cut and cleaned to remove all burrs or metal fragments. Apply a good grade of pipe dope. Use the dope liberally.

Always install valves in the inlet and the outlet lines. Install a drain line with the valve in the bottom (inlet) line.

Before opening the feeder, always close the inlet and the outlet valves. Open the drain valve to relieve the pressure and drain as much water as necessary. When adding crystals or chemicals molded in shapes (balls, briquettes, etc.), it is advisable to have the feeder about one-half filled with water.

If stirring in the feeder is necessary, use only a soft wood. Stir gently to avoid damaging the epoxy lining.

Fill the feeder to the level above the outlet line. Coat the top opening, the gasket, and the locking grooves of the cap with petroleum jelly or a heavier lubricant.

Open the outlet valve fully. Then slowly open the inlet valve. If throttled flow is desired for control of treatment feed rate, throttle with inlet valve only.

Once the chemicals have been properly introduced, operate the system in the normal manner. Check the scale-remover strength in the sump by observing the color of the solution when using the solid scale remover or using test papers. When chemical removers are used, a green solution indicates a very strong cleaner. A blue solution indicates normal cleaning strength. A purple solution indicates more cleaner is needed. If necessary, dip a sample of the sump solution in a glass to aid color check. Check with the maker of the chemicals and their suggested color chart for accurate work.

CLOSED-SYSTEM INSTALLATIONS

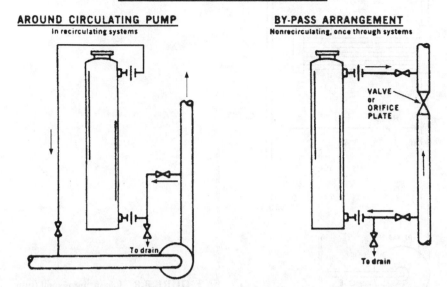

FIGURE 8.7 Feed through a bypass feeder. *(Virginia Chemicals)*

If, for instance, Virginia Chemicals scale remover is used in either solid or liquid form, use test papers to check for proper mixture and solution strength. Red test paper indicates there is enough cleaner. Inspection of the evaporative condenser tubes or lowering of head pressure to normal will indicate when the unit is clean. With shell and tube condensers, inspection of the inside of the water outlet pipe of the condenser will indicate the amount of scale in the unit.

After scale removal is completed, drain the spent solution to the sewer. Thoroughly rinse out the system with at least two fillings of water. Do not drain spent solutions to lawns or near valuable plants. The solution will cause plant damage, just as will any other strong salt solution. Do not drain to a septic tank. Refill the sump with fresh water and resume normal operation.

Shell (Tube or Coil) Condensers

Isolate the condenser to be cleaned from the cooling-tower system by an appropriate valve arrangement or by disconnecting the condenser piping. Pump in at the lowest point of the condenser. Venting the high points with tubing returning to the solution drum is necessary in some units to assure complete liquid filling of the waterside.

As shown in Figure 8.8, start circulating from a plastic pail or drum the minimum volume of water necessary to maintain circulation. After adding antifoam reagent or solid scale remover, slowly add liquid scale remover until the test strips indicate the proper strength for cleaning. Test frequently and observe the sputtering in the foam caused by carbon dioxide in the return line. Add scale remover as necessary to maintain strength. Never add more than 1 pound of solid scale remover per gallon of solution. Most condensers with moderate amounts of carbonate scale can be cleaned in about one hour. Circulation for 30 to 40 minutes without having to add cleaner to maintain cleaning strength usually indicates that action has stopped and that the condenser is clean. Empty and flush the condenser after cleaning with at least two complete fillings of water. Reconnect the condenser in the line.

If a condenser is completely clogged with scale, it is sometimes possible to open a passageway for the cleaning solution by using the standpipe method, as shown in Figure 8.9. Enough liquid scale remover is mixed with an equal volume of water to fill the two vertical pipes to a level slightly above the condenser. Some foaming will result from the action of the cleaner solution on the scale. Thus, some protective meas-

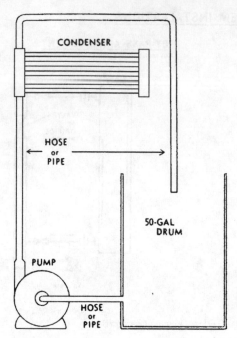

FIGURE 8.8   Cleansing a shell (tube or coil) condenser. *(Virginia Chemicals)*

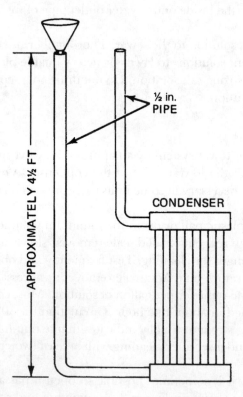

FIGURE 8.9   Opening a passageway for cleaning solution by using the stand-pipe method.

ures should be taken to prevent foam from injuring surrounding objects. The antifoam reagent supplied with each package will help control this nuisance. When the cleaning operation has been completed, drain the spent solution to the sewer. Rinse the condenser with at least two fillings of fresh water.

SAFETY

Most areas of the tower must be inspected for safe working and operating conditions. A number of items should be inspected yearly and repaired immediately in order to guarantee the safety of maintenance personnel:

- Scale remover contains acid and can cause skin irritation—avoid contact with your eyes, skin, and clothing; in case of contact, flush the skin or eyes with plenty of water for at least 15 minutes, and, if the eyes are affected, get medical attention

- Keep scale remover and other chemicals out of the reach of children

- Do not drain the spent solution to the roof or to a septic tank; always drain the spent solution in an environmentally safe way, not to the storm sewer

- Safety is all-important: all chemicals, especially acids, should be treated with great respect and handled with care; rubber gloves, acid-proof coveralls, and safety goggles should be worn when working with chemicals

Cleaning a system through the tower, although easier and faster than some of the other methods, presents one unique hazard—wind drift. Wind drift, even with the tower fan off, is a definite possibility. Wind drift will carry tiny droplets of acid that can burn eyes and skin. These acid droplets will also damage automobile finishes and buildings. Should cleaning solution contact any part of the person, it should be washed off immediately with soap and water.

Using forethought and reasonable precaution can prevent grief and expense.

SOLVENTS AND DETERGENTS

There are several uses for solvents and detergents in the ordinary maintenance schedule of air-cooled fin coil condensers, evaporator coils, permanent-type air filters, and fan blades. In most instances, a high-pressure spray washer is used to clean the equipment with detergent. Then a high-pressure spray rinse is used to clean the unit being scrubbed. The pump is usually rated at 2 gallons per minute at 500 pounds per square inch of pressure. The main function is to remove dirt and grease from fans and cooling surfaces. It takes about 10 to 15 minutes for the cleaning solution to do its job. It is then rinsed with clean water.

Use the dipping method to clean permanent-type filters. Prepare a cleaning solution of one part detergent to one part water. Use this solution as a bath in which the filters may be immersed briefly. After dipping, set the filter aside for 10 to 15 minutes. Flush with a stream of water. If water is not available, good results may be obtained by brisk agitation in a tank filled with fresh water.

When draining the solution used for cleaning purposes, be sure to follow the local codes on the use of storm sewers for disposal purposes. Proper disposal of the spent solution is critical for legal operation of this type of air-conditioning unit.

Another more recent requirement is the use of asbestos in the construction of the tower fill. If discovered when inspections are conducted, make sure it is replaced with the latest materials. The older towers can become more efficient with newer fill of more modern design.

Cooling-tower manufacturers design their units for a given performance standard and for conditions such as: the type of chiller used, ambient temperatures, location, and specifications.

As a system ages, it may lose efficiency. The cleanliness of the tower and its components are crucial to the success of the system. An unattended cold-water temperature will rise. This will send warmer water to the chiller. When the chiller kicks out on high head pressure, the system may shut down. Certain precautions should be taken to prevent shutdown from occurring.

MULTIPLE-CHOICE EXAM

1. The two types of refrigeration systems used for air-conditioning applications in buildings are:

    a. vapor compression and water-cooled

    b. vapor compression and absorption

    c. absorption and compression

    d. none of the above

2. Which compression system can use a wide variety of refrigerants?

    a. vapor   b. absorption

    c. motorized   d. natural

3. Where are higher-pressure refrigerants effectively used?

    a. in direct heat chillers   b. in water chillers

    c. in centripetal chillers   d. in centrifugal chillers

4. What type of compressor systems have primarily used HCFC-22 refrigerant?

    a. high displacement   b. negative displacement

    c. positive displacement   d. low displacement

5. Virtually all _____ compression systems require lubricants to allow efficient operation.

    a. liquid   b. vapor

    c. solid   d. water

6. In what unit are the viscosity grades measured in refrigeration systems?

    a. ounce/mm   b. ounce^3/cm

    c. pounds/gallon   d. centistokes (cS)

7. Some of the factors you should consider in process chillers are:

   a. type of refrigerant used

   b. correct sizing of expansion valve

   c. low-pressure switch

   d. low-temperature cutout

   e. oil separator

   f. none of the above

   g. all of the above

8. Packaged chillers used for HVAC are designed primarily for human comfort and are capable of producing a temperature difference of _____ degrees F.

   a. 20   d. 10

   c. 30   d. 40

9. Condensers are heat _____ designed to condense the high-pressure, high-temperature refrigerant discharged by the compressor.

   a. absorbers   b. spreaders

   c. collectors   d. exchangers

10. Compressors are cycled on and off for _____ reduction.

   a. unit   b. heat

   c. refrigerant   d. capacity

11. Condenser fans are of a _____ type and are statically and dynamically balanced.

   a. blade   b. propeller

   c. twisted   d. none of the above

12. Typical direct-driven fan motors on condensers are six-pole and operate at _____ rpm.

   a. 1,600   b. 3,475

   c. 1,100   d. 1,750

13. In most cases a ball-bearing motor is needed to allow operation at _____ speeds.

   a. low   b. high

   c. medium   d. very high

14. Fan guards are mounted around a fan venture and are made to meet the standards of the _____.

    a. International Plumbing Code   b. OSHA

    c. International Building Code   d. International Mechanical Code

15. In an air-cooled condenser, air is drawn in from the _____ of the condenser and exhausted upward.

    a. top   b. sides

    c. bottom   d. none of the above

16. In air-cooled condensers with flooded-heads, pressure control can hold back enough refrigerant in the condenser coils to render some of the coil surface _____.

    a. active   b. inactive

    c. useless   d. improved

17. Which of the following is not a water-cooled condenser?

    a. shell and tube   b. shell and coil

    c. tube and tube   d. none of the above

    e. all of the above

18. Condenser tubes can be cleaned mechanically and _____.

    a. chemically   b. legally

    c. spotlessly   d. electrically

19. Cooling water for condensers can be obtained from a cooling tower or _____ water.

    a. city   b. well

    c. pond   d. lake

    e. all of the above

20. The following are parts that make up a typical evaporative-cooled condenser:

    a. condensing coil   b. water-distribution system

    c. drift eliminator   d. water makeup and drains

    e. centrifugal or propeller fans   f. all of the above

    g. none of the above

21. An evaporative-cooled condenser operates at a much _____ condensing temperature than an air- or water-cooled system.

    a.　lower　　　　　　　　　　b.　higher

    c.　ambient　　　　　　　　　d.　longer

22. Absorption chillers are machines that use heat energy directly to chill the circulating medium, which is usually _____.

    a.　polyethylene glyco　　　　b.　Freon

    c.　water　　　　　　　　　　d.　R-22

23. Which compression cycle is the absorption cycle similar to?

    a.　water　　　　　　　　　　b.　vapor

    c.　liquid　　　　　　　　　　d.　refrigerant-12

24. In the absorption cycle, energy in the form of _____ is added to the first-stage generator.

    a.　water　　　　　　　　　　b.　air

    c.　cold　　　　　　　　　　　d.　heat

25. The direct-fired two-stage absorption chiller can double as a _____.

    a.　heater　　　　　　　　　　b.　freezer

    c.　cooler　　　　　　　　　　d.　condenser

26. Two-stage steam chillers do not usually have heating _____.

    a.　capability　　　　　　　　b.　tubes

    c.　exchangers　　　　　　　　d.　ability

27. Most centrifugal chillers have a(n) _____ -cooled condenser.

    a.　air　　　　　　　　　　　b.　water

    c.　refrigerant　　　　　　　　d.　none of the above

28. Most absorption equipment made today and used in air conditioning is _____ -cooled using the lithium-bromide water cycle.

    a.　water　　　　　　　　　　b.　steam

    c.　vapor　　　　　　　　　　d.　ice water

29. Absorption chillers can be installed almost anywhere in a building. They are particularly desirable where there is a need for little or no noise and _____.

a. vibration   b. water

c. air   d. wind

30. Absorption chillers will operate properly and produce maximum capacity only if they are installed _____.

a. level   b. at home

c. at the office   d. at the plant

31. In absorption chillers the capacity is directly proportional to solution concentration in the _____.

a. absorber   b. evaporator

c. compressor   d. condenser

32. Absorption chillers usually take from _____ minutes to 1 hour to reach full load from a cold start.

a. 15   b. 20

c. 30   d. 35

33. Two important maintenance concerns for the absorption chiller are the need to maintain corrosion control and _____ _____.

a. leak tightness   b. lithium bromide

c. clean water   d. clean pipes

34. Direct-fired absorption chillers should be checked and serviced _____ per season.

a. once   b. twice

c. every other month   d. once a month

35. What causes crystallization in an absorption chiller?

a. power failure   b. lack of water

c. air leak   d. dirt entrapment

TRUE-FALSE EXAM

1. The new lubricants, and particularly the POEs, are much stronger solvents than the mineral oils they replace.

   True   False

2. One of the most pressing problems associated with maintaining a refrigeration system today is the price of the refrigerant.

   True   False

3. A reciprocating compressor is a double-acting piston machine driven directly by a pin and connecting rod from its crankshaft.

   True   False

4. Electric motor-driven reciprocating compressors are either belt-driven or directly coupled to the compressor by a flexible coupling.

   True   False

5. Hermetic compressors are also known as sealed or welded or as cans.

   True   False

6. Screw compressors have almost linear capacity-control mechanisms.

   True   False

7. Scroll technology is based on three scrolls, the first one is fixed and the other two are orbits that rotate around the fixed scroll.

   True   False

8. Scroll compressors produce a vibrating machine.

   True   False

9. A scroll compressor has fewer parts than a reciprocating compressor.

   True   False

10. A liquid chiller system cools water, glycol, brine, alcohol, acids, chemicals, or other fluids.

    True   False

11. Motors used for scroll compressors are suction-gas-cooled.

    True   False

12. Screw compressors have replaced the traditional shipboard centrifugals.

     True   False

13. In packaged liquid chiller systems the high condenser-pressure switch opens if the compressor discharge pressure reaches a preset value.

     True   False

14. The most popular refrigerants for reciprocating liquid chillers are R-22 and R-134a.

     True   False

15. Heat-pump chillers utilize the same heat exchanger for cooling water as they do for heating water.

     True   False

MULTIPLE-CHOICE ANSWERS

1. B	8. B	15. C	22. C	29. A
2. A	9. D	16. B	23. B	30. A
3. D	10. D	17. E	24. D	31. A
4. C	11. B	18. A	25. A	32. A
5. B	12. C	19. A	26. A	33. A
6. D	13. A	20. F	27. B	34. A
7. G	14. B	21. A	28. A	35. A

TRUE-FALSE ANSWERS

1. T	5. T	9. T	13. T
2. T	6. T	10. T	14. T
3. F	7. F	11. T	15. T
4. T	8. F	12. T	

—NOTES—

Chapter 9
FILTERS AND AIR FLOW

AIR MOVEMENT

Convection, Conduction, and Radiation

Heat always passes from a warmer to a colder object or space. The action of refrigeration depends upon this natural law. The three methods by which heat can be transferred are convection, conduction, and radiation.

Convection is heat transfer that takes place in liquids and gases. In convection, the molecules carry the heat from one point to another. Convection can be used to remove heat from an area, and then it can be used to cool. Air or water can be cooled in one plan and circulated through pipes of radiators in another location. In this way the cool water or air is used to remove heat.

Conduction is heat transfer that takes place chiefly in solids. In conduction, the heat passes from one molecule to another without any noticeable movement of the molecules.

Radiation is heat transfer in wave form, such as light or radio waves. It takes place through a transparent medium such as air, without affecting that medium's temperature, volume, and pressure. Radiant heat is not apparent until it strikes an opaque surface, where it is absorbed. The presence of radiant heat is felt when it is absorbed by a substance or by your body.

COMFORT CONDITIONS

The surface temperature of the average adult's skin is 80 degrees F (26.7 degrees C). The body can either gain or lose heat according to the surrounding air. If the surrounding air is hotter than the skin temperature, the body gains heat and the person may become uncomfortable. If the surrounding air is cooler than the skin temperature, then the body loses heat. Again, the person may become uncomfortable. If the temperature is much higher than the skin temperature or much cooler than the body temperature, then the person becomes uncomfortable. If the air is about 70 degrees F (21.1 degrees C) then the body feels comfortable. Skin temperature fluctuates with the temperature of the surface air. The total range of skin temperature is between 40 and 105 degrees F (4.4 and 40.6 degrees C). However, if the temperature rises 10 degrees F (5.5 degrees C), the skin temperature rises only 3 degrees F (1.7 degrees C). Most of the time the normal temperature of the body ranges from 75 to 100 degrees F (23.9 to 37.8 degrees C). Both humidity and temperature affect the comfort of the human body. However, they are not the only factors that cause a person to be comfortable or uncomfortable. In heating or cooling a room, the air velocity, noise level, and temperature variation caused by the treated air must also be considered. When checking for room comfort, it is best to measure the velocity of the air at a distance of 4 to 72 inches from floor level. Velocity is measured with a velometer. Following is a range of air velocities and their characteristics:

- Slower than 15 feet per minute (fpm): stagnant air

- 20 fpm to 50 fpm: acceptable air velocities

- 25 fpm to 35 fpm: the best range for human comfort

- 35 fpm to 50 fpm: comfortable for cooling purposes

Velocities of 50 fpm or higher call for a very high speed for the air entering the room. A velocity of about 750 fpm or greater is needed to create a velocity of 50 fpm or more inside the room. When velocities greater than 750 fpm are introduced, noise will also be present.

Sitting and standing levels must be considered when designing a cooling system for a room. People will tolerate cooler temperatures at ankle level than at sitting level, which is about 30 inches from the floor.

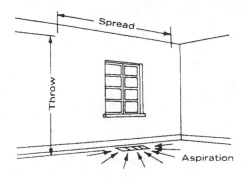

FIGURE 9.1 Aspiration, throw, and spread. *(Lima)*

Variations of 4 degrees F (2.2° C) are acceptable between levels. This is also an acceptable level for temperature variations between rooms.

To make sure that the air is properly distributed for comfort, it is necessary to look at the methods used to accomplish the job.

Terminology

The following terms apply to the movement of air. They are frequently used in referring to air-conditioning systems.

Aspiration is the induction of room air into the primary air stream. Aspiration helps eliminate stratification of air within the room. When outlets are properly located along exposed walls, aspiration also aids in absorbing undesirable currents from these walls and windows (see Figure 9.1).

Cubic feel per minute (cfm) is the measure of a volume of air. It is computed by multiplying the face velocity times the free area in square feet. For example, a resister with 144 square inches (1 square foot) of free area and a measured face velocity of 500 feet per minute would be delivering 500 cubic feet per minute (cfm).

Decibels (db) are units of measure of sound level. It is important to keep noise at a minimum. In most catalogs for outlets, there is a line dividing the noise level of the registers or diffusers. Lower total pressure loss provides a quieter system.

Diffusers are outlets that have a widespread, fan-shaped pattern of air.

Drop is generally associated with air that is discharged horizontally from high sidewall outlets. Since cool air has a natural tendency to drop, it will fall progressively as the velocity decreases. Measured at the point of terminal velocity, drop is the distance in feet that the air has fallen below the level of the outlet (see Figure 9.2).

Effective area is the smallest net area of an outlet utilized by the air stream in passing through the outlet passages. It determines the maximum, or jet, velocity of the air in the outlet. In many outlets, the effective area occurs at the velocity measuring point and is equal to the outlet area.

Face velocity is the average velocity of air passing through the face of an outlet or a return.

Feet per minute (fpm) is the measure of the velocity of an air stream. This velocity can he measured with a velocity meter that is calibrated in feet per minute.

Free area is the total area of the openings in the outlet or inlet through which air can pass. With gravity systems, free area is of prime importance. With forced-air systems, free area is secondary to total pressure loss, except in sizing return air grilles.

Noise criteria (nc) refer to an outlet sound rating in pressure level at a given condition of operation, based on established criteria and a specific room's acoustic absorption value.

Occupied zone is that interior area of a conditioned space that extends to within 6 inches of all room walls and to a height of 6 feet above the floor.

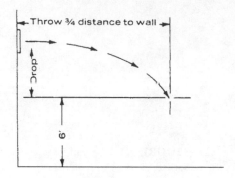

FIGURE 9.2 Drop. *(ARI)*

Outlet area is the area of an outlet utilized by the air stream at the point of the outlet velocity as measured with an appropriate meter. The point of measurement and type of meter must be defined to determine cfm accurately.

Outlet velocity (V_k) is the measured velocity at the starting point with a specific meter.

Perimeter systems are heating and cooling installations in which the diffusers are installed to blanket the outside walls. Returns are usually located at one or more centrally located places. High sidewall or ceiling returns are preferred, especially for cooling. Low returns are acceptable for heating. High sidewall or ceiling returns are highly recommended for combination heating and cooling installations.

Registers are outlets that deliver air in a concentrated stream into the occupied zone.

Residual velocity (V_R) is the average sustained velocity within the confines of the occupied zone, generally ranging from 20 to 70 fpm.

Sound power level (L_w) is the total sound created by an outlet under a specified condition of operation.

Spread is the measurement (in feet) of the maximum width of the air pattern at the point of terminal velocity (see Figure 9.4).

Static pressure (sp) is the outward force of air within a duct. This pressure is measured in inches of water. The static pressure within a duct is comparable to the air pressure within an automobile tire. A manometer measures static pressure.

Temperature differential (ΔT) is the difference between primary supply and room air temperatures.

Terminal velocity is the point at which the discharged air from an outlet decreases to a given speed, generally accepted as 50 feet per minute.

Throw is the distance (measured in feet) that the air stream travels from the outlet to the point of terminal velocity. Throw is measured vertically from perimeter diffusers and horizontally from registers and ceiling diffusers (see Figure 9.1).

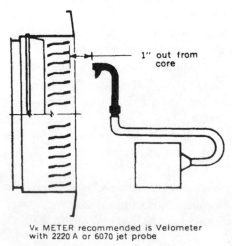

Vₖ METER recommended is Velometer
with 2220 A or 6070 jet probe

CFM = Aₖ × Vₖ

FIGURE 9.3 Air measurement at the grille. *(Lima)*

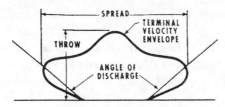

FIGURE 9.4 Typical air-stream pattern. *(Tuttle & Bailey)*

Total pressure (tp) is the sum of the static pressure and the velocity pressure. Total pressure is also known as impact pressure. This pressure is expressed in inches of water. The total pressure is directly associated with the sound level of an outlet. Therefore, any factor that increases the total pressure will also increase the sound level. The under-sizing of outlets or increases in the speed of the blower will increase total pressure and sound level.

Velocity pressure (vp) is the forward-moving force of air within a duct. This pressure is measured in inches of water. The velocity pressure is comparable to the rush of air from a punctured tire. A velometer is used to measure air velocity (see Figure 9.3).

DESIGNING A PERIMETER SYSTEM

After the heat loss or heat gain has been calculated, the sum of these heat losses or heat gains will determine the size of the duct systems and the heating and cooling unit.

The three factors that ensure proper delivery and distribution of air within a room are location, type, and size of outlet. Supply outlets, if possible, should always be located to blanket every window and every outside wall (see Figure 9.5). Thus, a register is recommended under each window. The outlet selected should be a diffuser whose air pattern is fan-shaped to blanket the exposed walls and windows.

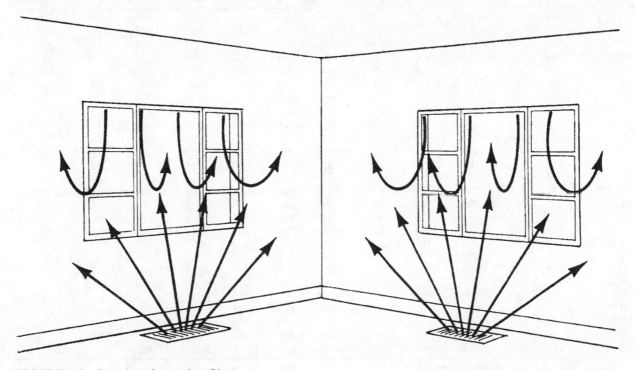

FIGURE 9.5 Location of an outlet. *(Lima)*

The American Society of Heating, Refrigeration, and Air Conditioning Engineers (ASHRAE) furnishes a chart with the locations and load factors needed for the climate of each major city in the United States. The chart should be followed carefully. The type of house, the construction materials, location, room sizes, and exposure to sun and wind are important factors. With such information, you can determine how much heat will be dissipated. You can also determine how much cooling will be lost in a building. The *ASHRAE Handbook of Fundamentals* lists the information needed to compute the load factors.

Calculate the heat loss or heat gain of the room; divide this figure by the number of outlets to be installed. From this you can determine the Btu/h required of each outlet. Refer to the performance data furnished by the manufacturer to determine the size the outlet should be. For residential application, the size selected should be large enough so that the Btu/h capacity on the chart falls to the side where the quiet zone is indicated. There is still a minimum vertical throw of 6 feet where cooling is involved.

Locating and Sizing Returns

Properly locating and sizing return air grilles is important. It is generally recommended that the returns be installed in high sidewalls or the ceiling. They should be in one or more centrally located places. This depends upon the size and floor plan of the structure. Although such a design is preferred, low returns are acceptable for heating.

To minimize noise, care must be taken to size the return air grille correctly. The blower in the equipment to be used is rated in cfm by the manufacturer. This rating can usually be found in the specification sheets. Select the grille or grilles necessary to handle this cfm.

The grille or grilles selected should deliver the necessary cfm for the air to be conditioned. The throw should reach approximately three-quarters of the distance from the outlet to the opposite wall (see Figure 9.2). The face velocity should not exceed the recommended velocity for the application (see Table 9.1). The

TABLE 9.1   Register or grille size related to air capacities (in cfm).

Register of Grille Size	Area in Ft²	Air Capacities in cfm										
		250 fpm	300 fpm	400 fpm	500 fpm	600 fpm	700 fpm	750 fpm	800 fpm	900 fpm	1,000 fpm	1,250 fpm
8 × 4	.163	41	49	65	82	98	114	122	130	147	163	204
10 × 4	.206	52	62	82	103	124	144	155	165	185	206	258
10 × 6	.317	79	95	127	158	190	222	238	254	285	317	396
12 × 4	.249	62	75	100	125	149	174	187	199	224	249	311
12 × 5	.320	80	96	128	160	192	224	240	256	288	320	400
12 × 6	.383	96	115	153	192	230	268	287	306	345	383	479
14 × 4	.292	73	88	117	146	175	204	219	234	263	292	365
14 × 5	.375	94	113	150	188	225	263	281	300	338	375	469
14 × 6	.449	112	135	179	225	269	314	337	359	404	449	561
16 × 5	.431	108	129	172	216	259	302	323	345	388	431	539
16 × 6	.515	129	155	206	258	309	361	386	412	464	515	644
20 × 5	.541	135	162	216	271	325	379	406	433	487	541	676
20 × 6	.647	162	194	259	324	388	453	485	518	582	647	809
20 × 8	.874	219	262	350	437	524	612	656	699	787	874	1,093
24 × 5	.652	162	195	261	326	391	456	489	522	587	652	815
24 × 6	.779	195	234	312	390	467	545	584	623	701	779	974
24 × 8	1.053	263	316	421	527	632	737	790	842	948	1,053	1,316
24 × 10	1.326	332	398	530	663	796	928	995	1,061	1,193	1,326	1,658
24 × 12	1.595	399	479	638	798	951	1,117	1,196	1,276	1,436	1,595	1,993
30 × 6	.978	245	293	391	489	587	685	734	782	880	978	1,223
30 × 8	1.321	330	396	528	661	793	925	991	1,057	1,189	1,371	1,651
30 × 10	1.664	416	499	666	832	998	1,165	1,248	1,331	1,498	1,664	2,080
30 × 12	2.007	502	602	803	1,004	1,204	1,405	1,505	1,606	1,806	2,007	2,509
36 × 8	1.589	397	477	636	795	953	1,112	1,192	1,271	1,430	1,589	1,986
36 × 10	2.005	501	602	802	1,003	1,203	1,404	1,504	1,604	1,805	2,005	2,506
36 × 12	2.414	604	724	966	1,207	1,448	1,690	1,811	1,931	2,173	2,414	3,018

* Based on LIMA registers of the 100 Series.

TABLE 9.2 Outlet velocity ratings.

Area	Rating (in. fpm)
Broadcast studios	500
Residences	500 to 750
Apartments	500 to 750
Churches	500 to 750
Hotel bedrooms	500 to 750
Legitimate theatres	500 to 1,000
Private offices, acoustically treated	500 to 1,000
Motion picture theatres	1,000 to 1,250
Private offices, not acoustically treated	1,000 to 1,250
General offices	1,250 to 1,500
Stores	1,500
Industrial buildings	1,500 to 2,000

drop should be such that the air stream will not fall into the occupied zone. The occupied zone is generally thought of as 6 feet above floor level.

The sound caused by an air outlet in operation varies in direct proportion to the velocity of the air passing through it. Air velocity depends partially on outlet size. Table 9.2 lists recommendations for outlet velocities within safe sound limits for most applications.

Air-Flow Distribution

Bottom or side outlet openings in horizontal or vertical supply ducts should be equipped with adjustable flow-equalizing devices. Figure 9.6 indicates the pronounced one-sided flow effect from an outlet opening. This is before the corrective effect of air-turning devices. A control grid is added in Figure 9.7 to equalize flow in the takeoff collar. A Vectrol® is added in Figure 9.8 to turn air into the branch duct and provide volume control. Air-turning devices are recommended for installation at all outlet collars and branch duct connections.

Square unvaned elbows are also a source of poor duct distribution and high-pressure loss. Non-uniform flow in a main duct, occurring after an unvaned ell, severely limits the distribution of air into branch ducts in the vicinity of the ell. One side of the duct may be void, thus starving a branch duct. Conversely, all flow may be stacked up on one side. This requires dampers to be excessively closed, resulting in higher sound levels.

Flow diagrams show the pronounced turbulence and piling up of airflow in an ell (see Figure 9.9). Duct-turns reduce the pressure loss in square elbows as much as 80 percent. Their corrective effect is shown in Figure 9.10.

FIGURE 9.6 This flow path diagram shows the pronounced one-sided flow effect from an outlet opening before the corrective effect of air-turning devices. *(Tuttle & Bailey)*

FIGURE 9.7 A control grid is added to equalize flow in the takoff collar. *(Tuttle & Bailey)*

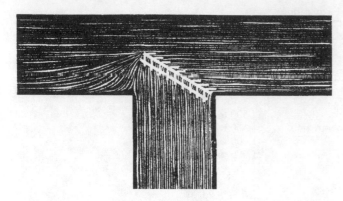

FIGURE 9.8   A Vectrol is added to turn air into the branch duct and provide volume control. *(Tuttle & Bailey)*

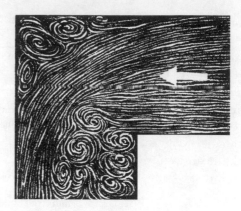

FIGURE 9.9   Note the turbulence and piling up of airflow in an ell. *(Tuttle & Bailey)*

FIGURE 9.10   A ducturn reduces the pressure loss in square elbows by as much as 80 percent. *(Tuttle & Bailey)*

Selection of Diffusers and Grilles

The selection of a linear diffuser or grille involves the job condition requirements, judgment, and performance-data analysis. Diffusers and grilles should be selected and sized according to the following characteristics:

- Type and style
- Function
- Air-volume requirement
- Throw requirement
- Pressure requirement
- Sound requirement

The air volume per diffuser or grille is that which is necessary for the cooling, heating, or ventilation requirements of the area served by the unit. The air volume required, when related to throw, sound, or pressure design limitations, determines the proper diffuser or grille size.

Generally, air volumes for internal zones of building spaces vary from 1 to 3 cfm per square foot of floor area. Exterior zones require higher air volumes of 2.5 to 4 cfm per square foot. In some cases, only the heating or cooling load of the exterior wall panel or glass surface is to be carried by the distribution center. Then, the air volume per linear foot of diffuser or grille will vary from 20 to 200 cfm, depending on heat-transfer coefficient, wall height, and infiltration rate.

Throw and occupied-area air location are closely related. Both could be considered in the analysis of specific area requirements. The minimum-maximum throw for a given condition of aeration is based upon a terminal velocity at that distance from the diffuser. The residual room velocity is a function of throw to terminal velocity. Throw values are based on terminal velocities ranging from 75 to 150 fpm with corresponding residual room velocities of 75 to 150 fpm. The diffuser or grille location, together with the air pattern selected, should generally direct the air path above the occupied zone. The air path then induces room air along its throw as it expands in cross-section. This equalizes temperature and velocity within the stream. With the throw terminating in a partition or wall surface, the mixed air path further dissipates energy.

Ceiling-mounted grilles and diffusers are recommended for vertical down patterns. Some locations in the room may need to be cooler than others. Also, some room locations may be harder to condition because of air-flow problems. These grilles are used in areas adjacent to perimeter wall locations that require localized spot conditioning. Ceiling heights of 12 feet or greater are needed. The throw for vertical projection is greatly affected by supply air temperature and proximity of wall surfaces.

Sidewall-mounted diffusers and grilles have horizontal values based on a ceiling height of 8 to 10 feet. The diffuser or grille is mounted approximately 1 foot below the ceiling. For a given throw, the room air motion will increase or decrease inversely with the ceiling height. For a given air-pattern setting and room air motion, the listed minimum-maximum throw value can be decreased by 1 foot for each 1-foot increase in ceiling height above 10 feet. Throw values are furnished by the manufacturer. When sidewall grilles are installed remote from the ceiling (more than 3 feet away), reduce rated throw values by 20 percent.

Sill-mounted diffusers or grilles have throw values based on an 8- to 10-foot ceiling height with the outlet installed in the top of a 30-inch-high sill. For a given throw, the room air motion will change with the ceiling height. For a given air-pattern setting and room air motion, the listed minimum-maximum throw value can be decreased by 2 feet for each 1-foot increase in ceiling height above 10 feet. Decrease 1 foot for each 1-foot decrease in sill height.

The minimum throw results in a room air motion higher than that obtained when utilizing the maximum throw. Thus, 50 fpm rather than 35 fpm is the air motion. The listed minimum throw indicates the minimum distance recommended. The minimum distance is from the diffuser to a wall or major obstruction, such as a structural beam. The listed maximum throw is the recommended maximum distance to a wall or major obstruction. Throw values and the occupied area velocity for sidewall grilles and ceiling diffusers are based on flush ceiling construction, providing an unobstructed air-stream path. The listed maximum throw times 1.3 is the complete throw of the air stream where the terminal velocity equals the room air velocity. Rated occupied-area velocities range from 25 to 35 fpm for maximum listed throws and 35 to 50 fpm for minimum listed throws.

Cooled air drop or heated-air rise are of practical significance when supplying heated or cooled air from a sidewall grille. If the throw is such that the air stream prematurely enters the occupied zone, considerable draft may be experienced. This is due to incomplete mixing. The total airdrop must be considered when the wall grille is located a distance from the ceiling. Cooled air drop is controlled by spacing the wall grille from the ceiling and adjusting the grilles upward 15 inches. Heated-air rise contributes significantly to temperature stratification in the upper part of the room.

The minimum separation between grille and ceiling must be 2 feet or more. The minimum mounting separation must also be 2 feet or more. The minimum mounting height should be 7 feet.

The diffuser or grille minimum pressure for a given air volume is reflected in ultimate system fan horsepower requirements. A diffuser or grille with a lower pressure rating requires less total energy than a unit with a higher pressure rating for a given air volume and effective area. Diffusers and grilles of a given size with lower pressure ratings usually have a lower sound level rating at a specified air volume.

TABLE 9.3   Recommend NC criteria.

NC Curve	Communication Environment	Typical Occupancy
Below NC 25	Extremely quiet environment, suppressed speech is quite audible, suitable for acute pickup of all sounds	Broadcasting studios, concert halls, music rooms.
NC 30	Very quiet office, suitable for large conferences; telephone use satisfactory.	Residences, theatres, libraries, executive offices, directors' rooms.
NC 35	Quiet office; satisfactory for conference at a 15 ft table; normal voice 10 to 30 ft telephone use satisfactory.	Private offices, schools, hotel rooms, courtrooms, churches, hospital rooms.
NC 40	Satisfactory for conferences at a 6 to 8 ft table; normal voice 6 to 12 ft; telephone use satisfactory.	General offices, labs, dining rooms.
NC 45	Satisfactory for conferences at a 4 to 5 ft; table; normal voice 3 to 6 ft; raised voice 6 to 12 ft; telephone use occasionally difficult.	Retail stores, cafeterias, lobby areas large drafting and engineering offices, reception areas.
Above NC 50	Unsatisfactory for conferences of more than two or three persons; normal voice 1 to 2 ft; raised voice 3 to 6 ft; telephone use slightly difficult.	Photocopy rooms, stenographic pools, print machine rooms, process areas.

Diffusers and grilles should be selected for the recommended noise criteria rating for a specific application. The data for each specific diffuser or grille type contains an NC rating. Table 9.3 lists recommended noise criteria (nc) and area of application.

High velocities in the duct or diffuser typically generate air noise. The flow turbulence in the duct and the excessive pressure reductions in the duct and diffuser system also generate noise. Such noise is most apparent directly under the diffuser. Room background levels of nc 35 and less provide little masking effect. Any noise source stands out above the background level and is easily detected.

Typically, air noise can be minimized by the following procedures:

- Limiting branch-duct velocities to 1,200 fpm

- Limiting static pressure in branch ducts adjacent to outlets to 0.15 inches H_2O

- Sizing diffusers to operate at outlet jet velocities up to 1,200 fpm (neck velocities limited to 500 to 900 fpm) and total pressures of 0.10 inches H_2O

- Using several small diffusers (and return grilles) instead of one or two large outlets or inlets that have a higher sound power

- Providing low-noise dampers in the branch duct where pressure drops of more than 0.20 inches of water must be taken

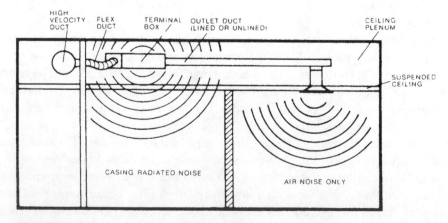

FIGURE 9.11   Casing noises. *(Tuttle & Bailey)*

- Internally lining branch ducts near the fan to quiet this noise source

- Designing background sound levels in the room to be a minimum of nc 35 or nc 40

Casing noise differs from air noise in the way it is generated. Volume controllers and pressure-reducing dampers generate casing noise. Inside terminal boxes are sound baffles, absorbing blankets, and orifice restrictions to eliminate line of sight through the box. All these work to reduce the generated noise before the air and air noise discharge from the box into the outlet duct. During this process, the box casing is vibrated by the internal noise. This causes the casing to radiate noise through the suspended ceiling into the room (see Figure 9.11).

Locating Terminal Boxes

In the past, terminal boxes and ductwork were separated from the room by dense ceilings. These ceilings prevented the system noise from radiating into the room. Plaster and taped sheetrock ceilings are examples of dense ceilings. Current architectural practice is to utilize lightweight (and low-cost) decorative suspended ceilings. These ceilings are not dense. They have only one-half the resistance to noise transmission that plaster and sheetrock ceilings have. Exposed tee-bar grid ceilings with 2×4 glass fiber pads and perforated metal pan ceilings are examples. The end result is readily apparent. Casing radiated noise in lightweight modern buildings is a problem.

Terminal boxes can sometimes be located over noisy areas (corridors, toilet areas, machine-equipment rooms) rather than over quiet areas. In quiet areas casing noise can penetrate the suspended ceiling and become objectionable. Enclosures built around the terminal box (such as sheetrock or sheet lead over a glass-fiber blanket wrapped around the box) can reduce the radiated noise to an acceptable level.

However, this method is cumbersome and limits access to the motor and volume controllers in the box. It depends upon field conditions for satisfactory performance and is expensive. Limiting static pressure in the branch ducts minimizes casing noise. This technique, however, limits the flexibility of terminal-box systems. It hardly classifies as a control.

Vortex Shedding

Product research in controlling casing noise has developed a new method of reducing radiated noise. The technique is known as vortex shedding. When applied to terminal boxes, casing radiated noise is dramatically lowered. Casing-radiation-attenuation vortex shedders (CRA) can be installed in all single- or dual-duct boxes up to 7000 cfm, both constant volume or variable volume, with or without reheat coils. CRA devices provide unique features and the following benefits:

- No change in terminal box size; box is easier to install in tight ceiling plenums to ensure minimum casing noise under all conditions

- Factory-fabricated box and casing noise eliminator, a one-piece assembly, reduces cost of installation: only one box is hung, and only one duct connection is made

- Quick-opening access door is provided in box to assure easy and convenient access to all operating parts without having to cut and patch field-fabricated enclosures

- Equipment is laboratory-tested and performance-rated, and engineering measurements are made in accordance with industry standards, thus, on-job performance is ensured; quiet rooms result and owner satisfaction is assured

Return Grilles

Return air grilles are usually selected for the required air volume at a given sound level or pressure value. The intake-air velocity at the face of the grille depends mainly on the grille size and the air volume.

The grille style and damper setting have a small effect on this intake velocity. The grille style, however, has a very great effect on the pressure drop. This, in turn, directly influences the sound level.

The intake velocity is evident only in the immediate vicinity of the return grille. It cannot influence room air distribution. Recent ASHRAE research projects have developed a scientific computerized method of relating intake-grille velocities, measured 1 inch out from the grille face, to air volume. Grille measuring factors for straight, deflected-bar, open, and partially closed dampers are in the engineering data furnished with the grille.

It still remains the function of the supply outlets to establish proper coverage, air motion, and thermal equilibrium. Because of this, the location of return grilles is not critical, and their placement can be largely a matter of convenience. Specific locations in the ceiling may be desirable for local heat loads or smoke exhaust, or a location in the perimeter sill or floor may be desirable for an exterior-zone intake under a window wall section. It is not advisable to locate large centralized return grilles in an occupied area. The large mass of air moving through the grille can cause objectionable air motion for nearby occupants.

Return air grilles should be selected for static pressures. These pressures will provide the required nc rating and conform to the return-system performance characteristics. Fan sound power is transmitted through the return-air system as well as the supply system. Fan silencing may be necessary or desirable in the return side. This is particularly so if silencing is being considered on the supply side.

Transfer grilles venting into the ceiling plenum should be located remotely from the plenum noise source. The use of a lined sheet-metal elbow can reduce transmitted sound. Lined elbows on vent grilles and lined common ducts on ducted return grilles can minimize "cross talk" between private offices.

The spread of an unrestricted air stream is determined by the grille-bar deflection. Grilles with vertical face bars at 0-degree deflection will have a maximum throw value. As the deflection setting of vertical bars is increased, the air stream covers a wider area and the throw decreases.

Registers are available with adjustable valves. An air-leakage problem is eliminated if the register has a rubber gasket mounted around the grille. When it pulls up tightly against the wall, an airtight seal is made. This helps to eliminate noise. The damper has to be cam-operated so that it will stay open and not blow shut when the air comes through.

On some registers, a simple tool can be used to change the direction of the deflection bars. This means that adjusting the bars in the register can result in a number of deflection patterns.

Fire and Smoke Dampers

Ventilating, air conditioning, and heating ducts provide a path for fire and smoke, which can travel throughout a building. The ordinary types of dampers that are often installed in these ducts depend on gravity-close action or spring-and-level mechanisms. When their releases are activated, they are freed to drop inside the duct.

A fusible link attachment to individual registers also helps control fire and smoke. Figure 9.12 shows a fusible-link register. The link is available with melting points of 160 degrees F (71.1 degrees C) or 212 degrees F (100 degrees C). When the link melts, it releases a spring that forces the damper to a fully closed position. The attachment does not interfere with damper operation.

Fire and smoke safety concepts in high-rise buildings are increasingly focusing on providing safety havens for personnel on each floor. This provision is to optimize air flow to or away from the fire floor or adjacent floors. Such systems require computer-actuated smoke dampers. Dampers are placed in supply and return ducts that are reliable. They must be tight-closing and offer minimum flow resistance when fully open.

Ceiling Supply Grilles, Registers, and Diffusers

Some ceiling grilles and registers have individually adjustable vanes. They are arranged to provide a one-way ceiling air pattern. They are recommended for applications in ceiling and sidewall locations for heating and cooling systems. They work best where the system has 0.75 to 1.75 cfm per square foot of room area (see Figure 9.13).

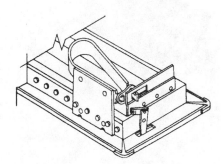

FIGURE 9.12 Register with fusible link for fire control. *(Lima)*

FIGURE 9.13 Ceiling grille. *(Tuttle & Bailey)*

Some supply ceiling grilles and registers have individually adjustable curved vanes. They are arranged to provide a three-way ceiling air pattern. The vertical face vanes are a three-way diversion for air. A horizontal pattern with the face vanes also produces a three-way dispersion of air. These grilles and registers are recommended for applications in ceiling locations for heating and cooling systems handling 1.0 to 2.0 cfm per square foot of room area.

Figure 9.14 shows a grille with four-way vertical face vanes. Horizontal face vanes are also available. They, too, are adjustable individually for focusing an air stream in any direction. Both the three-way and four-way pattern grilles can be adjusted to a full or partial down blow position. The curved streamlined vanes are adjusted to a uniform partially closed position. This deflects the air path while retaining an effective area capacity of 35 percent of the neck area. In the full down blow position, grille effective area is increased by 75 percent.

Perforated adjustable diffusers for ceiling installation are recommended for heating and cooling (see Figure 9.15). They are also recommended for jobs requiring on-the-job adjustment of air diffusion patterns.

Full-flow square or round necks have expanded metal air-pattern deflectors. They are adjustable for four-, three-, two-, or one-way horizontal diffusion patterns. This can be done without change in the air volume, pressure, or sound levels. These deflectors and diffusers have high diffusion rates. The result is rapid temperature and velocity equalization of the mixed air mass well above the zone of occupancy. They diffuse efficiently with six to eighteen air changes per hour.

There are other designs in ceiling diffusers. The type shown in Figure 9.16 is often used in a supermarket or other large store. Here, it is difficult to mount other means of air distribution. These round diffusers with

FIGURE 9.14 Vertical face vanes in a four-way ceiling supply grille. *(Tuttle & Bailey)*

FIGURE 9.15 Perforated-face adjustable diffuser for full flow and a deflector for ceiling installation. *(Tuttle & Bailey)*

FIGURE 9.16 Round diffusers with flush face and fixed pattern for ceiling installation. *(Tuttle & Bailey)*

FIGURE 9.17 Control grid with multiblade devices to control airflow in a diffuser collar. *(Tuttle & Bailey)*

a flush face and fixed pattern are for ceiling installation. They are used for heating, ventilating, and cooling. They are compact and simple flush diffusers. High induction rates result in rapid temperature and velocity equalization of the mixed air mass. Mixing is done above the zone of occupancy.

Grids are used and sold as an accessory to these diffusers. The grid (Figure 9.17) is a multi-blade device designed to ensure uniform airflow in a diffuser collar. It is individually adjustable. The blades can be moved to control the air stream precisely.

For maximum effect, the control grid should be installed with the blades perpendicular to the direction of approaching airflow. Where short collars are encountered, a double bank of control grids is recommended. The upper grid is placed perpendicular to the branch duct flow. The lower grid is placed parallel to the branch duct flow. The control grid is attached to the duct collar by means of mounting straps. It is commonly used with volume dampers.

The antismudge ring is designed to cause the diffuser discharge-air path to contact the ceiling in a thin-layered pattern. This minimizes local turbulence, the cause of distinct smudging (see Figure 9.18).

For the best effect, the antismudge ring must fit evenly against the ceiling surface. It is held in position against the ceiling by the diffuser margin. This eliminates any exposed screws.

Air-channel supply diffusers are designed for use with integrated air-handling ceiling systems. They are adaptable to fit between open parallel tee bars. They fit within perforated or slotted ceiling runners. The appearance of the integrated ceiling remains unchanged regardless of the size of the unit. They are painted out to be invisible when viewing the ceiling. These high-capacity diffusers provide a greater air-handling capability (see Figure 9.19).

FIGURE 9.18 Antismudge ring. (Tuttle & Bailey)

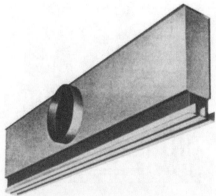

FIGURE 9.19 High-capacity air-channel diffuser with fixed pattern for suspended grid ceilings. *(Tuttle & Bailey)*

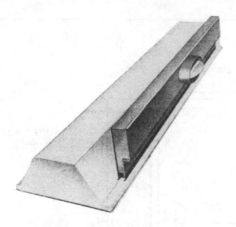

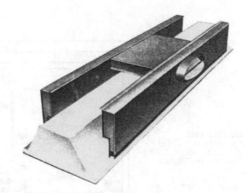

FIGURE 9.20 Single-side diffuser with side inlet. *(Tuttle & Bailey)*

FIGURE 9.21 Dual-side diffuser with side inlet. *(Tuttle & Bailey)*

The luminaire is a complete lighting unit. The luminaire diffuser fits close to the fluorescent lamp fixtures in the ceiling. The single-side diffuser with side inlet is designed to provide single-side concealed air distribution (see Figure 9.20). The diffuser is designed with oval-shaped side inlets and inlet dampers and provides effective single-point dampering.

Dual-side diffusers with side inlets are designed to provide concealed air distribution. Note the crossover from the oval side inlet to the other side of the diffuser. This type of unit handles more air and spreads it more evenly when used in large areas (see Figure 9.21). This type of diffuser is also available with an insulation jacket when needed.

Figure 9.22 illustrates the airflow from ceiling diffusers. The top view illustrates the motion from the diffuser. The side view shows how the temperature differential is very low. Note that the temperature is 68 degrees F near the ceiling and sidewall and 73 degrees F on the opposite wall near the ceiling.

Linear Grilles

Linear grilles are designed for installation in the sidewall, sill, floor, and ceiling. They are recommended for supplying heated, ventilated, or cooled air and for returning or exhausting room air (see Figure 9.23).

When installed in the sidewall near the ceiling, linear grilles provide a horizontal pattern above the occupied zone. Core deflections of 15 and 30 degrees direct the air path upward to overcome the drop effect resulting from cool primary air.

When installed in the top of a sill or enclosure, linear grilles provide a vertical up pattern. This is effective in overcoming uncomfortable cold downdrafts. It also offsets the radiant effect of glass surfaces. Core deflections of 0 and 15 degrees directed toward the glass surface provide upward airflow to the ceiling and along the ceiling toward the interior zone.

When installed in the ceiling, linear grilles provide a vertical downward air pattern. This pattern is effective in projection heating and in cooling the building perimeter from ceiling heights above 13 to 15 feet. Application of down-flow primary air should be limited to ensure against excessive drafts at the end of the throw. Core deflections of 0, 15, and 30 degrees direct the air path angularly downward as required. Debris screens can be integrally attached (see Figure 9.24).

Fans and Mechanical Ventilation

Mechanical ventilation differs from natural ventilation mainly in that the air circulation is performed by mechanical means (such as fans or blowers). In natural ventilation, the air is caused to move by natural forces. In mechanical ventilation the required air changes are affected partly by diffusion but chiefly by positive currents put in motion by electrically operated fans or blowers, as shown in Figure 9.25. Fresh air is usually circulated through registers connected with the outside and warmed as it passes over and through the intervening radiators.

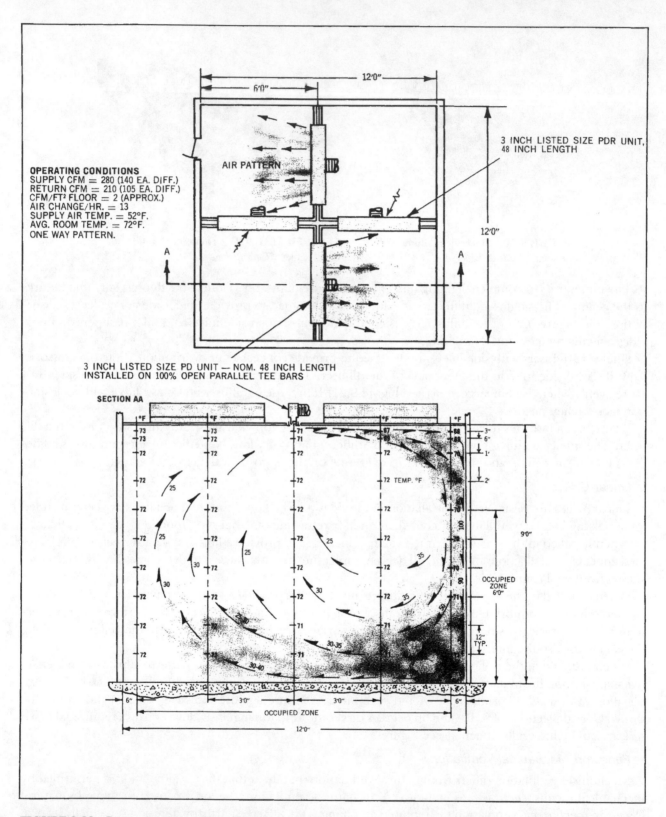

OPERATING CONDITIONS
SUPPLY CFM = 280 (140 EA. DIFF.)
RETURN CFM = 210 (105 EA. DIFF.)
CFM/FT² FLOOR = 2 (APPROX.)
AIR CHANGE/HR. = 13
SUPPLY AIR TEMP. = 52°F.
AVG. ROOM TEMP. = 72°F.
ONE WAY PATTERN.

3 INCH LISTED SIZE PDR UNIT, 48 INCH LENGTH

3 INCH LISTED SIZE PD UNIT — NOM. 48 INCH LENGTH INSTALLED ON 100% OPEN PARALLEL TEE BARS

SECTION AA

FIGURE 9.22   Room air motion. *(Tuttle & Bailey)*

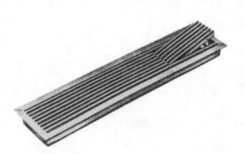

FIGURE 9.23 Linear grille with a hinged access door. *(Tuttle & Bailey)*

FIGURE 9.24 Debris screen for linear grilles. *(Tuttle & Bailey)*

The volume of air required is determined by the size of the space to be ventilated and the number of times per hour that the air in the space is to be changed. In many cases, existing local regulations or codes will govern the ventilating requirements. Some of these codes are based on a specified amount of air per person and others on the air required per square foot of floor area.

The various devices used to supply air circulation in air-conditioning applications are known as fans, blowers, exhausts, or propellers. The different types of fans may be classified with respect to their construction as follows:

- Propeller

- Tube axial

- Vane axial

- Centrifugal

A propeller fan consists essentially of a propeller or disk-type wheel within a mounting ring or plate and includes the driving-mechanism supports for either belt or direct drive. A tube axial fan consists of a propeller or disk-type wheel within a cylinder and includes the driving-mechanism supports for either belt drive or direct connection. A vane axial fan consists of a disk-type wheel within a cylinder and a set of air-guide vanes located before or after the wheel. It includes the driving-mechanism supports for either belt drive or direct connection. A centrifugal fan consists of a fan rotor or wheel within a scroll-type housing and includes the driving-mechanism supports for either belt drive or direct connection. Figure 9.26 shows the mounting arrangements.

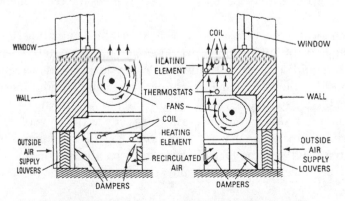

FIGURE 9.25 Typical mechanical ventilators for residential use. Note placement of fans and other details.

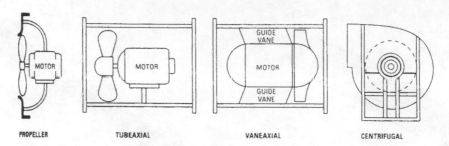

FIGURE 9.26 Fan classifications with proper mounting arrangement.

Fan performance may be stated in various ways, with the air volume per unit time, total pressure, static pressure, speed, and power input being the most important. The terms, as defined by the National Association of Fan Manufacturers, are as follows:

- Volume handled by a fan is the number of cubic feet of air per minute expressed as fan-outlet conditions

- Total pressure of a fan is the rise of pressure from fan inlet to fan outlet

- Velocity pressure of a fan is the pressure corresponding to the average velocity determination from the volume of airflow at the fan outlet area

- Static pressure of a fan is the total pressure diminished by the fan-velocity pressure

- Power output of a fan is expressed in horsepower and is based on fan volume and the fan total pressure

- Power input of a fan is expressed in horsepower and is measured as horsepower delivered to the fan shaft

- Mechanical efficiency of a fan is the ratio of power output to power input

- Static efficiency of a fan is the mechanical efficiency multiplied by the ratio of static pressure to total pressure

- Fan-outlet area is the inside area of the fan outlet

- Fan-inlet area is the inside area of the inlet collar

The volume of air required is determined by the size of the space to be ventilated and the number of times per hour that the air in the space is to be changed. Table 9.4 shows the recommended rate of air change for various types of spaces.

In many cases, existing local regulations or codes will govern the ventilating requirements. Some of these codes are based on a specified amount of air per person and on the air required per square foot of floor area. Table 9.4 should serve as a guide to average conditions. Where local codes or regulations are involved, they should be taken into consideration. If the number of persons occupying the space is larger than would be normal for such a space, the air should be changed more often than shown.

The horsepower required for any fan or blower varies directly as the cube of the speed, provided that the area of the discharge orifice remains unchanged. The horsepower requirements of a centrifugal fan generally decrease with a decrease in the area of the discharge orifice if the speed remains unchanged. The horsepower requirements of a propeller fan increase as the area of the discharge orifice decreases if the speed remains unchanged.

Whenever possible, the fan wheel should be directly connected to the motor shaft. This can usually be accomplished with small centrifugal fans and with propeller fans up to about 60 inches in diameter. The

TABLE 9.4   Volume of air required for various facilities.

Space to Be Ventilated	Air Changes per Hour	Minutes per Change
Auditoriums	6	10
Bakeries	20	3
Bowling alleys	12	5
Club rooms	12	5
Churches	6	10
Dining rooms (restaurants)	12	5
Factories	10	6
Foundries	20	3
Garages	12	5
Kitchens (restaurants)	30	2
Laundries	20	3
Machine shops	10	6
Offices	10	6
Projection booths	60	1
Recreation rooms	10	6
Sheet-metal shops	10	6
Ship holds	6	10
Stores	10	6
Toilets	20	3
Tunnels	6	10

deflection and the critical speed of the shaft, however, should be investigated to determine whether or not it is safe.

When selecting a motor for fan operation, it is advisable to select a standard motor one size larger than the fan requirements. It should be kept in mind, however, that direct-connected fans do not require as great a safety factor as do belt-driven units. It is desirable to employ a belt drive when the required fan speed or horsepower is in doubt, since a change in pulley size is relatively inexpensive if an error is made (see Figure 9.27).

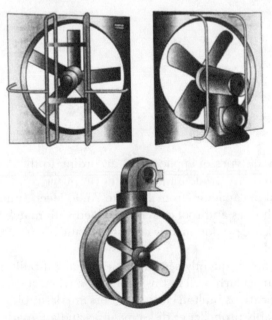

FIGURE 9.27   Various types of propeller fan drives and mounting arrangements.

Directly connected small fans for single-phase AC motors of the split-phase, capacitor, or shaded-pole type usually drive various applications. The capacitor motor is more efficient electrically and is used in districts where there are current limitations. Such motors, however, are usually arranged to operate at one speed. With such a motor, if it is necessary to vary the air volume or pressure of the fan or blower, the throttling of air by a damper installation is usually made.

In large installations (such as when mechanical draft fans are required), various drive methods are used, including: a slip-ring motor to vary the speed and a constant-speed, directly connected motor, which, by means of movable guide vanes in the fan inlet serves to regulate the pressure and air volume.

Most often, the service determines the type of fan to use. When operation occurs with little or no resistance, and particularly when no duct system is required, the propeller fan is commonly used because of its simplicity and economy in operation. When a duct system is involved, a centrifugal or axial type of fan is usually employed. In general, centrifugal and axial fans are comparable with respect to sound effect, but axial fans are somewhat lighter and require considerably less space. The following information is usually required for proper fan selection:

- Capacity requirement in cubic feet per minute

- Static pressure or system resistance

- Type of application or service

- Mounting arrangement of system

- Sound level or use of space to be served

- Nature of load and available drive

The various fan manufacturers generally supply tables or characteristic curves that ordinarily show a wide range of operating particulars for each fan size. The tabulated data usually include static pressure, outlet velocity, revolutions per minute, brake horsepower, tip or peripheral speed, and so on.

The numerous applications of fans in the field of air conditioning and ventilation are well known, particularly to engineers and air-conditioning repair and maintenance personnel. The various fan applications are as follows:

- Attic fans

- Circulating fans

- Cooling-tower fans

- Exhaust fans

- Kitchen fans

Exhaust fans are found in all types of applications, according to the American Society of Heating and Ventilating Engineers. Wall fans are predominantly of the propeller type, since they operate against little or no resistance. They are listed in capacities from 1,000 to 75,000 feet^3/minute. They are sometimes incorporated in factory-built penthouses and roof caps or provided with matching automatic louvers. Hood exhaust fans involving ductwork are predominantly centrifugal, and are used especially in handling hot or corrosive fumes.

Spray-booth exhaust fans are frequently centrifugal, especially if built into self-contained booths. Tube axial fans lend themselves particularly well to this application when suspension in a section of ductwork is advantageous. For such applications built-in cleanout doors are desirable.

Circulating fans are invariably propeller or disk-type units and are made in a vast variety of blade shapes and arrangements. They are designed for appearance as well as utility. Cooling-tower fans are predominantly of the propeller type. However, axial types are also used for packed towers, and occasionally a centrifugal fan is used to supply draft. Kitchen fans for domestic use are small propeller fans arranged for window or wall mounting and with various useful fixtures. They are listed in capacity ranges from 300 to 800 feet^3/minute.

Attic fans are used during the summer to draw large volumes of outside air through the house or building whenever the outside temperature is lower than that of the inside. It is in this manner that the relatively cool evening or night air is utilized to cool the interior in one or several rooms, depending on the location of the air-cooling unit. It should be clearly understood, however, that the attic fan is not strictly a piece of air-conditioning equipment since it only moves air and does not cool, clean, or dehumidify. Attic fans are used primarily because of their low cost and economy of operation, combined with their ability to produce comfort cooling by circulating air rather than conditioning it.

Fans may be centrally located in an attic or other suitable space (such as a hallway) and arranged to move air proportionally from several rooms. A local unit may be installed in a window to provide comfort cooling for one room only when desired. Attic fans are usually propeller types and should be selected for low velocities to prevent excessive noise. The fans should have sufficient capacity to provide at least 30 air changes per hour.

To decrease the noise associated with air-exchange equipment, the following rules should be observed:

- The equipment should be properly located to prevent noise from affecting the living area

- The fans should be of the proper size and capacity to obtain reasonable operating speed

- Equipment should be mounted on rubber or other resilient material to assist in preventing transmission of noise to the building

If it is unavoidable to locate the attic air-exchange equipment above the bedrooms, it is essential that every precaution be taken to reduce the equipment noise to the lowest possible level. Since high-speed AC motors are usually quieter than low-speed ones, it is often preferable to use a high-speed motor connected to the fan by means of an endless V-belt if the floor space available permits such an arrangement.

Because of the low static pressures involved (usually less than 1/8 inch of water) disk or propeller fans are generally used instead of the blower or housed types. It is important that the fans have quiet operating characteristics and sufficient capacity to give at least 30 air changes per hour. For example, a house with 10,000 feet^3 volume would require a fan with a capacity of 300,000 feet^3/hour or 5,000 feet^3/minute to provide 30 air changes per hour.

The two general types of attic fans in common use are boxed-in fans and centrifugal fans. The boxed-in fan is installed within the attic in a box or suitable housing located directly over a central ceiling grille or in a bulkhead enclosing an attic stair. This type of fan may also be connected by means of a direct system to individual room grilles. Outside cool air entering through the windows in the downstairs room is discharged into the attic space and escapes to the outside through louvers, dormer windows, or screened openings under the eaves.

Although an air-exchange installation of this type is rather simple, the actual decision about where to install the fan and where to provide the grilles for the passage of air up through the house should be left to a ventilating engineer. The installation of a multi-blade centrifugal fan is shown in Figure 9.28. At the suction side the fan is connected to exhaust ducts leading to grilles, which are placed in the ceilings of the two bedrooms. The air exchange is accomplished by admitting fresh air through open windows and up through the suction side of the fan; the air is finally discharged through louvers as shown.

Another installation is shown in Figure 9.29. This fan is a centrifugal curved-blade type, mounted on a light angle-iron frame, which supports the fan wheel, shaft, and bearings. The air inlet in this installation is placed close to a circular opening, which is cut in an airtight board partition that serves to divide the attic

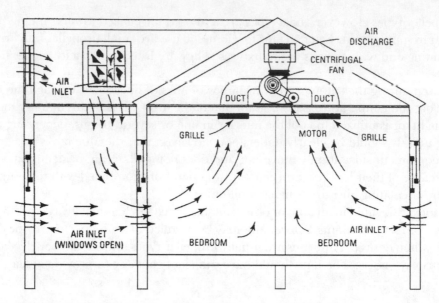

FIGURE 9.28 Installing a centrifugal fan in a one-family dwelling.

space into a suction and discharge chamber. The air is admitted through open windows and doors and is then drawn up the attic stairway through the fan into the discharge chamber.

The routine of operation to secure the best and most efficient results with an attic fan is important. A typical operating routine might require that, in the late afternoon when the outdoor temperature begins to fall, the windows on the first floor and the grilles in the ceiling or the attic floor be opened and the second-floor windows kept closed. This will place the principal cooling effect in the living rooms. Shortly before bedtime,

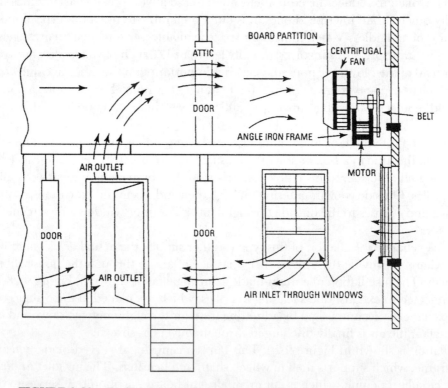

FIGURE 9.29 Typical attic installation of a belt-driven fan.

the first-floor windows may be closed and those on the second floor opened to transfer the cooling effect to the bedrooms. A suitable time clock may be used to shut the motor off before arising time.

Ventilation Methods

Ventilation is produced by two basic methods: natural and mechanical. Open windows, vents, or drafts obtain natural ventilation, whereas mechanical ventilation is produced by the use of fans.

Thermal effect is possibly better known as flue effect. Flue effect is the draft in a stack or chimney that is produced within a building when the outdoor temperature is lower than the indoor temperature. This is caused by the difference in weight of the warm column of air within the building and the cooler air outside.

Air may be filtered two ways: dry filtering and wet filtering. Various air-cleaning equipment (such as filtering, washing, or combined filtering and washing devices) is used to purify the air. When designing the duct network, ample filter area must be included so that the air velocity passing through the filters is sufficient. Accuracy in estimating the resistance to the flow of air through the duct system is important in the selection of blower motors. Resistance should be kept as low as possible in the interest of economy. Ducts should be installed as short as possible.

Competent medical authorities have properly emphasized the effect of dust on health. Air-conditioning apparatus removes these contaminants from the air. The apparatus also provides the correct amount of moisture so that the respiratory tracts are not dehydrated but are kept properly moist. Dust is more than just dry dirt. It is a complex, variable mixture of materials and, as a whole, is rather uninviting, especially the type found in and around human habitation. Dust contains fine particles of sand, soot, earth, rust, fiber, animal and vegetable refuse, hair, and chemicals.

MULTIPLE-CHOICE EXAM

1. According to the International Mechanical Code, heating and air conditioning systems of the central type must be provided with approved filters. Filters should be installed in the return air system, _____ from any heat exchanger or coil, in an approved convenient location.

   a. midstream
   b. away
   c. upstream
   d. downstream

2. Liquid adhesive coatings used on filters, according to the Code, shall have a flash point not lower than _____ degrees F.

   a. 175
   b. 325
   c. 425
   d. 575

3. Stratification is the separation of air into _____ having different temperatures.

   a. stratus
   b. streams
   c. units
   d. layers

4. Air filters used in hospitals, laboratories, and clean rooms are used for absolute filtration, where _____ percent of dust particles over 0.3 μm are removed.

    a. 90   b. 99

    c. 80   d. 50

5. It is possible to obtain very high-efficiency filters. HEPA filters are capable of obtaining an efficiency of _____ percent.

    a. 99.99   b. 90

    c. 95   d. 97

6. All HVAC systems should include the capability to introduce some _____ air.

    a. polluted   b. clean

    c. outside   d. inside

7. Heat always passes from a warmer to a _____ object or space.

    a. colder   b. moving

    c. stable   d. warmer

8. Convection is heat transfer that takes place in liquids and _____.

    a. gases   b. solids

    c. water   d. metal

9. Conduction is heat transfer that takes place chiefly in _____.

    a. metals   b. liquids

    c. solids   d. gases

10. The surface temperature of the average adult's skin is _____ degrees F.

    a. 100   b. 80

    c. 97.5   d. 96.8

11. Both temperature and _____ affect the comfort of the human body.

    a. atmosphere   b. altitude

    c. attitude   d. humidity

12. Velocities of _____ fpm or higher call for a very high speed for the air entering a room.

    a. 750   b. 100

    c. 50   d. 75

13. A velocity of air entering a room below _____ fpm is referred to as stagnant air.

    a. 25   b. 15

    c. 35   d. 50

14. What is the unit of measurement of a volume of air?

    a. feet per minute   b. feet^3/second

    c. cubic feet per minute   d. feet^3/hour

15. The outward force or static pressure of air within a duct is measured in _____ of water.

    a. inches   b. feet

    c. cubic feet   d. square feet

16. The three factors that ensure proper delivery and distribution of air within a room are location of outlet, type of outlet, and _____ of outlet.

    a. length   b. volume

    c. area   d. size

17. The sound caused by an air outlet in operation varies in direct proportion to the _____ of the air passing through it.

    a. pollutants   b. velocity

    c. color   d. humidity

18. A ducturn reduces the pressure loss in square elbows by as much as _____ percent.

    a. 90   b. 80

    c. 10   d. 100

19. The air volume per linear foot of a diffuser or grille will vary from 20 to _____ cfm, depending on heat-transfer coefficient, wall height, and infiltration rate.

    a. 100   b. 200

    c. 300   d. 400

20. Casing noise is different from air noise in the way that it is _____.

    a. eliminated   b. damped

    c. utilized   d. generated

21. Cross-talk between private offices can be minimized by:

   a. use of lined elbows on vent grilles

   b. use of simple sheet metal elbows on vent grilles

   c. use of plastic elbows on vent grilles

   d. none of the above

22. Fire and smoke in an air-supply system can be controlled by:

   a. use of a fusible link attachment to individual registers

   b. use of a straight line duct

   c. use of a sheet-metal screw on any attachment

   d. none of the above

23. For maximum effect, the control grid should be installed with the blades _____ to the direction of approaching airflow.

   a. slanted   b. flat

   c. perpendicular   d. none of the above

24. Linear grilles are designed for installation in the side wall, sill, floor, and _____.

   a. ceiling   b. stairways

   c. foyers   d. lights

25. The volume of air required is determined by the size of the space to be ventilated and the number of times per _____ that the air in the space is to be changed.

   a. day   b. minute

   c. second   d. hour

26. How is ventilation produced in an HVAC system?

   a. by opening a door

   b. by opening a window

   c. by natural and mechanical means

   d. by a water sprayer

27. Two general types of attic fans are in common use: boxed-in fans and _____ fans.

   a. centrifugal   b. centripetal

   c. direct-driven   d. louvered

28. Spray booth fans are of the _____ type.

    a. centrifugal   b. centripetal

    c. split-pole   d. shaded-pole

29. How many air changes per hour should be allowed for in a system for restaurant dining rooms?

    a. 12   b. 10

    c. 20   d. 30

30. The horsepower required for any fan or blower varies _____ as the cube of the speed.

    a. steadily   b. directly

    c. indirectly   d. intermittently

TRUE-FALSE EXAM

1. Radiation is heat transfer in wave form, such as light or radio waves.

    True   False

2. Cubic feet per minute is the unit of measurement for a volume of air.

    True   False

3. Sound level is measured in decibels (db).

    True   False

4. Perimeter systems refer to heating and cooling installations in which the diffusers are installed to blanket the outside walls.

    True   False

5. Velocity pressure is the backward force of air within a duct.

    True   False

6. A 24 × 12-inch register will handle a volume of 1,117 cfm at a rate of 700 fpm.

    True   False

7. Broadcast-studio-outlet velocity ratings are 500 fpm.

    True   False

8.   Both liquid and solid particles can be undesirable contaminants in gases.

True   False

9.   Ozone is very corrosive to rubber and plastics.

True   False

10.   The ASHRAE 55-1992 Standard requires a minimum "fresh" air supply of about 5 feet^3/minute for each occupant of a space to carry away carbon dioxide and supply oxygen.

True   False

11.   Fibrous filters have four general configurations.

True   False

12.   Membrane filters are made of very thin polymer sheets pierced by extremely small airflow passages.

True   False

13.   The charged particles in an electrostatic filter are called molecules.

True   False

14.   Electrostatic filters operate with very high voltages to produce charged particles.

True   False

15.   For ventilation applications, oxidation is the only catalytic process likely to be used in gaseous contaminant filters.

True   False

MULTIPLE-CHOICE ANSWERS

1.	C	7.	A	13.	B	19.	B	25.	D
2.	B	8.	A	14.	C	20.	D	26.	C
3.	D	9.	C	15.	A	21.	A	27.	A
4.	B	10.	B	16.	D	22.	A	28.	A
5.	A	11.	D	17.	B	23.	C	29.	A
6.	C	12.	C	18.	B	24.	A	30.	B

TRUE-FALSE ANSWERS

1.	T	5.	F	9.	T	13.	F
2.	T	6.	T	10.	T	14.	T
3.	T	7.	T	11.	T	15.	T
4.	T	8.	T	12.	T		

—NOTES—

Chapter 10

MAINTENANCE, SERVICING, AND SAFETY

One of the most important aspects of working around air-conditioning and refrigeration equipment is that of doing the job safely. Following incorrect procedures can be very painful both physically and mentally. Some of the suggestions that follow should aid in your understanding of careful work habits and use of the proper tools for the job

Safe practices are important in servicing refrigeration units. Such practices are common sense but must be reinforced to make one aware of the problems that can result when a job is done incorrectly.

GENERAL SAFETY PRECAUTIONS

Cylinders

Refrigeration and air-conditioning service persons must be able to handle compressed gases. Accidents occur when compressed gases are not handled properly. One of the first rules is that oxygen or acetylene must never be used to pressurize a refrigeration system. Oxygen will explode when it comes into contact with oil. Acetylene will explode under pressure, except when properly dissolved in acetone as used in commercial acetylene cylinders.

Dry nitrogen or dry carbon dioxide are suitable gases for pressurizing refrigeration or air-conditioning systems for leak tests or system cleaning. However, the following specific restrictions must be observed.

Commercial cylinders contain pressures in excess of 2,000 pounds per square inch of nitrogen (N_2) at normal room temperature. Commercial cylinders contain pressures in excess of 800 pounds per square inch of carbon dioxide (CO_2) at normal room temperature. Cylinders should be handled carefully. Do not drop them or bump them. Keep cylinders in a vertical position and securely fastened to prevent them from tipping over.

Do not heat the cylinder with a torch or other open flame. If heat is necessary to withdraw gas from the cylinder, apply heat by immersing the lower portion of the cylinder in warm water. Never heat a cylinder to a temperature over 110 degrees F (43 degrees C).

Pressurizing

Pressure-testing or cleaning refrigeration and air-conditioning systems can be dangerous! Extreme caution must be used in the selection and use of pressurizing equipment. Follow these procedures:

1. Never attempt to pressurize a system without first installing an appropriate pressure-regulating valve on the nitrogen or carbon dioxide cylinder discharge. This regulating valve should be equipped with two functioning pressure gauges. One gauge indicates cylinder pressure; the other indicates discharge or downstream pressure.

2. Always install a pressure-relief valve or frangible-disk type of pressure-relief device in the pressure supply line. This device should have a discharge port of at least 1/2 inch National Pipe Thread (NPT) size. This valve or frangible-disk device should be set to release at 175 pounds per square inch gauge (psig).

3. A system can be pressurized up to a maximum of 150 psig for leak testing or purging. Tecumseh hermetic-type compressors, for example, are low-pressure housing compressors. The compressor housings (cans or domes) are not normally subjected to discharge pressures. They operate instead at relatively low suction pressures. These Tecumseh compressors are generally installed on equipment where it is impractical to disconnect or isolate the compressor from the system during pressure testing. Therefore, do not exceed 150 psig when pressurizing such a complete system.

4. When flushing or purging a contaminated system, care must be taken to protect the eyes and skin from contact with acid-saturated refrigerant or oil mists. The eyes should be protected with goggles. All parts of the body should be protected by clothing to prevent injury by refrigerant. If contact with either skin or eyes occurs, flush the exposed area with cold water. Apply an ice pack if the burn is severe, and see a physician at once.

Refrigerants

R-12 refrigerants have effectively been replaced in modern air-conditioning equipment with R-134a or any of the other approved substitutes, and R-22 has some acceptable substitutes also. They are considered to be nontoxic and noninflammable. However, any gas under pressure can be hazardous. The latent energy in the pressure alone can cause damage. In working with R-12 and R-22 (or their substitutes), observe the same precautions that apply when working with other pressurized gases.

Never completely fill any refrigerant-gas cylinder with liquid. Never fill more than 80 percent with liquid. This will allow for expansion under normal conditions.

Make sure an area is properly ventilated before purging or evacuating a system that uses R-12, R-22, or their equivalents. In certain concentrations and in the presence of an open flame such as a gas range or a gas water heater, R-12 and R-22 may break down and form a small amount of harmful phosgene gas. This poison gas was used in World War I.

Lifting

Lifting heavy objects can cause serious problems. Strains and sprains are often caused by improper lifting methods. To avoid injury, learn to lift the safe way. Bend your knees, keep your back erect, and lift gradually with your leg muscles.

The material you are lifting may slip from your hands and injure your feet. To prevent foot injuries, wear the proper shoes.

Electrical Safety

Many Tecumseh single-phase compressors are installed in systems requiring off-cycle crankcase heating. This is designed to prevent refrigerant accumulation in the compressor housing. The power is on at all times. Even if the compressor is not running, power is applied to the compressor housing where the heating element is located.

Another popular system uses a run capacitor that is always connected to the compressor motor windings, even when the compressor is not running. Other devices are energized when the compressor is not running. That means there is electrical power applied to the unit even when the compressor is not running. This calls for an awareness of the situation and the proper safety procedures.

Be safe. Before you attempt to service any refrigeration system, make sure that the main circuit breaker is open and all power is off.

SERVICING THE REFRIGERATOR SECTION

The refrigerant cycle is a continuous cycle, which occurs whenever the compressor is operating. Liquid refrigerant is evaporated in the evaporator by the heat that enters the cabinet through the insulated walls and by product load and door openings. The refrigerant vapor passes from the evaporator through the suction line to the compressor dome, which is at suction pressure. From the top interior of the dome, the vapor passes down through a tube into the pump cylinder. The pressure and temperature of the vapor are raised in the cylinder by compression. The vapor is then forced through the discharge valve into the discharge line and the condenser. Air passing over the condenser surface removes heat from the high-pressure vapor, which then condenses to a liquid. The liquid refrigerant flows from the condenser to the evaporator through the small-diameter liquid line (capillary tube). Before it enters the evaporator, it is subcooled in the heat exchanger by the low-temperature suction vapor in the suction line.

Sealed Compressor and Motor

All models are equipped with a compressor with internal spring suspension. Some compressors have a plug-in magnetic starting relay, with a separate motor overload protector. Others have a built-in metallic motor overload protector. When ordering a replacement compressor, you should always give the refrigerator model number and serial number and the compressor part number. Every manufacturer has a listing available to servicepersons.

Condenser

Side-by-side and top-freezer models with a vertical natural-draft, wire-tube-type condenser have a water-evaporating coil connected in series with the condenser. The high-temperature, high-pressure, compressed refrigerant vapor passes first through the water-evaporating coil. There, part of the latent heat of evaporation and sensible heat of compression are released. The refrigerant then flows back through the oil cooling coil in the compressor shell. There, additional heat is picked up from the oil. The refrigerant then flows back to the main condenser, where sufficient heat is released to the atmosphere. This results in the condensation of refrigerant from a high-pressure vapor to high-pressure liquid.

Filter Drier

A filter drier is located in the liquid line at the outlet of the condenser. Its purpose is to filter or trap minute particles of foreign materials and absorb any moisture in the system. Fine mesh screens filter out foreign particles. The desiccant absorbs the moisture.

Capillary Tube

The capillary tube is a small-diameter liquid line connecting the condenser to the evaporator. Its resistance, or pressure drop, due to the length of the tube and its small diameter, meters the refrigerant flow into the evaporator.

The capillary tube allows the high-side pressure to unload, or balance out, with the low-side pressure during the off-cycle. This permits the compressor to start under a no-load condition.

The design of the refrigerating system for capillary feed must be carefully engineered. The capillary feed must be matched to the compressor for the conditions under which the system is most likely to operate. Both the high side (condenser) and the low side (evaporator) must be specifically designed for use with a capillary tube.

Heat Exchanger

The heat exchanger is formed by soldering a portion of the capillary tube to the suction line. The purpose of the heat exchanger is to increase the overall capacity and efficiency of the system. It does this by using the cold suction gas leaving the evaporator to cool the warm liquid refrigerant passing through the capillary tube to the evaporator. If the hot liquid refrigerant from the condenser were permitted to flow uncooled into the evaporator, part of the refrigerating effect of the refrigerant in the evaporator would have to be used to cool the incoming hot liquid down to evaporator temperature.

Freezer Compartment and Provision Compartment Assembly

Liquid refrigerant flows through the capillary and enters the freezer evaporator. Expansion and evaporation starts at this point.

COMPRESSOR REPLACEMENT

Replacement compressor packages are listed by the manufacturer. Check with the refrigerator manufacturer to be sure you have the proper replacement, Refer to the compressor number in the refrigerator under repair. Compare that number to the suggested replacement number. Replacement compressors are charged with oil and a holding charge of nitrogen. A replacement filter drier is packaged with each replacement compressor. It must be installed with the compressor. The A-line replacement compressor is used on Kelvinator® chest or upright freezers.

The new relay-overload protector assembly supplied with the N-line replacement compressor should always be used. The new motor overload protector supplied with the A-line replacement compressor should always be used. Transfer the relay, the relay cover, and the cover clamp from the original compressor to the replacement compressor. If a relatively small quantity of refrigerant is used, a major portion of it will be absorbed by the oil in the compressor when the refrigerator has been inoperative for a considerable length of time. When opening the system, use care to prevent oil from blowing out with the refrigerant.

TROUBLESHOOTING COMPRESSORS

There are several common compressor problems. Table 10.1 lists these problems and their solutions.

The illustrations following the tables show how a refrigerator is repaired in a sequential manner. This should aid in troubleshooting. Figure 10.1 shows the refrigerator's refrigeration system and where various pressures exist. Figures 10.2 through 10.21 relate the troubleshooting and repairing of the refrigerator. Figures 10.22 through 10.29 show important materials and devices as they are electrically connected. Figure 10.30 through 10.33 illustrate the compressor-condenser components and their locations in a home air conditioning unit.

TABLE 10.1 Compressor troubleshooting and service.

Complaint	Possible Cause	Repair
Compressor will not start. There is no hum.	1. Line disconnect switch open. 2. Fuse removed or blown. 3. Overload-protector tripped. 4. Control stuck in open position. 5. Control off due to cold location. 6. Wiring improper or loose.	1. Close start or disconnect switch. 2. Replace fuse. 3. Refer to electrical section. 4. Repair or replace control. 5. Relocate control. 6. Check wiring against diagram.
Compressor will not start. It hums, but trips on overload protector.	1. Improperly wired. 2. Low voltage to unit. 3. Starting capacitor defective. 4. Relay failing to close. 5. Compressor motor has a winding open or shorted. 6. Internal mechanical trouble in compressor. 7. Liquid refrigerant in compressor.	1. Check wiring against diagram. 2. Determine reason and correct. 3. Determine reason and replace. 4. Determine reason and correct, replace if necessary. 5. Replace compressor. 6. Replace compressor. 7. Add crankcase heater and/or accumulator.
Compressor starts, but does not switch off of start winding.	1. Improperly wired. 2. Low voltage to unit. 3. Relay failing to open. 4. Run capacitor defective. 5. Excessively high-discharge pressure. 6. Compressor motor has a winding open or shorted. 7. Internal mechanical trouble in compressor (tight).	1. Check wiring against diagram. 2. Determine reason and correct. 3. Determine reason and replace if necessary. 4. Determine reason and replace. 5. Check discharge shut-off valve, possible overcharge, or insufficient cooling of condenser. 6. Replace compressor. 7. Replace compressor.
Compressor starts and runs, but short cycles on overload protector.	1. Additional current passing through the overload protector. 2. Low voltage to unit (or unbalanced if three-phase). 3. Overload-protector defective. 4. Run capacitor defective. 5. Excessive discharge pressure. 6. Suction pressure too high. 7. Compressor too hot—return gas hot. 8. Compressor motor has a winding shorted.	1. Check wiring against diagram. Check added fan motors, pumps, and the like, connected to wrong side of protector. 2. Determine reason and correct. 3. Check current, replace protector. 4. Determine reason and replace. 5. Check ventilation, restrictions in cooling medium, restrictions in refrigeration system. 6. Check for possibility of misapplication. Use stronger unit. 7. Check refrigerant charge. (Repair leak.) Add refrigerant, if necessary. 8. Replace compressor.

Unit runs, but short cycles on.	1. Overload protector.
	2. Thermostat.
	3. High pressure cut-out due to insufficient air or water supply, overcharge, or air in system.
	4. Low pressure cut-out due to:
	  a. Liquid-line solenoid leaking.
	  b. Compressor valve leak.
	  c. Undercharge.
	  d. Restriction in expansion device.
		1. Check current. Replace protector.
		2. Differential set too close. Widen.
		3. Check air or water supply to condenser. Reduce refrigerant charge, or purge.
		4.
		  a. Replace.
		  b. Replace.
		  c. Repair leak and add refrigerant.
		  d. Replace expansion device.

Unit operates long or continuously.	1. Shortage of refrigerant. 2. Control contacts stuck or frozen closed. 3. Refrigerated or air conditioned space has excessive load or poor insulation. 4. System inadequate to handle load. 5. Evaporator coil iced. 6. Restriction in refrigeration system. 7. Dirty condenser. 8. Filter dirty.	1. Repair leak. Add charge. 2. Clean contacts or replace control. 3. Determine fault and correct. 4. Replace with larger system. 5. Defrost. 6. Determine location and remove. 7. Clean condenser. 8. Clean or replace.

TABLE 10.1 Compressor troubleshooting and service. *(Continued)*

Complaint	Possible Cause	Repair

Start capacitor open, shorted, or blown.	1. Relay contacts not operating properly.
	2. Prolonged operation on start cycle due to:
	   a. Low voltage to unit.
	   b. Improper relay.
	   c. Starting load too high.
	
	3. Excessive short cycling.
	
	4. Improper capacitor.
		1. Clean contacts or replace relay if necessary.
		2.
		   a. Determine reason and correct.
		   b. Replace.
		   c. Correct by using pump-down arrangement if necessary.
		3. Determine reason for short cycling as mentioned in previous complaint.
		4. Determine correct size and replace.

Run capacitor open, shorted, or blown.	1. Improper capacitor. 2. Excessively high line voltage (110% of rated maximum).	1. Determine correct size and replace. 2. Determine reason and correct.
Relay defective or burned out.	1. Incorrect relay. 2. Incorrect mounting angle. 3. Line voltage too high or too low. 4. Excessive short cycling. 5. Relay being influenced by loose vibrating mounting. 6. Incorrect run capacitor.	1. Check and replace. 2. Remount relay in correct position. 3. Determine reason and correct. 4. Determine reason and correct. 5. Remount rigidly. 6. Replace with proper capacitor.
Space temperature too high.	1. Control setting too high. 2. Expansion valve too small. 3. Cooling coils too small. 4. Inadequate air circulation.	1. Reset control. 2. Use larger valve. 3. Add surface or replace. 4. Improve air movement.
Suction-line frosted or sweating.	1. Expansion-valve oversized or passing excess refrigerant. 2. Expansion valve stuck open. 3. Evaporator fan not running. 4. Overcharge of refrigerant.	1. Readjust valve or replace with smaller valve. 2. Clean valve of foreign particles. Replace if necessary. 3. Determine reason and correct. 4. Correct charge.
Liquid-line frosted or sweating.	1. Restriction in dehydrator or strainer. 2. Liquid shutoff (king valve) partially closed.	1. Replace part. 2. Open valve fully.
Unit noisy.	1. Loose parts or mountings. 2. Tubing rattle. 3. Bent fan blade causing vibration. 4. Fan motor bearings worn.	1. Tighten. 2. Reform to be free of contact. 3. Replace blade. 4. Replace motor.

TABLE 10.2 PSC compressor motor troubles and corrections.

Causes	Corrections
Low Voltage	
1. Inadequate wire size.	1. Increase wire size.
2. Watt-hour meter too small.	2. Call utility company.
3. Power transformer too small or feeding too many homes.	3. Call utility company.
4. Input voltage too low.	4. Call utility company.
	(Note: Starting torque varies as the square of the input voltage.)
Branch Circuit Fuse or Circuit Breaker Tripping	
1. Rating too low.	1. Increase size to a minimum of 175% of unit FLA (Full Load Amperes) to a maximum of 225% of FLA.
System Pressure High or Not Equalized	
1. Pressure not equalizing within 3 min.	1. a. Check metering device (capillary tube or expansion valve). b. Check room thermostat for cycling rate. Off cycle should be at least 5 min. Also check for "chattering." c. Has some refrigerant dryer or some other possible restriction been added?
2. System pressure too high.	2. Make sure refrigerant charge is correct.
3. Excessive liquid in crankcase (split-system applications).	3. Add crankcase heater and suction line accumulator.
Miscellaneous	
1. Run capacitor open or shorted.	1. Replace with new, properly-sized capacitor.
2. Internal overload open.	2. Allow two hours to reset before changing compressor.

TABLE 10.3 Dimensional data for the unit shown in Figure 10.31.

Condensing	(-)AKA-	018, 024	030	036, 042	048, 060	
unit	(-)ALB-		018, 024	030, 036	042, 048, 060	
model	(-)AMA-				018, 024, 030	
					036, 042, 048	060
Length "H"		16 3/4	20 3/4	20 3/4	26 3/4	34 3/4
Length "L"		33 11/16	33 11/16	38 11/16	42 9/18	43
Width "W"		23 1/4	23 1/4	27 1/8	31	31

TABLE 10.4 Refrigerant line size information.

CONDENSING UNIT REFRIGERANT TUBING DATA																						
MODEL	(-)AKA-	018		024		030		036			042							048		060			
	(-)ALB-		018		024		030					036		042			048				060		
	(-)AMA-								018	024	030						036		042	048			060
FACTORY CHARGE, OZ ①		46	61	51	63	68	74	75	88	85	96	79	78	100		96		192	113	100	172	112	173	192
REFRIGERANT VAPOR CONNECTION SIZE ON UNIT		5/8" I.D. SWEAT	3/4" I.D. SWEAT										7/8" I.D. SWEAT						1-1/8" I.D. SWEAT④				
REFRIGERANT LIQUID CONNECTION SIZE ON UNIT			5/16" I.D. SWEAT③																3/8" I.D. SWEAT				

RECOMMENDED VAPOR AND LIQUID LINE SIZES (O.D.) FOR VARIOUS LENGTH RUNS

LENGTHS UP TO 30 FEET	SUCT.	5/8" O.D.	3/4" O.D.	7/8" O.D.	1-1/8" O.D
	LIQ.		1/4" O.D.	5/16" O.D.	3/8" O.D.
LENGTHS 31 TO 45 FEET	SUCT.	5/8" O.D.	3/4" O.D.	7/8" O.D.	1-1/8" O.D.
	LIQ.		20'-1/4" O.D. BALANCE 5/16" O.D.	25'-5/16" O.D. BALANCE 3/8" O.D.	25'-3/8" O.D. BALANCE 1/2" O.D.
LENGTHS 46 TO 60 FEET	SUCT.	3/4" O.D.	3/4" or 7/8" O.D. SEE CAPACITY MULTIPLIER②	7/8" or 1-1/8" O.D. SEE CAPACITY MULTIPLIER②	1-1/8" O.D
	LIQ.		15'-1/4" O.D. BALANCE 5/16" O.D.	20'-5/16" O.D. BALANCE 3/8" O.D.	20 -3/8" O.D. BALANCE 1/2" O.D.
LENGTHS 61 TO 90 FEET	SUCT.	3/4" O.D.	3/4" or 7/8" O.D. SEE CAPACITY MULTIPLIER②	7/8" or 1-1/8" O.D. SEE CAPACITY MULTIPLIER②	1-1/8" O.D.
	LIQ.		5/16" O.D.	10'-5/16" O.D. BALANCE 3/8" O.D.	10'-3/8" O.D. BALANCE 1/2" O.D.
LENGTHS 91 TO 120 FEET	SUCT.	3/4" O.D.	3/4" or 7/8" O.D. SEE CAPACITY MULTIPLIER②	7/8" or 1-1/8" O.D. SEE CAPACITY MULTIPLIER②	1-1/8" O.D.
	LIQ.		80'-5/16" O.D. BALANCE 3/8" O.D.	3/8" O.D.	1/2" O.D.
LENGTHS 121 TO 150 FEET	SUCT.	3/4" O.D.	3/4" or 7/8" O.D. SEE CAPACITY MULTIPLIER②	7/8" or 1-1/8" O.D. SEE CAPACITY MULTIPLIER②	1-1/8" O.D.
	LIQ.		60'-5/16" O.D. BALANCE 3/8" O.D.	100'-3/8" O.D. BALANCE 1/2" O.D.	1/2" O.D.

CAPACITY MULTIPLIER FOR VARYING VAPOR LINE SIZES AND LENGTH RUNS②

VAPOR TUBING SIZE O.D.	5/8 3/4	5/8 3/4 7/8	3/4 7/8	3/4 7/8	3/4 7/8 1-1/8	3/4 7/8 1-1/8	7/8 1-1/8	7/8 1-1/8 1-3/8	7/8 1-1/8 1-3/8
30 FT. LENGTH	1.00 1.01	.98 1.00 1.01	1.00 1.01	1.00 1.01	99 1.00 1.01	.99 1.00 1.01	1.00 1.01	99 1.00 1.01	.99 1 00 1.01
60 FT. LENGTH	.98 1.00	96 .98 1.00	98 1.00	.98 1.00	.97 .99 1.01	.97 .99 1.01	.98 1.00	97 1.00 1.01	97 1.00 1.01
90 FT. LENGTH	.97 .99	.94 .97 .99	.97 99	96 .99	.96 .98 1.00	.94 98 1.00	.96 1 00	96 99 1 00	.95 99 1 00
120 FT. LENGTH	.95 .99	.91 .95 .99	.95 .99	.94 .98	.94 .98 1.00	.92 .97 1.00	.94 .99	.94 .99 1.00	.93 .99 1.00
150 FT. LENGTH	.94 .98	.88 .94 .98	.94 .98	.93 .97	.92 .97 1.00	.89 .96 1.00	.92 .98	93 .99 1.00	91 .98 1.00

① Factory charge is sufficient for 25 Ft. of recommended liquid line and matching evaporator. For different lengths, adjust charge accordingly:

1/4" Liquid Line ± .3 oz per foot
5/16" Liquid Line ± .4 oz per foot
3/8" Liquid Line ± .6 oz per foot
1/2" Liquid Line ± 1.2 oz per foot

② Capacity multiplier × rated capacity = actual capacity
Example -024 with 60' of 3/4" O.D. Vapor Line
24.000 × 98 = 23.520 BTUH.

③ Approx. 5/16" I.D. will accept 1/4" O.D. field liquid line.
④ Requires adapter supplied with unit (packed inside).

TABLE 10.5 Line sizing with condensing unit over 10 feet above evaporator.

Nominal Tons	Liquid Line Sizing (O.D.)		
	Horizontal Run*	Vertical Run†	
1.1/2	1/4″	100% · 3/16″	
2	1/4″	80% · 1/4″ & 20% · 5/16″	
2.1/2	1/4″	40% · 1/4″ & 60% · 5/16″	
3	5/16″	20% · 1/4″ & 80% · 5/16″	
3.1/2	5/16″	100% · 5/16″	
4	3/8″	80% · 5/16″ & 20% · 3/8″	
5	3/8″	40% · 5/16″ & 60% · 3/8″	

* See Table 12.4 if horizontal run exceeds 30 feet.
† The smaller size tubing must be at the bottom of the run. The combination shown will result in approximately zero net pressure drop for vertical run.

TABLE 10.6 Condensing-unit approved application matches with flow-check piston sizes required.

Condensing Unit Model and Size	Evaporator and Model Number and Size	Piston Size Required	*Coil Code		System Chg. OZ
			Elec. Furn.	HP AH	
(-)AKA-018	RCBA-2453	53	B	B	46
(-)AKA-024	RCBA-2457	57	C	D	51
(-)AKA-030	RCBA-3665	65	B	D	68
(-)AKA-036	RCBA-3673	73	C	C	75
(-)AKA-042	RCBA-4878	78	B	B	79
(-)AKA-048	RCBA-4876	76	C	D	113
(-)AKA-060	RCBA-6089	89	B	B	112

* Coil code in electric furnace or air-handler model number.

TABLE 10.7 Condensing-unit approved application matches with TXV and piston sizes required.

Condensing Unit Model and Size	Evaporator Model Number and Size	TXV Size (TON)	Piston Size	Coil Slabs
(-)ALB-018 (-)AMA-018	RCGA-24A1	1.5	120	4
(-)ALB-024 (-)AMA-024	RCGA-24A2	2.0	172	4
(-)ALB-030 (-)AMA-030	RCGA-36A1	2.5	157	6
(-)ALB-036 (-)AMA-036	RCGA-36A2	3.0	157	6
(-)ALB-042 (-)AMA-042	RCGA-48A1	4.0	172	8
(-)ALB-048 (-)AMA-048	RCGA-48A1	4.0	172	8
(-)ALB-060	RCGA-60A1	5.0	172	10
(-)AMA-060	RCGA-60A1	5.0	172	10

TABLE 10.8 Electrical and physical data.

AWG Copper Wire Size	AWG Aluminum Wire Size	Connector Type & Size (or equivalent)
12	10	T & B Wire Nut PT2
10	8	T & B Wire Nut PT3
8	8	Sherman Split Bolt TSP6
6	4	Sherman Split Bolt TSP4
4	2	Sherman Split Bolt TSP2

TABLE 10.9 Copper wire size AWG.

Supply wire length (feet)	(1% Voltage Drop)							
200	6	4	4	4	3	3	2	2
150	8	6	6	4	4	4	3	3
100	10	8	8	6	6	6	4	4
50	14	12	10	10	8	8	6	6
	15	20	25	30	35	40	45	50
	Supply-circuit ampacity							

TABLE 10.10 Field wire sizes for 24-volt thermostat circuits.

Thermostat Load-Amps	Solid Copper Wire-Awg.					
3.0	16	14	12	10	10	10
2.5	16	14	12	12	10	10
2.0	18	16	14	12	12	10
	50	100	150	200	250	300
	Length of Run-Feet*					

*Wire length equals twice the run distance.
NOTE: Do not use control wiring smaller than No. 18 AWG between thermostat and outdoor unit.

TABLE 10.11   Troubleshooting chart.

Symptom	Possible Cause	Remedy
High head-low suction	a. Restriction in liquid line or capillary tube or filter drier	a. Remove or replace defective component
High head-high or normal suction	a. Dirty condenser coil b. Overcharged c. Condenser fan not running	a. Clean coil b. Correct system charge c. Repair or replace
Low head-high suction	a. Incorrect capillary tube b. Defective compressor valves	a. Replace evaporator assembly b. Replace compressor
Unit will not run	a. Power off or loose electrical connection b. Thermostat out of calibration-set too high c. Defective contactor d. Blown fuses e. Transformer defective f. High-pressure control open g. Compressor overload contacts open	a. Check for unit voltage at contactor in condensing unit b. Reset c. Check for 24 volts at contactor coil replace if contacts are open d. Replace fuses e. Check wiring-replace transformer f. Reset-also see high head pressure remedy. The high-pressure control opens at 430 PSI g. If external overload-replace OL. If internal replace compressor NOTE: Wait at least 2 h for overload to reset
Condenser fan runs. compressor does not	a. Run or start capacitor defective b. Start relay defective c. Loose connection d. Compressor stuck, grounded or open motor winding, open internal overload. e. Low-voltage condition	a. Replace b. Replace c. Check for unit voltage at compressor-check and tighten all connections d. Wait at least 2 h for overload to reset If still open, replace the compressor. e. Add start kit components
Low suction-cool compressor Iced Evaporator Coil	a. Low-indoor airflow b. Operating unit at temperatures below 65° outdoor temperature.	a. Increase speed of blower or reduce restriction-replace air filter b. Add low ambient kit
Compressor short cycles	a. Defective overload protector	a. Replace-check for correct voltage
Registers sweat	a. Low airflow	a. Increase speed of furnace blower or reduce restriction replace air filter.
High-suction pressure	a. Excessive load b. Defective compressor	a. Recheck load calculation b. Replace
Insufficient cooling	a. Improperly sized unit b. Improper airflow c. Incorrect refrigerant charge d. Incorrect voltage	a. Recalculate load b. Check-should be approximately 400 CFM per ton c. Charge per procedure attached to unit service panel d. At compressor terminals, voltage must be within 10% of nameplate volts when unit is operating

WARNING: Disconnect all power to unit before servicing. Contactor may break only one side of line.

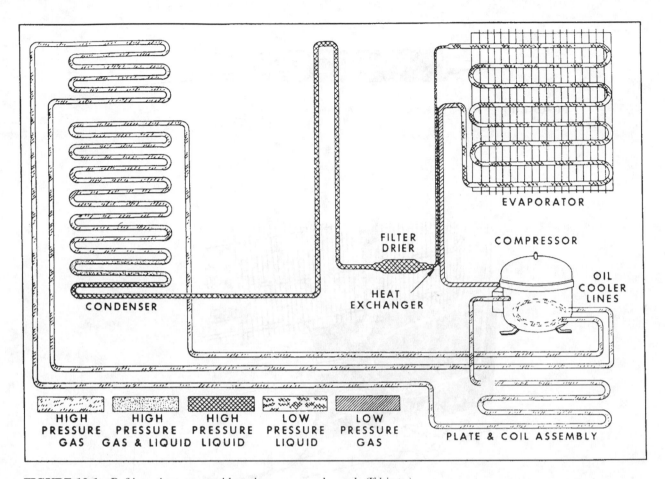

FIGURE 10.1 Refrigerating system with various pressures located. *(Kelvinator)*

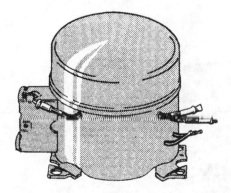

FIGURE 10.2 N-line replacement compressor for top-freezer models. *(Kelvinator)*

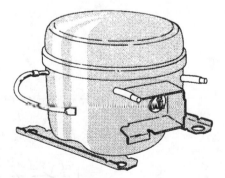

FIGURE 10.3 A-line replacement compressor. *(Kelvinator)*

FIGURE 10.4   Replacement filter drier. *(Kelvinator)*

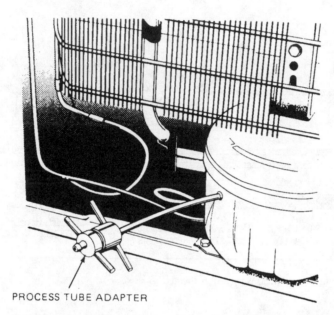

PROCESS TUBE ADAPTER

FIGURE 10.5   Process-tube adapter. *(Kelvinator)*

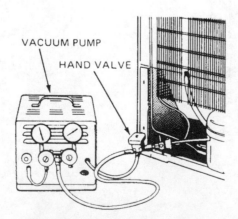

FIGURE 10.6   Vacuum pump and hand valve. *(Kelvinator)*

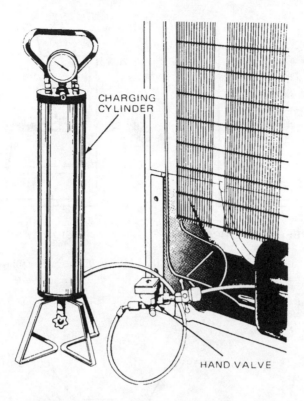

FIGURE 10.7 Charging cylinder. *(Kelvinator)*

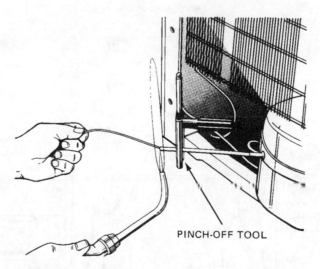

FIGURE 10.8 Brazing process tube with pinch-off tool in place. *(Kelvinator)*

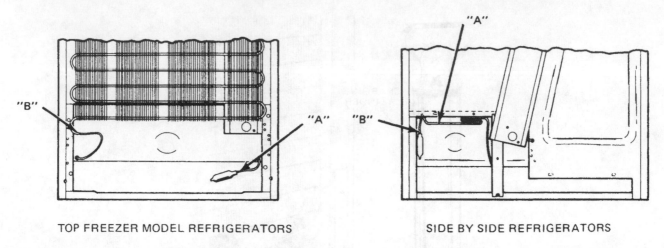

TOP FREEZER MODEL REFRIGERATORS   SIDE BY SIDE REFRIGERATORS

FIGURE 10.9   Repairing the perimeter tube. Cut and deburr the 3/16 in. OD perimeter tube at "A" and "B." *(Kelvinator)*

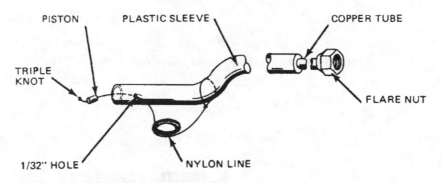

FIGURE 10.10   Repair tool made for repair of the perimeter tube. *(Kelvinator)*

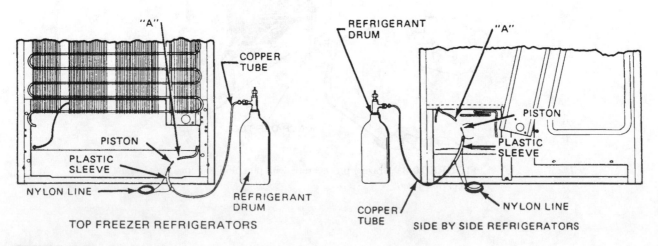

TOP FREEZER REFRIGERATORS   SIDE BY SIDE REFRIGERATORS

FIGURE 10.11   Repairing the perimeter tube. *(Kelvinator)*

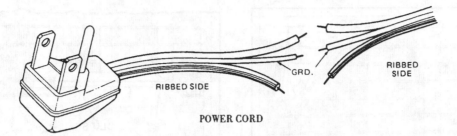

RIBBED SIDE

GRD.

RIBBED
SIDE

POWER CORD

FIGURE 10.12 Splicing a new power-cord plug onto the existing power-cord line. *(Kelvinator)*

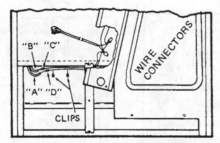

SIDE BY SIDE REFRIGERATORS

FIGURE 10.13 Power-cord location and splices, side-by-side models. *(Kelvinator)*

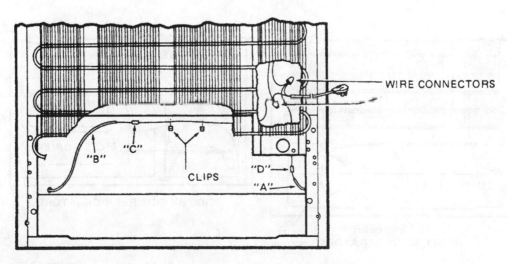

WIRE CONNECTORS

FIGURE 10.14 Power-cord splicing on top-freezer models. *(Kelvinator)*

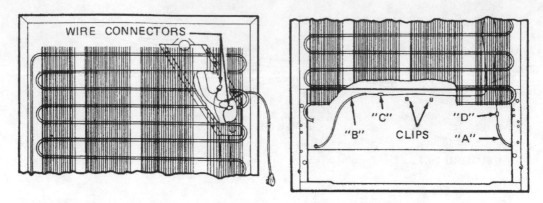

FIGURE 10.15 Another top-freezer refrigerator power-cord splice. *(Kelvinator)*

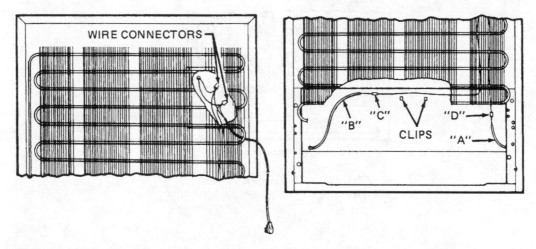

FIGURE 10.16 Top-freezer refrigerator-cord splice location. *(Kelvinator)*

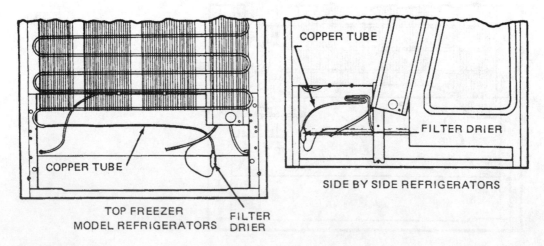

FIGURE 10.17 Side-by-side and top-freezer model refrigerator filter-drier location. *(Kelvinator)*

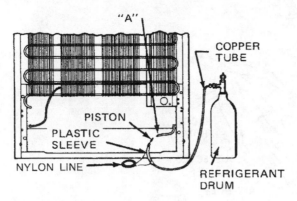

FIGURE 10.18 An older technique using vapor pressure from refrigerant drum to blow the piston and nylon assembly through the hot tube. *(Kelvinator)*

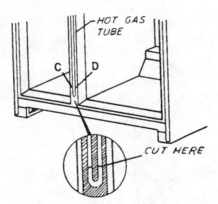

FIGURE 10.19 Where to cut the hot-gas tube for repair. *(Kelvinator)*

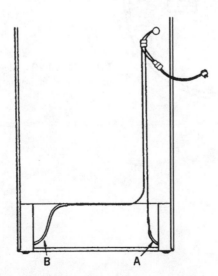

FIGURE 10.20 Location of the leads and power cord. *(Kelvinator)*

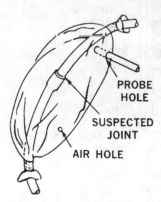

FIGURE 10.21   Leak-detection envelope. *(Kelvinator)*

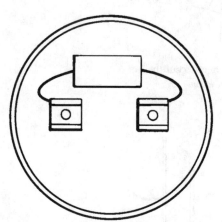

FIGURE 10.22   Bleeder resistor across the capacitor terminals. *(Tecumseh)*

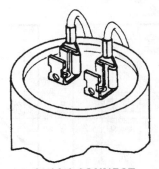

QUICK CONNECT

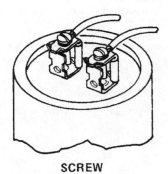

SCREW

FIGURE 10.23 There are three ways to attach terminals to an electrolytic capacitor. *(Tecumseh)*

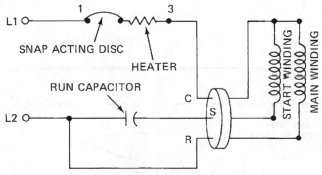

MAIN WINDING IS RUN WINDING

FIGURE 10.24 The PSC motor eliminates the need for potentially troublesome and costly extra electrical components, such as start capacitors and potential motor-starting relays. *(Tecumseh)*

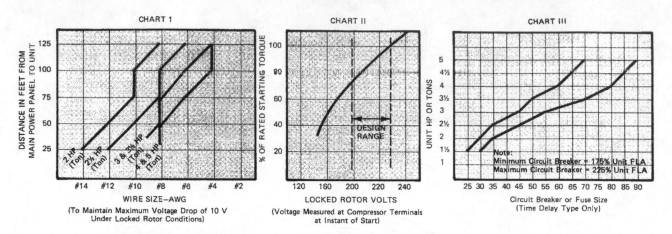

FIGURE 10.25   Wire size, locked rotor volts, and circuit breaker or fuse size requirements for compressors. *(Tecumseh)*

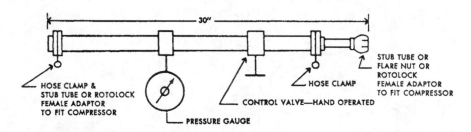

FIGURE 10.26   Compressor test stand. *(Tecumseh)*

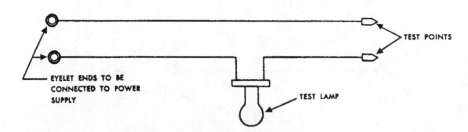

FIGURE 10.27   Continuity test cord and lamp. *(Tecumseh)*

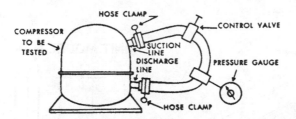

FIGURE 10.28　Bypass line from suction to discharge line. *(Tecumseh)*

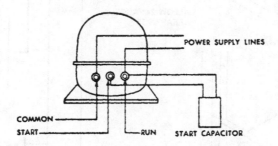

FIGURE 10.29　Location of start capacitor in a circuit. *(Tecumseh)*

FIGURE 10.30　Air-cooled condensing unit. *(Rheem)*

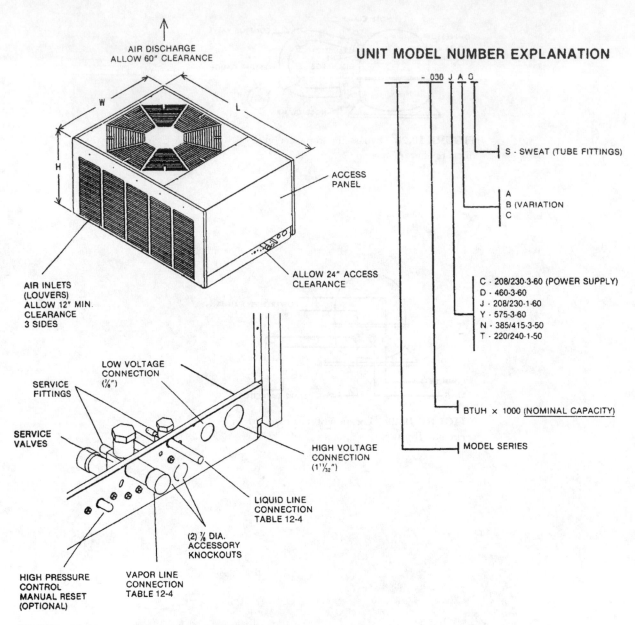

FIGURE 10.31 Dimensions and model number explanation. *(Rheem)*

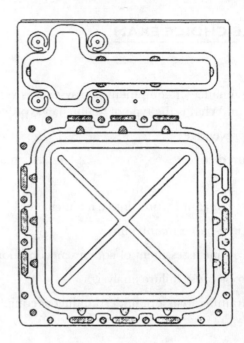

BOTTOM VIEW SHOWING
DRAIN OPENINGS (SHADED AREAS).

FIGURE 10.32 Base pan. *(Rheem)*

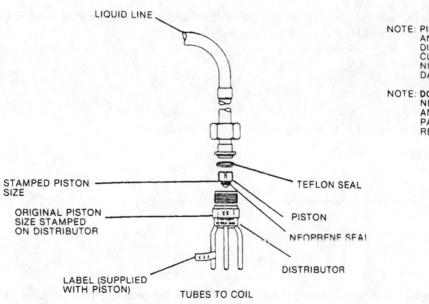

LIQUID LINE

NOTE: PISTON, PISTON SEAL
AND INSIDE OF
DISTRIBUTOR MUST BE
CLEAN AND FREE OF
NICKS, BURRS OR OTHER
DAMAGE.

NOTE: **DO NOT** REPLACE
NEOPRENE SEAL WITH
ANY "O" RING. CONTACT
PARTS DEPT. FOR EXACT
REPLACEMENT.

STAMPED PISTON
SIZE

TEFLON SEAL

ORIGINAL PISTON
SIZE STAMPED
ON DISTRIBUTOR

PISTON

NEOPRENE SEAL

DISTRIBUTOR

LABEL (SUPPLIED
WITH PISTON)

TUBES TO COIL

FIGURE 10.33 Piston and distributor assembly. *(Rheem)*

MULTIPLE-CHOICE EXAM

1. Maintenance of equipment and systems is very important in any air-conditioning and refrigeration system. What is the type of maintenance procedure most often discussed and utilized?

    a. preventive   b. proactive

    c. predictive   d. reactive

2. What is meant by preventive maintenance?

    a. time-based maintenance

    b. routine assessment of equipment condition

    c. root-cause failure analysis

    d. training to assure understanding of equipment

3. What questions about output results should be asked about the computerized-maintenance-management system?

    a. Can the CMMS system produce an accurate labor report to be utilized during budget preparation?

    b. Can repeat work orders be flagged to indicate a potential system problem?

    c. Can these reports be utilized when dealing with an oversight authority such as the Joint Commission on Accreditation of Healthcare Organizations?

    d. all of these

    e. none of these

4. Why are safe practices important in servicing refrigeration equipment?

    a. they prevent accidents

    b. they save lives

    c. they help keep equipment and maintenance personnel from danger

    d. all of the above

    e. none of he above

5. Cylinders of compressed gas must be handled carefully. What is the pressure of a cylinder of nitrogen at room temperature in pounds per inch squared:

    a. 1,000   b. 500

    c.   740   d. 2,000

6. When flushing or purging a contaminated system, care must be taken to protect the eyes and _____ from contact with acid-saturated refrigerant or oil mists.

    a. fingers   b. toes

    c. skin   d. all of these

    e. none of these

7. A frangible disk is a:

    a. breakable disk   b. sheet-metal disk

    c. cast-iron disk   d. ceramic disk

8. Never fill any refrigerant gas cylinder with liquid at more than _____ percent in order to allow for expansion.

    a. 90   b. 80

    c. 99   d. 90

9. One of the most often repeated admonishments in safety is:

    a. lift with the legs and not the back

    b. wear safety toed shoes

    c. wear goggles

    d. wear gloves

10. Why should you NOT handle an electrolytic capacitor when charged?

    a. it still may have a charge   b. it is still warm

    c. it explodes easily   d. it can shock you

11. When ordering a new compressor you should always give the refrigerator number, the compressor part number, and the _____ number.

    a. motor   b. serial

    c. capacitor   d. refrigerant

12. The capillary tube is a small _____ line connecting the condenser to the evaporator.

    a. water   b. vapor

    c. gas   d. liquid

13. A heat exchanger can be formed by soldering a portion of the capillary tube to the _____ line.

    a. suction   b. water

    c. refrigerator   d. pressurized

14. Replacement compressor packages are listed by the _____.

    a. manufacturer   b. serial number

    c. unit part number   d. bar code

15. When the refrigerator has been inoperative for a considerable length of time and you open the system, use care to prevent oil from blowing out the _____.

    a. water   b. air

    c. refrigerant   d. vacuum

16. When troubleshooting a unit and the compressor runs too much or 100 percent of the time, the cause is probably one or more of the following:

    a. low pumping-capacity compressor

    b. erratic thermostat or thermostat set too cold

    c. refrigerator is exposed to unusual heat

    d. all of the above

    e. none of the above

17. What type of gas is produced when refrigerant is exposed to fire or flames?

    a. phosgene   b. chlorine

    c. fluorine   d. oxygen

18. Which of the following causes motor burnout in a compressor?

    a. low line voltage   b. loss of refrigerant

    c. high head pressure   d. moisture

    e. all of the above   f. none of the above

19. Do not remove caps from the replacement filter dryer until all the _____ tubes have been processed for installation of the filter dryer.

    a. capillary   b. refrigerant

    c. filter   d. drier

20. What is used as a filler material when brazing for replacing the filter dryer?

    a.   silver solder   b.   40/60 solder

    c.   solder   d.   brass rod

21. What is the color of the ground wire in a three-wire cord for a refrigerator?

    a.   green   b.   white

    c.   black   d.   red

22. Why should you use care when stripping insulation from refrigerator doors?

    a.   you may damage the defroster

    b.   you may damage the capillary tube

    c.   you may damage the door seal

    d.   you may damage the heater-resistance wire

23. Always introduce refrigerant into the system in a _____ state.

    a.   solid   b.   free

    c.   vapor   d.   liquid

24. Various types of leak detectors are available. Which of the following is not one of them?

    a.   liquid detectors (bubbles)

    b.   halide torches

    c.   halogen-sensing electronic detectors

    d.   electronic transistor pressure-sensing detectors

    e.   vacuum-tube-operated detectors

25. In order to service refrigeration equipment properly, the service person must possess the following:

    a.   a thorough understanding of the theory of refrigeration

    b.   a working knowledge of the purpose, design, and operation of the various mechanical parts of the refrigerator

    c.   the ability to diagnose and correct any trouble that may develop

    d.   all of the above

    e.   none of the above

26. To check the cut-out and cut-in temperatures of a thermostat, use a refrigeration tester or a
    _____.

    a. ohmmeter   b. transistorized meter

    c. voltmeter   d. recording meter

27. Freezer and provision-compartment temperatures are affected by the following:

    a. improper door seal   b. service load

    c. ambient temperature   d. compressor efficiency

    e. none of the above   f. all of the above

28. What is the true measure of power?

    a. resistance   b. amperage

    c. voltage   d. wattage

29. The voltage rating and the _____ rating are important for an electrolytic capacitor during re-
    placement.

    a. resistance   b. amperage

    c. microfarad   d. ohm

30. Bleeder resistors are used across the _____ terminals.

    a. capacitor   b. condenser

    c. compressor   d. power-plug

TRUE-FALSE EXAM

1. A bleeder resistor is placed across the electrolytic capacitor to prevent shocks.

   True   False

2. Branch circuit fuses or circuit breakers sized too small will cause nuisance tripping.

   True   False

3. Too large or excessive refrigerant charge will cause a starting load too great to allow the compressor
   to operate properly.

   True   False

4.   A defective electrolytic capacitor must be replaced by one of the same voltage rating and same number of microfarads.

True   False

5.   The run (main) windings of a single-phase hermetic compressor motor consist of large-diameter wire having very high resistance.

True   False

6.   Avoid having lawn-sprinkler heads spray directly on the compressor cabinet of a residential air-conditioning unit.

True   False

7.   Factory-fresh air-conditioning units are shipped from the manufacturer with a full charge of refrigerant.

True   False

8.   There is no absolute fixed limit as to how high the condensing unit may be above the evaporator.

True   False

9.   Erratic operating pressure can result if piping is not properly sized.

True   False

10.   The flow-check piston is a multipurpose device.

True   False

11.   Do not use oxygen to purge lines of a pressure system for a leak test.

True   False

12.   Air in a system causes high-condensing temperatures and pressure, resulting in increased power input and reduced performance.

True   False

13.   The superheat charging method is used for charging systems when a flow-check piston or capillary tubes are used on the evaporator as a metering device.

True   False

14. The liquid-pressure method is used for charging systems in the cooling mode when an expansion valve is used on the evaporator.

True   False

15. If the low-voltage control wiring is run in conduit with the power supply, Class-I insulation is required. Class II insulation is required if it is run separately.

True   False

MULTIPLE-CHOICE ANSWERS

1. A	7. A	13. A	19. B	25. D
2. A	8. B	14. A	20. A	26. D
3. D	9. A	15. C	21. A	27. F
4. D	10. A	16. D	22. D	28. D
5. D	11. B	17. A	23. C	29. C
6. D	12. D	18. E	24. E	30. A

TRUE-FALSE ANSWERS

1. F	5. F	9. T	13. T
2. T	6. T	10. T	14. T
3. T	7. T	11. T	15. T
4. T	8. T	12. T	

—NOTES—

Chapter 11

HUMIDIFICATION, DEHUMIDIFICATION, AND PSYCHROMETRICS

HVAC Licensing Study Guide

HUMIDIFICATION, DEHUMIDIFICATION, AND PSYCHROMETRICS

Most refrigerated cooling processes also include dehumidification. The process has been studied from many angles. There are some very interesting approaches that are available from a variety of sources. A good starting point is a quick review of what you know about temperature and its measurement.

For air-conditioning and heating for comfort purposes, it is necessary to also know something about humidification. The word "sensible" implies that the heating or cooling takes place at a constant humidity. Although the humidity ratio remains constant, there can be a change in the relative humidity. As the dry-bulb temperature increases, the air will hold more moisture at saturation.

If an air stream is passed through a water spray in such a way that the departing air is saturated adiabatically (without heat loss or gain), then the process can be shown on charts. A constant wet-bulb process and the final wet and dry bulb temperatures are equal. In practice, the process is called evaporative cooling, and saturation is achieved. The efficiency of an air washer or evaporative cooler is the ratio of the dry-bulb temperature difference from point 1 to point 2 to the initial difference between the dry- and wet-bulb temperatures.

Psychrometry (from the Greek *psychro-*, meaning "cold") is the science and practice of air mixtures and their control. The science deals mainly with dry air, water-vapor mixtures, and the specific heat of dry air and its volume. It also deals with the heat of water, the heat of vaporization or condensation, and the specific heat of steam in reference to moisture mixed with dry air. Psychrometry is a specialized area of thermodynamics.

TEMPERATURE

Temperature is defined as the thermal state of matter. Matter receives or gives up heat as it is contacted by another object. If no heat flows upon contact, there is no difference in temperature. Figure 11.1 shows the different types of dry-bulb thermometers. The metric centigrade scale is now referred to as degrees Celsius. The Celsius scale is divided into 100 degrees, from the freezing point of water to the boiling point.

American industry and commerce still use the Fahrenheit scale for temperature measurement. However, the metric scale is becoming more accepted. The Fahrenheit scale divides into 180 parts the temperature range from the freezing point of water to its boiling point. The Fahrenheit temperature scale measures water at its freezing point of 32 degrees and its boiling point of 212 degrees. The pressure reference is sea level, or 14.7 pounds per square inch.

Absolute temperatures are measured from absolute zero. This is the point at which there is no heat. On the Fahrenheit scale, absolute zero is -460 degrees. Temperatures on the absolute Fahrenheit scale (Rankine) can be found by adding 460 degrees to the thermometer reading. On the Celsius scale, absolute zero is -273 degrees. Any temperatures on the absolute Celsius scale (Kelvin) can be found by adding 273 to the thermometer reading (see Figure 11.1).

Absolute-zero temperature is the base point for calculations of heat. For example, if air or steam is kept in a closed vessel, the pressure will change roughly in direct proportion to its absolute temperature. Thus, if 0-degree F air (460 degrees absolute) is heated to 77 degrees F (537 degrees absolute) without increasing the volume, the pressure will increase to roughly 537/460 times the original pressure. A more formal statement of the important physical law involved states that at a constant temperature, as the absolute temperature of a perfect gas varies, its absolute pressure will vary directly. Or, at a constant pressure, as the absolute tem-

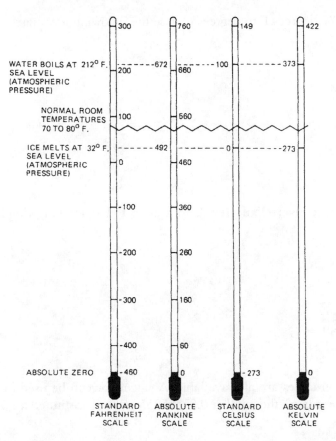

WATER BOILS AT 212° F.
SEA LEVEL
(ATMOSPHERIC
PRESSURE)

NORMAL ROOM
TEMPERATURES
70 TO 80° F.

ICE MELTS AT 32° F.
SEA LEVEL
(ATMOSPHERIC
PRESSURE)

ABSOLUTE ZERO

STANDARD
FAHRENHEIT
SCALE

ABSOLUTE
RANKINE
SCALE

STANDARD
CELSIUS
SCALE

ABSOLUTE
KELVIN
SCALE

FIGURE 11.1 Standard dry-bulb thermometer scales. *(Johnson Controls)*

perature of a perfect gas varies, the volume of the gas will vary directly. This statement is known as the Perfect Gas Law. It can be expressed mathematically by the following equation:

$$PV = TR$$

$P =$ absolute pressure

$V =$ volume

$T =$ absolute temperature

$R =$ a constant, depending on the units selected
for P, V, and T.

It is sometimes necessary to convert from one temperature scale to another. In converting from the Fahrenheit to the Celsius scale, 5/9 degree Fahrenheit is equal to one degree Celsius. Or, 9/5 degree Celsius is equal to one degree Fahrenheit. Equations facilitate converting from one scale to the other.

For example, convert 77 degrees F to degrees C. Use the following formula:

°C = 5/9 (°F − 32)

5/9 (77 − 32)

5/9 (45)

$\dfrac{5\,(45)}{9}$

225/9 = 25

77° F = 25° C

To use another example, convert 25 degrees C to degrees F. Use the following formula:

°F = 9/5(°C) + 32

9/5(25) + 32

$\dfrac{9(25) + 32}{5}$

225/5 = 45 + 32 = 77

25° C = 77° F

Temperature-conversion tables are also available. A calculator can be used for the above temperature conversions. If a calculator is used, the number 0.55555555 can be substituted for 5/9. The number 1.8 can be substituted for 9/5.

PRESSURES

All devices that measure pressure must be exposed to two pressures. The measurement is always the difference between two pressures, such as gauge pressure and atmospheric pressure.

Gauge Pressure

On an ordinary pressure gauge, one side of the measuring element is exposed to the medium under pressure. The other side is exposed to the atmosphere. Atmospheric pressure varies with altitude and climatic conditions.

Thus, it is obvious that gauge pressure readings will not represent a precise, definite value unless the atmospheric pressure is known. Gauge pressures are usually designated as psig (pounds per square inch gauge). Pressure values that include atmospheric pressure are designated psia (pounds per square inch absolute).

Atmospheric Pressure

A barometer is used to measure atmospheric pressure. A simple mercury barometer may be made with a glass tube slightly more than 30 inches in length. It should be sealed at one end and filled with mercury. The open end should be inverted in a container of mercury (see Figure 11.2). The mercury will drop in the tube until the weight of the atmosphere on the surface of the mercury in the pan just supports the weight of the mercury column in the tube. At sea level and under certain average climatic conditions, the height of mercury in the tube will be 29.92 inches. The space above the mercury in the tube will be an almost perfect vacuum, except for a slight amount of mercury vapor.

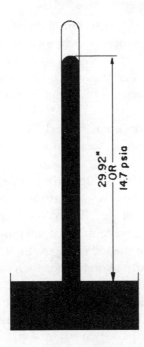

FIGURE 11.2   Simple mercury barometer.
(Johnson Controls)

A mercury manometer is an accurate instrument for measuring pressure. The mercury is placed in a glass U-tube. With both ends open to the atmosphere, the mercury will stand at the same level in both sides of the tube (see Figure 11.3). A scale is usually mounted on one of the tubes with its zero point at the mercury level.

Pressure-Measuring Devices

Low pressures, such as in an air-distribution duct, are measured with a manometer using water instead of mercury in the U-tube (see Figure 11.3). The unit of measurement, one inch of water, is often abbreviated as 1 H_2O or 1 w.g. (water gauge).

Water is much lighter than mercury. Thus, a water column is a more sensitive gauge of pressure than a mercury column. Figure 11.4 shows a well-type manometer. It indicates 0 inches water gauge. Figure 11.5 shows an incline manometer. It spreads a small range over a longer scale for accurate measurement

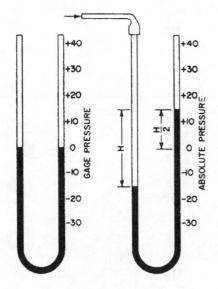

FIGURE 11.3   Mercury manometer at rest *(left)* and with pressure applied *(right)*. H is the height from the top of one tube to the top of the other; H/2 is one half of H.

of low pressures. Most manometers of this type use a red, oily liquid in place of water to provide a more practical and useful instrument. Figure 11.6 shows a magnahelic mechanical manometer. It is designed to eliminate the liquid in the water gauge. It is calibrated in hundredths of an inch water gauge for very sensitive measurement. Figure 11.7 shows a standard air-pressure gauge used for adjusting control instruments. It is calibrated in inches of water and psig. It is apparent from the scale that one psig is equal to 27.6 inches water gauge.

Bourdon (spring) tube gauges and metal diaphragm gauges are also used for measuring pressure. These gauges are satisfactory for most commercial uses. They are not as accurate as the manometer or barometer because of the mechanical methods involved.

Another unit of pressure measurement is the atmosphere. Zero-gauge pressure is equal to one atmosphere (14.7 pounds per square inch) at sea level. For rough calculations, one atmosphere can be considered 15 psig. A gauge pressure of 15 psia is approximately two atmospheres. The volume of a perfect gas varies inversely with its pressure as long as the temperature remains constant. Thus, measuring pressures in atmospheres is convenient in some cases, as may be observed from the following example: a 30-gallon tank, open to the atmosphere, contains 30 gallons of free air at a pressure of one atmosphere. If the tank is closed and air pumped in until the pressure equals two atmospheres, the tank will contain 60 gallons of free air. The original 30 gallons now occupies only one-half of the volume it originally occupied.

Conversion charts for pressures (psi to inches of Hg) may be found in various publications. The reference section of an engineering data book is a good source.

Hygrometer

A hygrometer is an instrument used to measure the amount of moisture in the air. If a moist wick is placed over a thermometer bulb, the evaporation of moisture from the wick will lower the thermometer reading (temperature). If the air surrounding a wet-bulb thermometer is dry, evaporation from the moist wick will

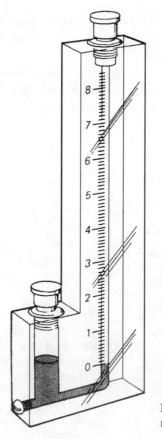

FIGURE 11.4 Well-type manometer. *(Johnson Controls)*

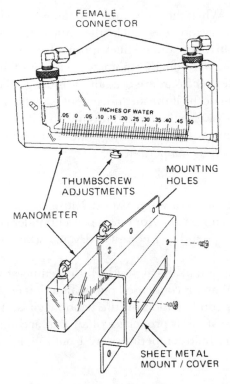

FIGURE 11.5 Incline manometer. *(Dwyer)*

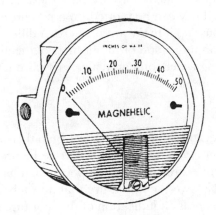

FIGURE 11.6 Magnahelic manometer.

FIGURE 11.7 Inches of water gauge. *(Johnson Controls)*

be more rapid than if the bulb thermometer is wet. When the air is saturated, no water will evaporate from the cloth wick, and the temperature of the wet-bulb thermometer will be the same as the reading on the dry-bulb thermometer. However, if the air is not saturated, the water will evaporate from the wick, causing the tempeature reading to be lower. The accuracy of the wet-bulb thermometer depends on how fast air passes over the bulb. Speeds of 5,000 feet/minute (60 miles per hour) are best, but it is dangerous to move a thermometer at that speed. Errors up to 15 percent can occur if the air movement is too slow or if there is too much radiant heat (sunlight, for example) present. A wet-bulb temperature taken with air moving at about 1 to 2 meters/second is referred to as a screen temperature, whereas a temperature taken with air moving about 3.5 meters/second or more is referred to as a sling temperature.

Properties of Air

Air is composed of nitrogen, oxygen, and small amounts of water vapor. Nitrogen makes up 77 percent, while oxygen accounts for 23 percent. Water vapor can account for 0 to 3 percent under certain conditions. Water vapor is measured in grains or, in some cases, pounds per pound of dry air. Seven thousand grains of water equal one pound.

Temperature determines the amount of water vapor that air can hold. Hotter temperatures mean that air has a greater capacity to hold water in suspension. Water is condensed out of air as it is cooled. Outside, water condensation becomes rain. Inside, it becomes condensation on the window glass. Thus, dry air acts somewhat like a sponge. It absorbs moisture. There are four properties of air that account for its behavior under varying conditions. These properties are dry-bulb temperature, wet-bulb temperature, dew-point temperature, and relative humidity.

Dry-Bulb Temperature

Dry-bulb temperature is the air temperature that is determined by an ordinary thermometer. There are certain amounts of water vapor per pound of dry air. They can be plotted on a psychrometric chart. A psychrometer is an instrument for measuring the aqueous vapor in the atmosphere. A difference between a wet-bulb thermometer and a dry-bulb thermometer is an indication of the dryness of the air. A psychrometer, then, is a hygrometer, which is a device for measuring water content in air. A psychrometric chart indicates the different values of temperature and water moisture in air.

Wet-Bulb Temperature

Wet-bulb temperature reflects the cooling effect of evaporating water. A wet-bulb thermometer is the same as a dry-bulb thermometer, except that it has a wet cloth around the bulb (see Figure 11.8A). The thermometer is swung around in the air. The temperature is read after this operation. The wet-bulb temperature is lower than the dry-bulb temperature. It is the lowest temperature that a water-wetted body will attain when exposed to an air current. The measurement is an indication of the moisture content of the air.

The Bacharach sling psychrometer, shown in Figure 11.8B, is a compact sling type that determines the percentage of relative humidity. It has a built-in slide-rule calculator that correlates wet- and dry-bulb temperatures to relative humidity. The dual-range, high and low temperature, scales are designed for better resolution. The thermometers telescope into the handle for protection when not in use. They are available in both red-spirit-filled and the mercury-filled thermometers as well as in degrees F or C. There is a built-in water reservoir that holds sufficient water for several hours of testing. It is designed for portability, ease of use, and ruggedness. Accuracy is within ±5 percent relative humidity. The thermometers are constructed of shock-resistant glass. The stems have deep-etched numbers and 1-degree scale divisions for easy reading. The mercury-filled and spirit-filled thermometers have a range of +25 F to +120 degrees F. They can be obtained in Celsius with a range of -5 to +50 degrees.

Dew-Point Temperature

Dew-point temperature is the temperature below which moisture will condense out of air. The dew point of air is reached when the air contains all the moisture it can hold. The dry-bulb and wet-bulb temperatures

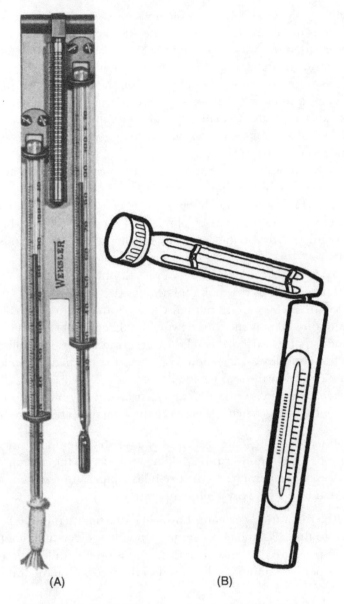

FIGURE 11.8 (A) Wet-bulb and dry-bulb thermometers mounted together. Note the knurled and ringed rod between the two at the top of the scales. It is used to hold the unit and twirl it in the air. (B) Bacharach sling psychrometer. *(Weksler)*

are the same at this point. The air is said to be at 100 percent relative humidity when both thermometers read the same. Dew point is important when designing a humidifying system for human comfort. If the humidity is too high in a room, the moisture will condense and form on the windows.

Relative Humidity

Relative humidity is a measure of how much moisture is present compared to how much moisture the air could hold at that temperature. Relative humidity (rh) is based on the percentage of humidity contained in the air relative to the saturation condition of the air. A reading of 70 percent means that the air contains 70 percent of the moisture it can hold. Relative-humidity lines on the psychrometric chart are sweeping curves, as shown in Figure 11.13.

To keep the home comfortable in winter, it is sometimes necesary to add humidity. Hot-air heat will in time remove most of the moisture in the living space. The addition of moisture is accomplished in a number of ways. Humidifiers are used to spray water into the air, or large areas of water are made available to evaporate. Showers and running water also add moisture to a living space.

In summer, however, the amount of moisture per pound on the outside is greater than on the inside, especially when the room is air-conditioned. This means the vapor pressure is greater on the outside than the inside. Under these conditions, moisture will enter the air-conditioned space by any available route. It will enter through cracks, around doors and windows, and through walls. In winter, the moisture moves the other way—from the inside to the outside.

The percentage of relative humidity is never more than 100 percent. When the air is not saturated, the dry-bulb temperature will always be higher than the wet-bulb temperature. The dew-point temperature will always be the lowest reading. The greater the difference between the dew-point temperature and the dry-bulb temperature, the lower will be the percentage of relative humidity. The wet-bulb reading can never be higher than the dry-bulb reading. Nor can the dew-point reading be higher than the dry-bulb reading. Consider the following example:

With saturated air (100 percent humidity) temperature is dry bulb 90 degrees F, wet bulb 90 degrees F, and relative humidity 100 percent, thus dew point 90 degrees F.

With unsaturated air (less than 100 percent humidity) temperature is dry bulb 80 degrees F, wet bulb 75 degrees F, and relative humidity 80 percent, thus dew point 73 degrees F.

Manufacturers of humidifiers furnish a dial similar to a thermostat for controlling the humidity. A chart on the control tells what the humidity setting should be when the temperature outside is at a given point. Table 11.1 gives an example of what the settings should be.

The relationship between humidity, wet-bulb temperature, and dry-bulb temperature has much to do with the designing of air-conditioning systems. There are three methods of controlling the saturation of air:

1. Keep the dry-bulb temperature constant. Raise the wet-bulb temperature and the dew-point temperature to the dry-bulb temperature. Adding moisture to the air can do this. In turn, it will raise the dew-point temperature to the dry-bulb temperature, which automatically raises the wet-bulb temperature to the dry-bulb temperature.

2. Keep the wet-bulb temperature constant. Lower the dry-bulb temperature. Raise the dew-point temperature to the wet-bulb temperature. Cooling the dry-bulb temperature to the level of the wet-bulb temperature does this. The idea here is do it without adding or removing any moisture. The dew point is automatically raised to the wet-bulb temperature.

TABLE 11.1 Permissible relative humidity (in the winter).

Outside Temperature		Brick Wall 12 in. Thick Plastered Inside	Single Glass	Double Glass
°F	°C			
			Percentage	
-20	-29.0	45	7	35
-10	-23.0	50	10	40
0	-17.8	60	18	45
10	-12.2	64	25	50
20	-6.7	67	30	55
30	-1.1	74	38	60
40	4.4	80	45	65
50	10.0	85	50	70
60	15.6	90	55	75

TABLE 11.2 Activity-heat relationships.

Activity	Total Heat in Btu Per Hour
Person at rest	385
Person standing	430
Tailor	480
Clerk	600
Dancing	760
Waiter	1,000
Walking	1,400
Bowling	1,500
Fast walking	2,300
Severe exercise	2,600

3. Keep the dew-point temperature constant and the wet-bulb temperature at the dew-point temperature. Cooling the dry-bulb and wet-bulb temperatures to the dew-point temperature can do this.

People and Moisture

People working inside a building or occupied space give off moisture as they work. They also give off heat. Such moisture and heat must be considered in determining air-conditioning requirements. Table 11.2 indicates some of the heat given off by the human body when working.

Psychrometric Chart

The psychrometric chart holds much information (see Figure 11.9). However, it is hard to read. It must be studied for some time. The dry-bulb temperature is located in one place and the wet-bulb in another. If the two are known, it is easy to find the relative humidity and other factors relating to checking air. Both customary and metric psychrometric charts are available.

Air contains different amounts of moisture at different temperatures. Table 11.3 shows the amounts of moisture that air can hold at various temperatures.

TABLE 11.3 Saturated vapor per pound of dry air (barometer reading at 29.92 inches per square inch).

Temperature		Weight in Grains
°F	°C	
25	-3.9	19.1
30	-1.1	24.1
35	1.7	29.9
40	4.4	36.4
45	7.2	44.9
50	10.0	53.5
55	12.8	64.4
60	15.6	77.3
65	18.3	92.6
70	21.1	110.5
75	23.9	131.4
80	26.7	155.8
85	29.4	184.4
90	32.2	217.6
95	35.0	256.3
100	37.8	301.3
105	40.6	354.0
110	43.3	415.0

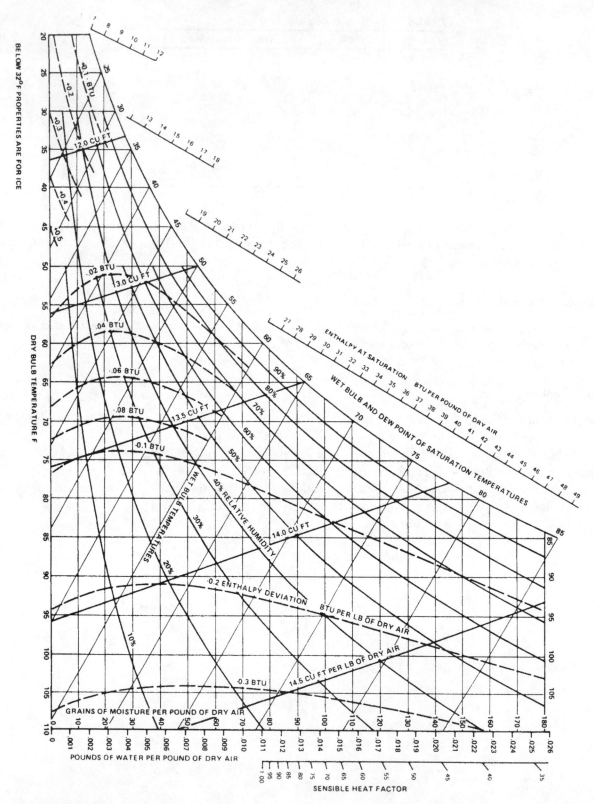

FIGURE 11.9 Psychrometric chart. *(Carrier)*

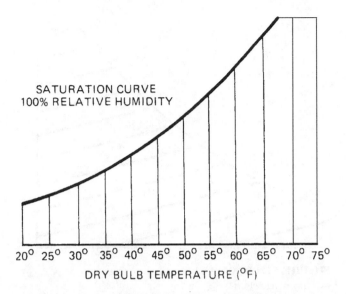

FIGURE 11.10   Temperature lines (dry-bulb) on
psychrometric chart. Only a portion of the chart is shown.

An explanation of the various quantities shown on a psychrometric chart will enable you to understand the chart. The different quantities on the chart are shown separately on the following charts. These charts will help you see how the psychrometric chart is constructed (see Figure 11.10).

Across the bottom, the vertical lines are labeled from 25 to 110 degrees F in increments of 5 degrees. These temperatures indicate the dry-bulb temperature (see Figure 11.11). The horizontal lines are labeled

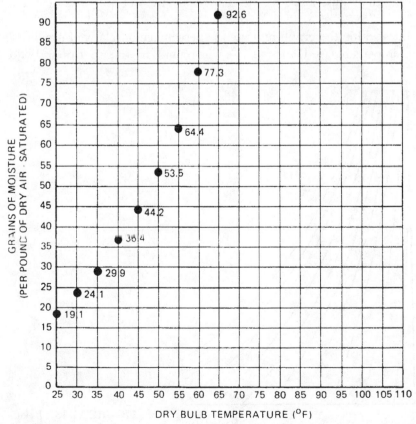

FIGURE 11.11   The moisture content in kilograms per kilogram of dry air is
measured on the vertical column. Here, only a portion of the chart is shown.

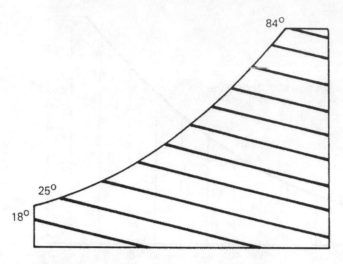

FIGURE 11.12 Wet-bulb temperature lines on a psychrometric chart.

from 0 to 180 degrees F. This span of numbers represents the grains of moisture per pound of dry air (when saturated). The outside curving line on the left side of the graph indicates the wet-bulb, dew-point, or saturation temperature (see Figure 11.12).

At 100 percent saturation the wet-bulb temperatures are the same as the dry-bulb and the dew-point temperatures. This means the wet-bulb lines start from the 100 percent saturation curve. Diagonal lines represent the wet-bulb temperatures. The point where the diagonal line of the wet bulb crosses the dry bulb's vertical line is the dew point. The temperature of the dew point will be found by running the horizontal line to the left and reading the temperature on the curve, since the wet-bulb and dew-point temperatures are on the curve.

The curving lines within the graph indicate the percentage of relative humidity. These lines are labeled 10, 20, and so on in percent (see Figure 11.13). The pounds of water per pound of dry air are shown in the

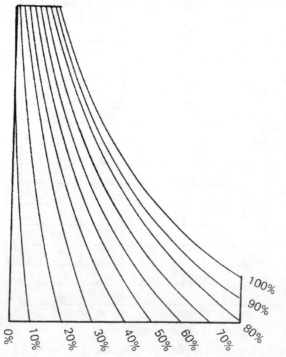

FIGURE 11.13 Relative-humidity lines on a psychrometric chart.

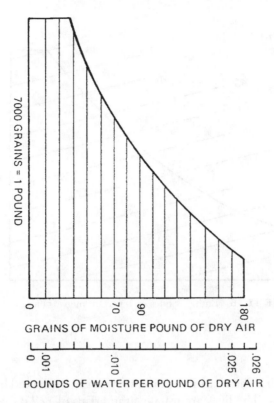

FIGURE 11.14 The sensible heat factor on a psychrometeric chart.

bottom column of numbers. The grains of moisture per pound of dry air are shown in the middle column of the three columns of numbers (see Figure 11.14).

Table 11.3 shows that one pound of dry air will hold 19.1 grains of water at 25 degrees F (-3.9 degrees C). One pound of dry air will hold 415 grains of water at 110 degrees F. It can be seen that the higher the temperature, the more moisture the air can hold. This is one point that should be remembered. To find the weight per grain, divide 1 by 7000 to get 0.00014 pounds per grain. Therefore, on the chart 0.01 pounds corresponds to about 70 grains. The volume of dry air (cubic feet per pound) is represented by diagonal lines (see Figure 11.15). The values are marked along the lines. They represent the cubic feet of the mixture of

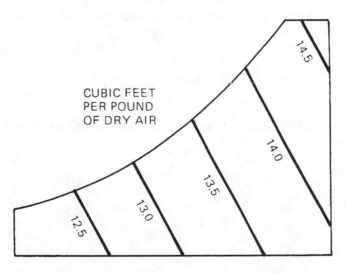

FIGURE 11.15 Air-volume lines on a psychrometric chart.

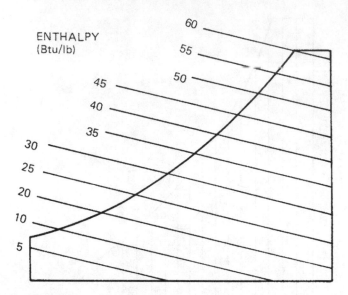

FIGURE 11.16   Enthalpy lines on a psychrometric chart.

vapor and air per pound of dry air. The chart indicates that volume is affected by temperature relationships of the wet- and dry-bulb readings. The lines are usually at intervals of 1/2 cubic foot per pound.

Enthalpy is the total amount of heat contained in the air above 0 degrees F (-17.8 degrees C) (see Figure 11.16). The lines on the chart that represent enthalpy are extensions of the wet-bulb lines. They are extended and labeled in Btu per pound. This value can be used to help determine the load on an air-conditioning unit.

MULTIPLE-CHOICE EXAM

1. The sorption of water vapor in a desiccant raises the temperature of the desiccant itself, which in turn _____ the surrounding air.

    a.   expands   b.   cools

    c.   warms   d.   heats

2. Common _____ desiccants include water solutions of lithium chloride or glycol.

    a.   solid   b.   liquid

    c.   paste   d.   gel

3. Lower vapor pressure at the surface of a desiccant means that it can attract moisture from air that has a _____ vapor pressure.

    a. higher

    b. very much higher

    c. lesser

    d. lower

4. In practice, dehumidifiers exhibit process-air dry-bulb temperature increases that typically range from slightly over one to two times that associated with the heat of condensation alone. The entire heat, including the heat of condensation, is termed the heat of _____.

    a. sorption

    b. adsorption

    c. absorption

    d. exothermic

5. In a solid desiccant dehumidifier, moist process air is passed through a rotating wheel that resembles a ceramic honeycomb. The desiccant in the wheel absorbs moisture. Then the wheel slowly rotates into a second, heated air stream. The hot reactivation air removes moisture from the desiccant so it can absorb more _____ when the wheel rotates back into the process air stream.

    a. vapor

    b. humidity

    c. water

    d. heat

6. Granular desiccants are placed in some form of bed to _____ air.

    a. humidify

    b. dehumidify

    c. clear

    d. unpollute

7. In order to be able to manufacture pharmaceuticals, the relative humidity has to be held to as low as _____ percent.

    a. 1

    b. 5

    c. 10

    d. 20

8. Desiccants are exceptionally effective at maintaining _____ humidity.

    a. low

    b. high

    c. relative

    d. excessive

9. Desiccant systems begin to have advantages over conventional systems when the proportion of fresh air in a given system goes above _____ percent.

    a. 5

    b. 10

    c. 15

    d. 20

10. Direct evaporative cooling has the advantages of low first cost and _____ operating costs.

    a. light   b. heavy

    c. low   d. high

11. Temperature is defined as the _____ state of matter.

    a. thermal   b. solid

    c. liquid   d. frozen

12. The Fahrenheit scale has a boiling temperature for water of 212 degrees and a freezing point of _____ degrees.

    a. 10   b. 0

    c. 32   d. 100

13. The Celsius scale has a boiling temperature of 100 degrees for water and a freezing point of _____ degrees.

    a. 0   b. 10

    c. 1,000   d. -10

14. The absolute Rankine scale has a freezing point of water at _____ degrees.

    a. 50   b. 100

    c. 672   d. 492

15. The absolute Kelvin scale has a freezing point of water at _____ degrees.

    a. 32   b. 273

    c. 100   d. 212

16. Psychrometry is the science and practice of air _____ and their control.

    a. pressures   b. mixtures

    c. distributions   d. diffusions

17. Psychrometry is a specialized area of _____.

    a. air conditioning   b. air humidifying

    c. thermodynamics   d. dehumidification

18. A barometer is used to measure _____ pressure.

    a. condenser   b. compressor

    c. refrigerant   d. atmospheric

19. A mercury manometer is an accurate instrument for measuring _____.

    a. pressure

    b. a vacuum

    c. compression

    d. condensation

20. Bourdon (spring) gauges and metal diaphragm gauges are used for measuring _____.

    a. compression

    b. vacuum

    c. pressure

    d. humidity

21. A hygrometer is described as an instrument used to measure the amount of _____ in the air.

    a. water

    b. ice droplets

    c. fog

    d. moisture

22. Air is composed of nitrogen, oxygen, and small amounts of _____ vapor.

    a. hydrogen

    b. water

    c. helium

    d. gas

23. What determines the amount of water vapor that air can hold?

    a. pressure

    b. a vacuum

    c. temperature

    d. sunshine

24. A wet-bulb temperature reflects the cooling effect of _____ water.

    a. evaporating

    b. boiling

    c. frozen

    d. condensing

25. How is the dry-bulb temperature obtained?

    a. by using a Celsius-scale thermometer

    b. by using a manometer

    c. by using a Kelvin-scale thermometer

    d. by using an ordinary thermometer

26. The percentage of relative humidity is never more than _____ percent.

    a. 75

    b. 100

    c. 25

    d. 50

27. What is used in the winter months in northern areas to spray water into the hot air in the furnace plenum that is used for heating the house?

    a. a humidifier   b. a dehumidifer

    c. a desiccant   d. an atomizer

28. What is the permissible relative humidity, in the winter, with an outside temperature of 10 degrees F and a single glass pane with a 12-inch brick wall that is plastered inside:

    a. 25   b. 10

    c. 40   d. 65

29. How much heat is given off by a person when working (standing):

    a. 1,400 Btu   b. 2,300 Btu

    c. 430 Btu   d. 385 Btu

30. Enthalpy is the total amount of _____ contained in the air above 0 degrees F.

    a. ice   b. heat

    c. fog   d. humidity

TRUE-FALSE EXAM

1. The centigrade scale is now referred to as Celsius.

   True   False

2. Absolute temperatures are measured from absolute zero.

   True   False

3. At a constant temperature, as the absolute temperature of a perfect gas varies, its absolute pressure will vary indirectly.

   True   False

4. Psychrometry is the science and practice of air mixtures and their control.

   True   False

5. Absolute zero is the point at which there is no heat.

   True   False

6. A barometer is used to measure the compressor's pressure.

　　True　　False

7. All devices that measure pressure must be exposed to two pressures. The measurement is always the difference between two pressures, such as gauge pressure and atmospheric pressure.

　　True　　False

8. Pressures that include atmospheric pressure are designated pounds per square inch absolute or psia.

　　True　　False

9. A mercury manometer is an accurate instrument for measuring pressure.

　　True　　False

10. When used for low-pressure measurements, such as in an air-distribution duct, the manometer can use water instead of mercury in the U-tube.

　　True　　False

11. The Bourdon-spring tube gauge is as sensitive and accurate as the manometer or barometer.

　　True　　False

12. A hygrometer is an instrument used to measure the amount of moisture in the air.

　　True　　False

13. Seven thousand grains of water equal 1 pound.

　　True　　False

14. The amount of water vapor that air can hold is determined by the temperature.

　　True　　False

15. Dew-point temperature is the temperature below which moisture will condense out of air.

　　True　　False

MULTIPLE-CHOICE ANSWERS

1. D	7. C	13. A	19. A	25. D
2. B	8. A	14. D	20. C	26. B
3. D	9. C	15. B	21. D	27. A
4. A	10. C	16. B	22. B	28. A
5. B	11. A	17. C	23. C	29. C
6. B	12. C	18. D	24. A	30. B

TRUE-FALSE ANSWERS

1. T	5. T	9. T	13. T
2. T	6. F	10. T	14. T
3. F	7. T	11. F	15. T
4. T	8. T	12. T	

Part E

CODES AND CERTIFICATIONS

HVAC Licensing Study Guide

Chapter 12
EPA REFRIGERATION HANDLERS

The EPA listing of reclaimers is updated when additional refrigerant reclaimers are approved. Reclaimers appearing on this list are approved to reprocess used refrigerant to at least the purity specified in appendix A to 40 CFR part 82, subpart F (based on ARI Standard 700, "Specifications for Fluorocarbon and Other Refrigerants"). Reclamation of used refrigerant by an EPA-certified reclaimer is required in order to sell used refrigerant not originating from and intended for use with motor-vehicle air conditioners.

The EPA encourages reclaimers to participate in a voluntary third-party reclaimer-certification program operated by the Air-Conditioning and Refrigeration Institute (ARI). The volunteer program offered by the ARI involves quarterly testing of random samples of reclaimed refrigerant. Third-party certification can enhance the attractiveness of a reclaimer program by providing an objective assessment of its purity.

Since the world has become aware of the damage the Freon® refrigerants can do to the ozone layer, there has been a mad scramble to obtain new refrigerants that can replace all those now in use. There are some problems with adjusting the new and especially existing equipment to the properties of new refrigerant blends.

It is difficult to directly replace R-12, for instance. It has been the mainstay in refrigeration equipment for years. However, the automobil air-conditioning industry has been able to reformulate R-12 to produce an acceptable substitute, R-134a. There are others now available to substitute in the more sophisticated equipment with large amounts of refrigerants. Some of these will be covered here.

FREON® REFRIGERANTS

The Freon® family of refrigerants has been one of the major factors responsible for the impressive growth of not only the home-refrigeration and air-conditioning industry but also the commercial-refrigeration industry. The safe properties of these products have permitted their use under conditions where flammable or more toxic refrigerants would be hazardous. Following are descriptions of commonly used Freon® refrigerants.

Freon-11 has a boiling point of 74.8 degrees F and has wide usage as a refrigerant in indirect industrial and commercial air-conditioning systems employing single or multistage centrifugal compressors with capacities of 100 tons and above. Freon-11 is also employed as brine for low-temperature applications. It provides relatively low operating pressures with moderate displacement requirements.

The boiling point of Freon-12 is -21.7 degrees F. It is the most widely known and used of the Freon refrigerants. It is used principally in household and commercial refrigerators, frozen-food cabinets, ice-cream cabinets, food-locker plants, water coolers, room and window air-conditioning units, and similar equipment. It is generally used in reciprocating compressors, ranging in size from fractional to 800 horsepower. Rotary compressors are useful in small units. The use of centrifugal compressors with Freon-12 for large air-conditioning and process-cooling applications is increasing.

The boiling point of Freon-13 is -144.6 degrees F. It is used in low-temperature specialty applications employing reciprocating compressors and generally in cascade with Freon-12 or Freon-22.

Freon-21 has a boiling point of 48 degrees F. It is used in fractional-horsepower household refrigerating systems and drinking-water coolers employing rotary vane-type compressors. Freon-21 is also used in comfort-cooling air-conditioning systems of the absorption type where dimethyl ether or tetraethylene glycol is used as the absorbent.

The boiling point of Freon-22 is -41.4 degrees F. It is used in all types of household and commercial refrigeration and in air-conditioning applications with reciprocating compressors. The outstanding thermodynamic properties of Freon-22 permit the use of smaller equipment than is possible with similar refrigerants, making it especially suitable where size is a problem.

The boiling point of Freon-113 is 117.6 degrees F. It is used in commercial and industrial air-conditioning and process-water and brine cooling with centrifugal compression. It is especially useful in small-tonnage applications.

The boiling point of Freon-114 is 38.4 degrees F. It is used as a refrigerant in fractional-horsepower household refrigerating systems and drinking-water coolers employing rotary vane-type compressors. It is also used in indirect industrial and commercial air-conditioning systems and in industrial process-water and brine cooling to -70 degrees F, employing multistage centrifugal-type compressors in cascades of 100 ton' refrigerating capacity and larger.

The boiling point of Freon-115 is -37.7 degrees F. It is especially stable, offering a particularly low discharge temperature in reciprocating compressors. Its capacity exceeds that of Freon-12 by as much as 50 percent in low-temperature systems. Its potential applications include household refrigerators and automobile air conditioning.

Freon-502 is an azeotropic mixture composed of 48.8 percent Freon-22 and 51.2 percent Freon-115 by weight. It boils at -50.1 degrees F. Because it achieves the capacity of Freon-22 with discharge temperatures comparable to Freon-12, it is finding new reciprocating compressor applications in low-temperature display cabinets and in storing and freezing of food.

Properties of Freons®

The Freon® refrigerants are colorless and almost odorless, and their boiling points vary over a wide range of temperatures. Those Freon® refrigerants that are produced are nontoxic, non-corrosive, nonirritating, and nonflammable under all conditions of usage. They are generally prepared by replacing chlorine or hydrogen with fluorine. Chemically, Freon® refrigerants are inert and thermally stable up to temperatures far beyond conditions found in actual operation. However, Freon® is harmful when allowed to escape into the atmosphere. It can deplete the ozone layer and cause more harmful ultraviolet rays to reach the surface of the Earth.

The pressures required in liquefying the refrigerant vapor affect the design of the system. The refrigerating effect and specific volume of the refrigerant vapor determine the compressor displacement. The heat of vaporization and specific volume of the liquid refrigerant affect the quantity of refrigerant to be circulated through the pressure-regulating valve or other system device.

Flammability

Freon® is nonflammable and noncombustible under conditions where appreciable quantities contact flame or hot metal surfaces. It requires an open flame at 1,382 degrees F to decompose the vapor. Even at this temperature, only the vapor decomposes to form hydrogen chloride and hydrogen fluoride, which are irritating but are readily dissolved in water. Air mixtures are not capable of burning and contain no elements that will support combustion. For this reason, Freon® is considered nonflammable.

Circulation

It should be noted that the Freon® refrigerants have relatively low heat values, but this must not be considered a disadvantage. It simply means that a greater volume of liquid must be circulated per unit of time to produce the desired amount of refrigeration. It does not concern the amount of refrigerant in the system. Actually, it is a decided advantage (especially in the smaller or low-tonnage systems) to have a refrigerant with low heat values. This is because the larger quantity of liquid refrigerant to be metered through the liquid-regulating device will permit the use of more accurate and more positive operating and regulating mechanisms of less sensitive and less critical adjustments. Table 12.1 lists the quantities of liquid refrigerant metered or circulated per minute under standard ton conditions.

TABLE 12.1 Quantities of refrigerant circulated per minute under standard ton conditions.

Refrigerant	Pounds Expanded per Minute	Ft^3/lb Liquid 86°F	In.^{3} Liquid Expanded per Minute	Specific Gravity Liquid 86°F (water-1)
Freon-22	2.887	0.01367	67.97	1.177
Freon-12	3.916	0.0124	83.9	1.297
Freon-114	4.64	0.01112	89.16	1.443
Freon-21	2.237	0.01183	45.73	1.360
Freon-11	2.961	0.01094	55.976	1.468
Freon-113	3.726	0.01031	66.48	1.555

Volume (Piston) Displacement

For reason of compactness, cost of equipment, reduction of friction, and compressor speed, the volume of gas that must be compressed per unit of time for a given refrigerating effect generally should be as low as possible. Freon-12 has a relatively low volume displacement, which makes it suitable for use in reciprocating compressors ranging from the smallest size up to 800-ton capacity, including compressors for household and commercial refrigeration. Freon-12 also permits the construction of compact rotary compressors in the commercial sizes. Generally, low-volume displacement (high-pressure) refrigerants are used in reciprocating compressors; high-volume displacement (low-pressure) refrigerants are used in large-tonnage centrifugal compressors; intermediate-volume (intermediate-pressure) refrigerants are used in rotary compressors. There is no standard rule governing this usage.

Condensing Pressure

Condensing (high-side) pressure should be low to allow construction of lightweight equipment, which affects power consumption, compactness, and installation. High pressure increases the tendency toward leakage on the low side as well as the high side when pressure is built up during idle periods. In addition, pressure is very important from the standpoint of toxicity and fire hazard.

In general, a low-volume displacement accompanies a high condensing pressure, and a compromise must usually be drawn between the two in selecting a refrigerant. Freon-12 presents a balance between volume displacement and condensing pressure. Extra-heavy construction is not required for this type of refrigerant, and so there is little or nothing to be gained from the standpoint of weight of equipment in using a lower-pressure refrigerant.

Evaporating Pressure

Evaporating (low-side) pressures above atmospheric are desirable to avoid leakage of moisture-laden air into the refrigerating systems and permit easier detection of leaks. This is especially important with open-type units. Air in the system will increase the head pressures, resulting in inefficient operations, and may adversely affect the lubricant. Moisture in the system will cause corrosion and may also freeze out and stop operation of the equipment.

In general, the higher the evaporating pressure, the higher the condensing pressure under a given set of temperatures. Therefore, to keep head pressures at a minimum and still have positive low-side pressures, the refrigerant selected should have a boiling point at atmospheric pressure as close as possible to the lowest temperature to be produced under ordinary operating conditions. Freon-12, with a boiling point of -21.7 degrees F, is close to ideal in this respect for most refrigeration applications. A still lower boiling point is of some advantage only when lower operating temperatures are required.

Freezing Point

The freezing point of a refrigerant should be below any temperature that might be encountered in the system. The freezing point of all refrigerants, except water (32 degrees F) and carbon dioxide (-69.9 degrees F, triple point), are far below the temperatures that might be encountered in their use. Freon-12 has a freezing point of -247 degrees F. (See Appendix A for more details on newer refrigerants.)

Critical Temperature

The critical temperature of a refrigerant is the highest temperature at which it can be condensed to a liquid, regardless of a higher pressure. It should be above the highest condensing temperature that might be encountered. With air-cooled condensers, in general, this would be above 130 degrees F. Loss of efficiency caused by superheating of the refrigerant vapor on compression and by throttling expansion of the liquid is greater when the critical temperature is low.

All common refrigerants have satisfactorily high critical temperatures except carbon dioxide (87.8 degrees F) and ethane (89.8 degrees F). These two refrigerants require condensers cooled to temperatures below their respective critical temperatures, and thus generally need water.

There are some hydrofluorocarbon refrigerants (such as R-134a) that are made to eliminate the problems with refrigerants in the atmosphere caused by leaks in systems. R-134a is a non-ozone-depleting refrigerant used in vehicle air-conditioning systems. DuPont's brand name is Suva, and the product is produced in a plant located in Corpus Christi, Texas, as well in Chiba, Japan. According to DuPont's Web site, R-134a was globally adopted by all vehicle manufacturers in the early 1990s as a replacement for CFC-12. The transition to R-134a was completed by the mid-1990s for most major automobile manufacturers. Today, there are more than 300 million cars with air conditioners using the newer refrigerant.

Latent Heat of Evaporation

A refrigerant should have a high latent heat of evaporation per unit of weight so that the amount of refrigerant circulated to produce a given refrigeration effect may be small. Latent heat is important when considering its relationship to the volume of liquid required to be circulated. The net result is the refrigerating effect. Since other factors enter into this determination, they are discussed separately.

The refrigerant effect per pound of refrigerant under standard ton conditions determines the amount of refrigerant to be evaporated per minute. The refrigerating effect per pound is the difference in Btu content of the saturated vapor leaving the evaporator (5 degrees F) and the liquid refrigerant just before passing through the regulating valve (86 degrees F). While the Btu refrigerating effect per pound directly determines the number of pounds of refrigerant to be evaporated in a given length of time to produce the required results, it is much more important to consider the volume of the refrigerant vapor required rather than the weight of the liquid refrigerant. By considering the volume of refrigerant necessary to produce standard ton conditions, it is possible to make a comparison between Freon-12 and other refrigerants so as to provide for the reproportioning of the liquid orifice sizes in the regulating valves, sizes of liquid refrigerant lines, and so on.

A refrigerant must not be judged only by its refrigerating effect per pound; the volume per pound of the liquid refrigerant must also be taken into account to arrive at the volume of refrigerant to be vaporized. Although Freon-12 has relatively low refrigerating effect, this is not a disadvantage, because it merely indicates that more liquid refrigerant must be circulated to produce the desired amount of refrigeration. Actually, it is a decided advantage to circulate large quantities of liquid refrigerant, because the greater volumes required will permit the use of less sensitive operating and regulating mechanisms with less critical adjustment.

Refrigerants with high Btu refrigerating effects are not always desirable, especially for household and small commercial installations because of the small amount of liquid refrigerant in the system and the difficulty encountered in accurately controlling its flow through the regulating valve. For household and small commercial systems, the adjustment of the regulating-valve orifice is most critical for refrigerants with high Btu values.

Specific Heat

A low specific heat of the liquid is desirable in a refrigerant. If the ratio of the latent heat to the specific heat of a liquid is low, a relatively high proportion of the latent heat may be used in lowering the temperature of the liquid from the condenser temperature to the evaporator temperature. This results in a small net refrigerating effect per pound of refrigerant circulated and, assuming other factors remain the same, reduces the capacity and lowers the efficiency. When the ratio is low, it is advantageous to precool the liquid before evaporation by heat interchange with the cool gases leaving the evaporator.

In the common type of refrigerating systems, expansion of the high-pressure liquid to a lower-pressure, lower-temperature vapor and liquid takes place through a throttling device such as an expansion valve. In this process, energy available from the expansion is not recovered as useful work. Since it performs no external work, it reduces the net refrigerating effect.

Power Consumption

In a perfect system operating between 5- and -86-degree F conditions, 5.74 Btu is the maximum refrigeration obtainable per Btu of energy used to operate the refrigerating system. This is the theoretical maximum coefficient of performance on cycles of maximum efficiency (for example, the Carnet cycle). The minimum horsepower would be 0.821 horsepower/ton of refrigeration. The theoretical coefficient of performance would be the same for all refrigerants if they could be used on cycles of maximum efficiency.

However, because of engineering limitations, refrigerants are used on cycles with a theoretical maximum coefficient of performance of less than 5.74. The cycle most commonly used differs in its basic form from the Carnet cycle, as already explained, in employing expansion without loss or gain of heat from an outside source, and in compressing adiabatic ally (compression without gaining or losing heat to an outside source) until the gas is superheated above the condensing medium temperature. These two factors, both of which increase the power requirement, vary in importance with different refrigerants. But it so happens that when expansion loss is high, compression loss is generally low, and vice versa. All common refrigerants (except carbon dioxide and water) show about the same overall theoretical power requirement on a 5- to -86-degree F cycle. At least the theoretical differences are so small that other factors are more important in determining the actual differences in efficiency.

The amount of work required to produce a given refrigerating effect increases as the temperature level to which the heat is pumped from the cold body is increased. Therefore, on a 5- to 86-degree F cycle, when gas is superheated above 86 degrees F on compression, efficiency is decreased and the power requirement increased unless the refrigerating effect caused by superheating is salvaged through the proper use of a heat interchanger.

Volume of Liquid Circulated

Volumes of liquid required to circulate for a given refrigerant effect should be low in order to avoid fluid-flow (pressure-drop) problems and to keep down the size of the required refrigerant change. In small-capacity machines, the volume of liquid circulated should not be so low as to present difficult problems in accurately controlling its flow through expansion valves or other types of liquid-metering devices.

With a given net refrigerating effect per pound, a high density of liquid is preferable to a low volume. However, a high density tends to increase the volume circulated by lowering the net refrigerating effect.

HANDLING REFRIGERANTS

One of the requirements of an ideal refrigerant is that it must be nontoxic. In reality, however, all gases (with the exception of pure air) are more or less toxic or asphyxiating. It is therefore important that wherever gases or highly volatile liquids are used, adequate ventilation be provided, because even nontoxic gases in air produce a suffocating effect.

Vaporized refrigerants (especially ammonia and sulfur dioxide) bring about irritation and congestion of the lungs and bronchial organs, accompanied by violent coughing, vomiting, and, when breathed in sufficient quantity, suffocation. It is of the utmost importance, therefore, that service workers subjected to a refrigerant gas find access to fresh air at frequent intervals to clear their lungs. When engaged in the repair of ammonia and sulfur-dioxide machines, approved gas masks and goggles should be used. Carrene, Freon® (R-12), and carbon-dioxide fumes are not irritating and can be inhaled in considerable concentrations for short periods without serious consequences.

It should be remembered that liquid refrigerant will refrigerate or remove heat from anything it meets when released from a container. In the case of contact with refrigerant, the affected or injured area should be treated as if it has been frozen or frostbitten.

Storing and Handling Refrigerant Cylinders

Refrigerant cylinders should be stored in a dry, sheltered, and well-ventilated area. The cylinders should be placed in a horizontal position if possible and held by blocks or saddles to prevent rolling. It is of the utmost importance to handle refrigerant cylinders with care and to observe the following precautions:

- Never drop the cylinders or permit them to strike each other violently

- Never use a lifting magnet or a sling (rope or chain) when handling cylinders; a crane may be used when a safe cradle or platform is provided to hold the cylinders

- Caps provided for valve protection should be kept on the cylinders at all times except when the cylinders are actually in use

- Never overfill the cylinders; whenever refrigerant is discharged from or into a cylinder, weigh the cylinder and record the weight of the refrigerant remaining in it

- Never mix gases in a cylinder

- Never use cylinders for rollers, supports, or for any purpose other than to carry gas

- Never tamper with the safety devices in valves or on the cylinders

- Open the cylinder valves slowly; never use wrenches or tools except those provided or approved by the gas manufacturer

- Make sure that the threads on regulators or other unions are the same as those on the cylinder valve outlets; never force a connection that does not fit

- Regulators and gauges provided for use with one gas must not be used on cylinders containing a different gas

- Never attempt to repair or alter the cylinders or valves

- Never store the cylinders near highly flammable substances (such as oil, gasoline, or waste)

- Cylinders should not be exposed to continuous dampness, salt water, or salt spray

- Store full and empty cylinders apart to avoid confusion

- Protect the cylinders from any object that will produce a cut or other abrasion on the surface of the metal

LUBRICANTS[1]

Lubricant properties can be evaluated to determine if the product is right for the job. Three basic properties are viscosity, lubricity, and chemical stability. They must be satisfactory to protect the compressor. The correct viscosity is needed to fill the gaps between parts and flow correctly where it is supposed to go. Generally speaking, smaller equipment with smaller gaps between moving parts requires a lighter viscosity, and larger equipment with bigger parts needs heavier viscosity oils. Lubricity refers to the lubricant's ability to protect the metal surfaces from wear.

Good chemical stability means that the lubricant will not react to form harmful chemicals such as acids, sludges, and so forth that may block tubing, or there may be carbon deposits. The interaction of lubricant and refrigerant can cause potential problems as well.

Miscibility defines the temperature region where refrigerant and oil mix or separate. If there is separation of the oil from the refrigerant in the compressor, it is possible that the oil is not getting to metal parts that need it. If there is separation in the evaporator or other parts of the system, it is possible that the oil does not return to the compressor and eventually there is not enough oil to protect it.

Solubility determines if the refrigerant will thin the oil too much. That would cause it to lose its ability to protect the compressor. The thinning effect also influences oil return.

Once you mix a blend at a given composition, the pressure-temperature relationships follow the same general rules as for pure components. For example, the pressure goes up when the temperature goes up. For three blends containing different amounts of A and B, the pressure curve is similarly shaped, but in the result pressure will be higher for the blend that contains more of the A or higher pressure component.

Some refrigerant blends are intended to match some other product. R-12 is a good example. It will rarely match the pressure at all points in the desired temperature range. What is more common is that the blend will match in one region and the pressures will be different elsewhere.

In Figure 12.1, the blend with concentration C1 matches the CFC at cold evaporator temperatures, but the pressures run higher at condenser conditions. The blend with composition C2 matches closer to room temperature. And, it may show the same pressure in a cylinder being stored, for example. The operation pressures at evaporator and condenser temperatures, however, will be somewhat different. Finally, the blend at C3 will generate the same pressures at hot condenser conditions. but the evaporator must run at lower pressures to get the same temperature.

It will be seen later that the choice of where the blend matches the pressure relationship can solve or cause certain retrofit-related problems.

Refrigerant Blends

Generally speaking, the R-12 retrofit blends have higher temperature glide. They do not match the pressure/temperature/capacity of R-12 across the wide temperature application range that R-12 has. In other words, one blend does not fit all. Blends that match R-12 at colder evaporator temperatures may generate higher pressures and discharge temperatures when used in warmer applications or in high ambient temperatures. These are called refrigeration blends.

In refrigeration it is often an easier and cheaper retrofit job if you can match evaporator pressures to R-12 and split the glide. That is because you can get similar box temperatures in similar run times. And, you would probably not need to change controls or the thermostatic expansion valves (TXVs), which are sensitive to pressure.

Blends that match R-12 properties in hot conditions, such as in automotive AC condensers, may lose capacity or require lower suction pressures when applied at colder evaporator temperatures. These are called automotive blends.

---

[1] *Courtesy of National Refrigerants.*

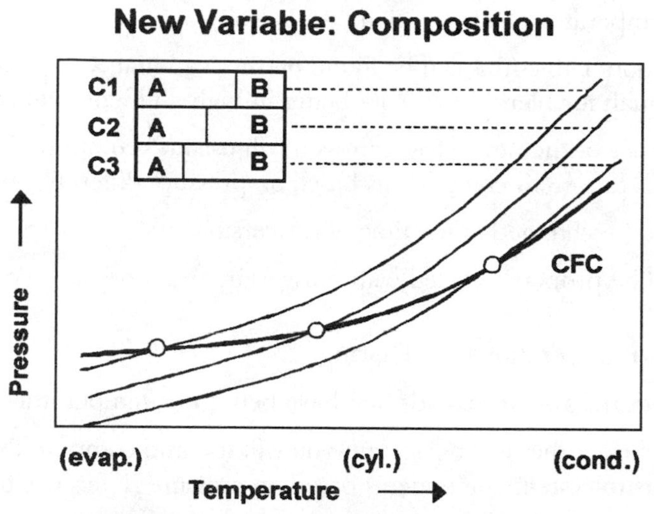

New Variable: Composition

FIGURE 12.1 New refrigerants of variable composition.
(National Refrigerants)

For automotive air conditioning many of the controls and safety switches are related to the highside pressure. If the blend generates higher discharge pressures, you could short cycle more often and lose capacity in general. It is better to pick the high side to match R-12 and let the low side run a little lower pressure.

R-134a Refrigerant

The blended refrigerant R-134a is a long-term HFC alternative with similar properties to R-12. It has become the new industry-standard refrigerant for automotive air-conditioning and refrigerator/freezer appliances. R-134a refrigerating performance will suffer at lower temperatures (below -10 degrees F). Some traditional R-12 applications have used alternatives other than 134a for lower temperatures.

R-134a requires polyolester (POE) lubricants. Traditional mineral oils and alkyl benzenes do not mix with HFC refrigerants, and their use with 134a may cause operation problems or compressor failures. In addition, automotive A/C systems may use polyalkaline glycols (PAGs), which are typically not seen in stationary equipment.

Both POEs and PAGs will absorb moisture and hold onto it to a much greater extent than traditional lubricants. The moisture will promote reactions in the lubricant as well as the usual problems associated with water—corrosion and acid formation. The best way to dry a wet HFC system is to rely on the filter dryer. Deep vacuum will remove "free" water but not the water that has absorbed into the lubricant.

Appliances, both commercial and self-contained refrigeration, centrifugal chillers, and automotive air conditioning utilize R-134a. Retrofitting equipment with a substitute for R-12 is sometimes difficult; there are a number of considerations to be examined before undertaking the task:

1. For centrifugal compressors it is recommended that the manufacturer's engineering staff become involved in the project—special parts or procedures may be required. This will ensure proper capacity and reliable operation after the retrofit.

2. Most older direct expansion systems can be retrofitted to R-401A, R-409A, R-414B, or R-416A (R-500 to R-401B or R-409A), so long as there are no components that will cause fractionation within the system to occur.

3. Filter driers should be changed at the time of conversion.

4. The system should be properly labeled with refrigerant and lubricant type.

R-12 Medium/High Temperature Refrigeration (>OF evap)

1. See Recommendation Table (this can be found on the National Refrigerants Web site—click on Technical Manual) for blends that work better in high ambient-heat conditions.

2. Review the properties of the new refrigerant you will use and compare them to R-12. Prepare for any adjustments to system components based on pressure difference or temperature glide.

3. Filter dryers should be changed at the time of conversion.

4. The system should be properly labeled with refrigerant and lubricant type.

R-12 Low Temperature Refrigeration (<20F evap)

1. See Recommendation Table for blends that have better low-temperature capacity.

2. Review the properties of the new refrigerant you will use and compare them to R-12. Prepare for any adjustments to system components based on pressure difference or temperature glide.

3. Filter dryers should be changed at the time of conversion.

4. The system should be properly labeled with refrigerant and lubricant type.

Another blended refrigerant that can be used to substitute for R-12 is 401A . It is a blend of R-22, 152a, and 124. The pressure and system capacity match R-12 when the blend is running an average evaporator temperature of 10 to 20 degrees F. Applications for this refrigerant are as a direct expansion refrigerate for R-12 in air-conditioning systems and in R-500 systems.

R-401B

This blend refrigerant is similar to R-401A except that it is higher in R-22 content. This blend has higher capacity at lower temperatures and matches R-12 at -20 degrees F. It also provides a closer match to R-500 at air-conditioning temperatures.

Applications for R-401B are in normally lower-temperature R-12 refrigeration locations, in transport refrigeration, and in R-500 as a direct expansion refrigerant in air-conditioning systems.

R-402A

R-402A is a blend of R-22 and R-125 with hydrocarbon R-290 (propane) added to improve mineral-oil circulation. This blend is formulated to match R-502 evaporator pressures, yet it has higher discharge pressure than 502. Although the propane helps with oil return, it is still recommended that some mineral oil be replaced with alkyl benzene.

Applications are in low-temperature (R-502) refrigeration locations. In retrofitting it is used to substitute for R-502.

R-402B

R-402B is similar to R-402A but with less R-125 and more R-22. This blend will generate higher discharge temperatures, which makes it work particularly well in ice machines. Applications are in ice machines where R-502 was used extensively.

RECLAIMING REFRIGERANT

One of the means available for reclaiming refrigerant is called the TOTALCLAIM® system. It is furnished to the trade by Carrier, long known for its dominance in the field of refrigeration and air conditioning.

The information in this section is designed to aid the service technician in understanding the construction and operation of the TOTALCLAIM system. A thorough understanding of the system is the most effective tool for troubleshooting.

Description

TOTALCLAIM extracts refrigerant from an air-conditioning or refrigeration system, removes contaminants from the refrigerant, and stores the charge until it is returned to the original system or another system. TOTALCLAIM can determine the level of acid and moisture contamination in the refrigerant through the use of TOTALTEST.

In recovery operations (Figure 12.2) refrigerant is extracted from an air-conditioning or refrigeration system and temporarily stored in the TOTALCLAIM storage cylinder. In recovery mode, the target system is evacuated to a pressure less than 0 psig. In recovery plus mode, the target is evacuated to a negative pressure of approximately 20 inches Hg (4 psia).

In the recycle mode (Figure 12.3) refrigerant already stored in the storage cylinder is reprocessed through the TOTALCLAIM unit to remove additional contaminants.

In the recharge mode (Figure 12.4), the refrigerant stored in the TOTALCLAIM storage cylinder is returned to the target air-conditioning or refrigeration system.

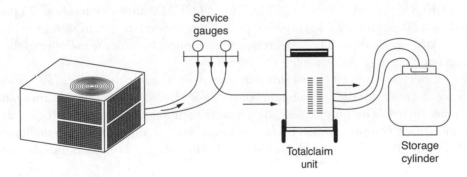

FIGURE 12.2 Recovery operations. *(Carrier)*

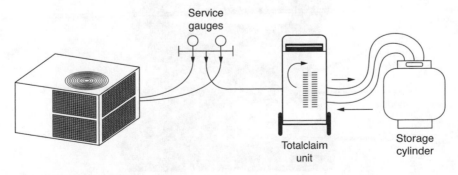

FIGURE 12.3 Recycle operations. *(Carrier)*

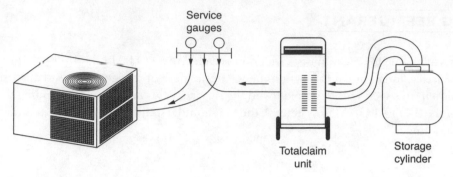

FIGURE 12.4   Recharge operations. *(Carrier)*

In the service mode, the internal solenoid valves are positioned so that the TOTALCLAIM system can be evacuated. Service mode would be used when a different refrigerant is to be recovered or when piping connections within TOTALCLAIM must be opened to permit repair.

The test mode permits the service technician to energize individual solenoid valves for the purpose of checking out the energizing paths. This mode is intended solely for control-circuit troubleshooting.

In all modes, the pattern of refrigerant flow is determined by solenoid valves, which are controlled by the microprocessor-based control.

Description and Component Location

The TOTALCLAIM unit is approximately 35 inches (90 cm) high, including the handle. It is 16 inches (40.7 cm) wide and 10.5 inches (26.5 cm) deep. The TOTALCLAIM unit weighs about 75 pounds (34 kg). It is accompanied by a 50-pound (22.7 kg) capacity D.O.T.-approved refrigerant storage cylinder modified for the TOTALCLAIM application. The hoses required to connect the storage cylinder to the unit are also provided. An external filter dryer is available as an accessory.

In Figure 12.5 all electrical and electronic controls, except for the solenoid valves, are located in the upper section. This section contains the control-panel display board, microprocessor control (standard control module), and a relay board. The only replaceable discrete components in the electronics section are the power switch, transformer, compressor/fan motor contactor, and circuit breaker. If a malfunction is traced to the electronic controls, the entire control module, display board, or relay board must be replaced.

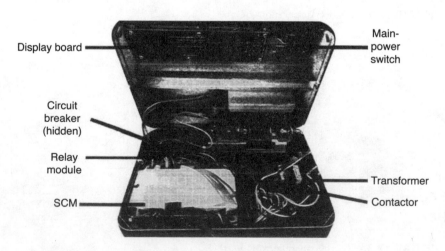

FIGURE 12.5   Electrical section. *(Carrier)*

Compressor

In Figure 12.6 the TOTALCLAIM uses a rotary compressor to pump refrigerant. The compressor is equipped with an external, automatic-reset overload device that trips on excess current or temperature. A discharge temperature thermistor (TDIS) senses the compressor discharge temperature. From here the data is sent to the microprocessor. Both the suction and discharge sides of the compressor are monitored by pressure transducers. These transducers send pressure data to the microprocessor.

Oil Separator

The oil separator collects lubricating oil that escapes with the compressor-discharge gas. A float-valve arrangement inside the oil separator returns oil to the compressor when it reaches a predetermined level.

Condenser

The condenser fan blows air across the condenser, which is mounted to the rear wall. It is a copper-tube/aluminum-fin condenser. The ambient air thermistor (TAMB) senses the temperature at the condenser and sends that data to the microprocessor.

Filter Dryer

The primary filter-dryer is located behind the access door on the side of the unit. The reset switch for the circuit breaker is also located in this section. Knurled, quick-connect fittings permit the filter dryer to be removed and installed without the need for tools. The filter-dryer shutoff valve must be turned off to allow the filter dryer to be replaced while the system is under pressure (see Figure 12.7).

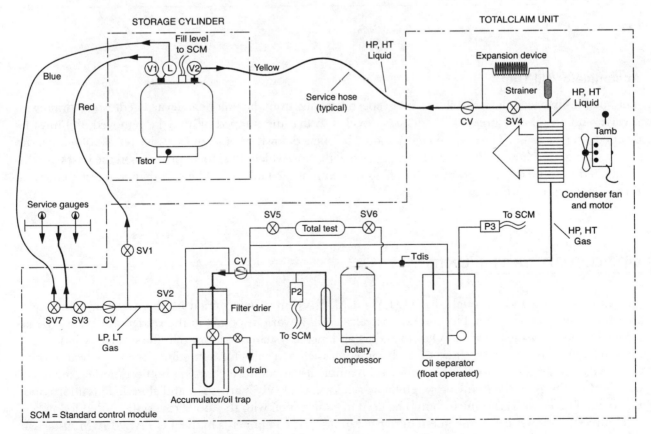

FIGURE 12.6 Liquid recovery-functional flow. *(Carrier)*

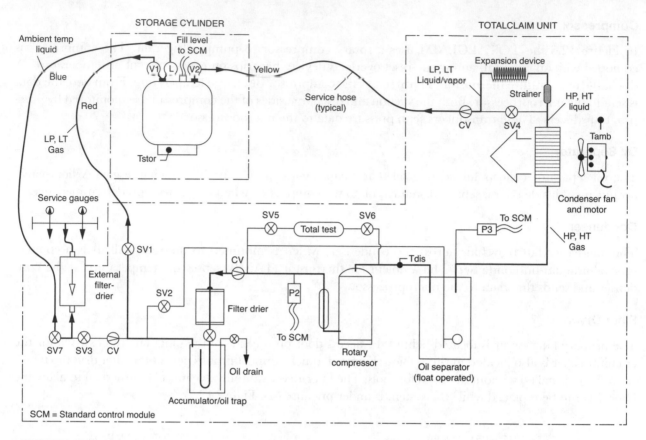

FIGURE 12.7   Vapor recovery-functional flow. *(Carrier)*

Accumulator/Oil Trap

The suction-line accumulator/oil trap intercepts oil coming from the unit being evacuated. An oil drain with a valve and an oil-measurement bottle are provided so that the trapped oil may be removed. Oil must be drained after each use. The oil-drain valve should be opened slowly to prevent excessive release of refrigerant.

The refrigerant hose connections are equipped with caps, which must be in place when the hoses are disconnected. The hoses have positive shutoff connections at the tank end. The end that connects to the unit is equipped with standard service fittings.

OPERATION OF THE UNIT

The flow of refrigerant through the TOTALCLAIM system is determined by the mode or submode in which the unit is operating. The state of the refrigerant at various points in the refrigerant cycle is determined by the mode or submode. A key to understanding the system is to know that the storage cylinder plays an important role in the refrigerant cycle. It sometimes acts as a collector, sometimes as an evaporator, other times as a condenser, and still other times as an ambient-temperature charging bottle. Another key to understanding the system is knowing when the various solenoid valves are open and closed. This information is provided in Table 12.1, which should be used in conjunction with the mode descriptions that follow.

Figure 12.6 shows the arrangement of refrigeration-cycle components in the TOTALCLAIM system. In a normal recovery operation, the unit will extract liquid refrigerant first. Table 12.1 shows that SV-7 is open and SV-3 is closed during liquid recovery. These conditions would exist if the operator selected the "LIQUID" option at the keyboard.

Given the solenoid-valve conditions shown, liquid at about ambient temperature is extracted from the target system and flows into the storage cylinder. Low-pressure, low-temperature gas is drawn from the cylinder through SV-1. The high-temperature, high-pressure gas leaving the compressor discharge is condensed to a high-temperature, high-pressure liquid. Note that SV-4 is closed in liquid-recovery mode, while in vapor recovery mode it is open. With SV-4 closed, the refrigerant flows through the expansion device. Because of the pressure drop, the refrigerant returns to the storage cylinder as a low-temperature, low-pressure liquid/vapor mixture. This process cools the storage cylinder. In the recovery plus and recovery modes, the system will automatically enter a storage "cylinder cooling" cycle after 2 minutes of liquid recovery. Storage-cylinder cooling is described later.

When R-22 or R-502 is being processed, the microprocessor will open SV-4, which acts as a parallel expansion device for these refrigerants at higher ambient temperatures.

In vapor-recovery operations, the flow is changed significantly, as shown in Figure 12.6. SV-7 and SV-1 are closed and SV-3 is open, bypassing the storage cylinder. Therefore, ambient temperature, low-pressure vapor extracted from the target system is pulled directly into the TOTALCLAIM unit. The other significant difference from liquid-recovery operations is that SV-4 is open. Thus, the relatively high-temperature, high-pressure liquid leaving the condenser will enter the storage cylinder in that state.

In the recovery mode, one complete recovery cycle is performed. The cycle ends when the pressure in the system being evacuated reaches 0 psig or below.

In the recovery-plus mode, multiple recovery cycles are performed. Refrigerant is extracted as shown in Figures 12.6 and 12.7. First it is liquid, then vapor, unless vapor recovery is selected at the control panel. During both the liquid and vapor recovery cycles, storage-cylinder cooling will be initiated as determined by time, temperature, and/or pressure conditions.

If TOTALTEST is selected (Figure 12.8), the microprocessor will open SV-5 and SV-6 for 1 to 4 minutes, depending on the refrigerant type, at the end of the recovery or recycle operation. The refrigerant will be sampled during that period.

Storage Cylinder Cooling

Low-suction pressures are created at the TOTALCLAIM compressor. As the target system approaches a vacuum, it causes reduced refrigerant flow. This, in turn, causes high temperatures that would eventually damage the TOTALCLAIM compressor.

To avoid compressor damage, the microprocessor automatically switches TOTALCLAIM into the "storage cylinder cooling" submode as needed to maintain proper cooling of the compressor. In this submode, connections to the target unit are closed and SV-1 is opened. (See Figure 12.9 and Table 12.1.) Now the TOTALCLAIM functions in a closed loop, SV-4 is closed, placing the capillary-tube expansion device in the loop. The storage cylinder now acts as a flooded evaporator to cool the refrigerant. The cooling period lasts from 90 seconds to 15 minutes, depending on the recovery mode and the state of the refrigerant being processed (liquid or vapor). Over a period of 10 to 15 minutes, the cylinder temperature is reduced from 60 to 70 degrees F (15.6-21.1 degrees C) below surrounding ambient temperature. Thus, during subsequent recovery cycles, the storage cylinder acts as a low-temperature condenser in addition to the higher-temperature air-cooled condenser.

Recycle Operation

In the recycle mode, shown in Figure 12.10, the microprocessor closes SV-3 and SV-7 and opens SV-1. Under those conditions, refrigerant vapor is drawn from the storage cylinder and cycled through the TOTALCLAIM unit to remove additional contaminants. The operator sets the recycle time at the control panel and

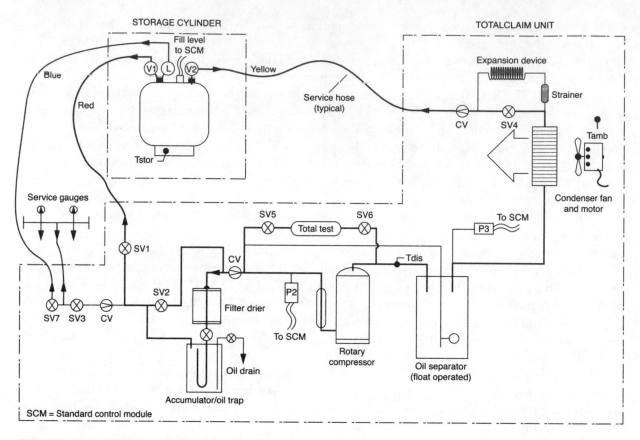

FIGURE 12.8   TOTAL TEST-functional flow. *(Carrier)*

the microprocessor stops the cycle at the end of that time. If the operator does not select a run time, a default time of 1 hour is automatically selected.

Recharge Operation

In the recharge mode, shown in Figure 12.11, TOTALCLAIM is basically a charging cylinder. All solenoid valves except SV-7 are closed. The compressor and condenser fan are turned off. The target unit draws liquid refrigerant from the TOTALCLAIM storage cylinder. In applications where vapor recharging is required, the blue hose must be moved from valve L (blue-handled) to valve V1 (red-handled) on the storage cylinder.

Service Operation

The service mode is selected when it is necessary to evacuate the TOTALCLAIM system. The compressor and condenser fan are shut off, and all solenoid valves, except SV-5, SV-6, and SV-7, are open to permit refrigerant to be drawn from the TOTALCLAIM system.

Test Operation

The test mode permits the technician to energize individual solenoid valves in order to simplify troubleshooting of the control circuits. This mode is selected by pressing the RESET and MODE keys for 5 seconds. The test mode takes priority over all other modes. When the test mode is turned on, all solenoids are deenergized. Then, using the arrow keys, the technician can energize individual solenoids and trace the energizing signal along the path. If there is a malfunction in the path, it can be isolated to the solenoid valve,

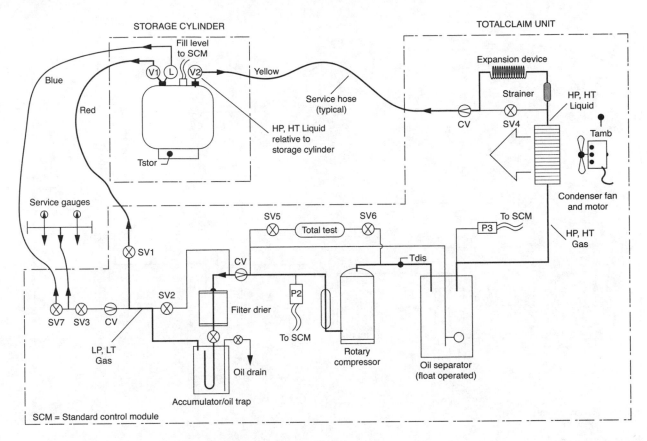

FIGURE 12.9 Storage cylinder cooling-functional flow. *(Carrier)*

relay module, or standard control module (SCM). SV-5 and SV-6 are energized at the same time. All the others are operated individually. The START/STOP key is used to exit the test mode.

Control Circuits

The compressor and fan both operate from 115-volt single-phase power. The contactor, C, which is energized by the SCM, controls both components. The relay is energized under the "compressor/condenser fan on" conditions shown in Table 12.1. An external overload device disables the compressor in the event of a current overload or excessively high temperature. The device will reset automatically when internal temperature drops to a safe level (see Figure 12.12).

The SCM receives data from several sources within the unit. Temperature data is supplied by thermistors located in the compressor discharge (T_{DIS}), the storage cylinder (T_{STOR}), and the ambient-air intake (T_{AMB}). These are identified as T1, T2, and T3 on the schematic diagram. The unit will shut down, and an error code will appear on the display if any of these thermistors fail, either open or shorted.

Error Code Thermistor
E03 Ambient air
E04 Compressor discharge
E05 Storage cylinder

A level-sensing device inside the storage cylinder allows the SCM to monitor the contents of the tank. When the tank reaches 80 percent full, 50 pounds or 22.7 kg, the SCM will stop the recovery process, and error code E09 will appear on the display.

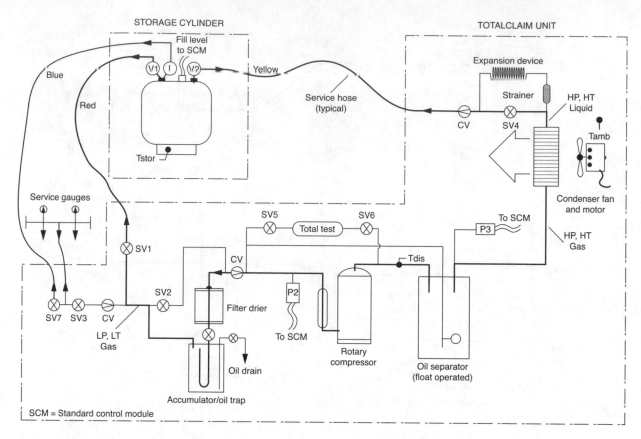

FIGURE 12.10   Recycle-functional flow. *(Carrier)*

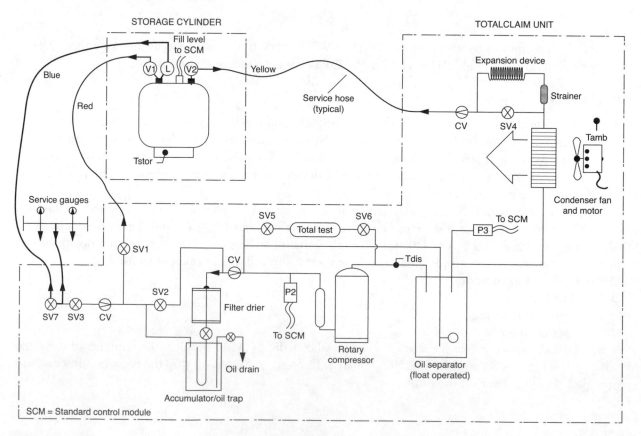

FIGURE 12.11   Recharge (liquid)-functional flow. *(Carrier)*

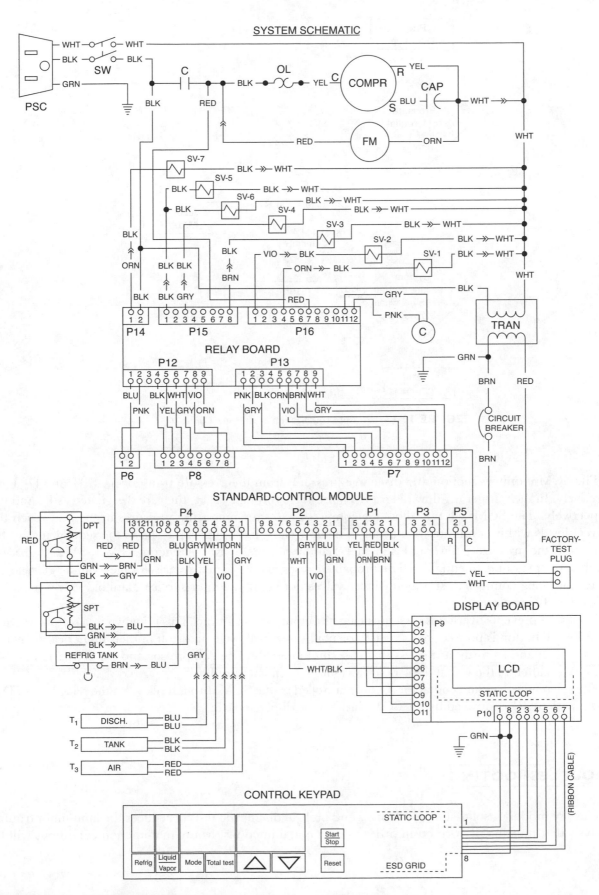

FIGURE 12.12 TOTALCLAIM® wiring diagram. *(Carrier)*

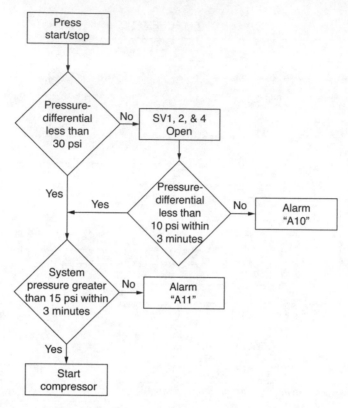

FIGURE 12.13   Pressure equalization at startup. *(Carrier)*

The SCM monitors suction and discharge pressures from the pressure transducers, SPT and DPT respectively. In the refrigerant-flow diagram (Figures 12.6 through 12.11) they are designated as P2 and P3, respectively. The SCM will not allow the compressor to start unless the pressures in the unit reach the correct levels within 3 minutes of starting. The flow chart in Figure 12.13 shows the sequencing of this process. If the pressure differential is greater than 30 psi, the SCM will energize SV-1, SV-2, and SV-4 to allow the pressures to equalize. The probable causes of an A11 alarm are storage-cylinder refrigerant hoses not being connected, storage-cylinder valves not being open, or service manifold valves or target unit valves being closed.

The SCM prevents overloads by sequencing its outputs one at a time. This is known as a soft start. Once the START button is pressed, the selected solenoid valves will cycle on or off in their numerical sequence (SV-1, SV-2, and so forth) at no less than 1-second intervals. The compressor and condenser fan are sequenced on after all the selected solenoid valves have been energized or deenergized.

The solenoid valves and contactor (C) are controlled by the SCM through relays on the relay board. Display functions are controlled by the SCM through the display board.

TROUBLESHOOTING

The use of modular, solid-state electronics and built-in diagnostic testing reduces the amount of trouble analysis that must be performed in order to isolate malfunctions. Many malfunction conditions will be diagnosed by the system, and an error message will be displayed to tell the service technician what component has failed. In some cases, however, it will be necessary to isolate the fault by using standard troubleshooting methods, supported by the built-in test capability.

The troubleshooting diagram at the front of the unit's manual contains most of the information needed to troubleshoot the TOTALCLAIM system. The flow diagram provides a good starting point if you have no idea what the problem is. If you have an error message on the display panel, you should go directly to the errors-and-alarms table and perform the indicated action or troubleshoot the failed component.

A test mode is provided as an aid in troubleshooting the relay/solenoid-valve logic. This approach is discussed in detail in the manual that is available on the Internet. For more information on this unit contact Carrier at www.carricr.com.

MULTIPLE-CHOICE EXAM

1. Why is Freon being replaced as a refrigerant?

   a. It is poisonous

   b. It is too expensive

   c. It is damaging to the environment

   d. It is not suited for safe operation in confined places

2. Freon 11 has a boiling point of:

   a. -100 degrees F   b. 32 degrees F

   c. 212 degrees F   d. 74.8 degrees F

3. Where is R-12 most often used as a refrigerant?

   a. household and commercial refrigeration

   b. automobile air conditioners

   c. processing equipment

   d. none of these

4. What type of compressors make good use R-12?

   a. reciprocating compressors   b. rotary compressors

   c. screw compressors   d. desiccant compressors

5. What is the boiling point of Freon-13?

   a. 100 degrees F   b. -144.6 degrees F

   c. 74.8 degrees F   d. 32 degrees F

6.   What is the boiling point of Freon-21?

    a. 74.8 degrees F   b. 100 degrees F

    c. 48 degrees F   d. 32 degrees F

7.   With what type of compressor does the Freon-21 refrigerant work best?

    a. reciprocating type   b. rotary-vane type

    c. piston type   d. absorption

8.   Freon-22 has a boiling point of _____.

    a. 48 degrees F   b. 212 degrees F

    c. 32 degrees F   d. -41.4 degrees F

9.   What type of compressor works best with R-22?

    a. rotary-vane type   b. reciprocating

    c. absorption system   d. screw type

10.   Where is R-22 used today?

    a. in azeotropic compressors   b. in piston-type compressors

    c. in rotary-vane compressors   d. in home air conditioners

11.   The boiling point of Freon 113 is:

    a. 48 degrees F   b. -41.4 degrees F

    c. 100 degrees F   d. 117.6 degrees F

12.   Where is R-113 most frequently used?

    a. in small-tonnage applications

    b. in large-tonnage applications

    c. in outdoor compressors

    d. in indoor compressors

13.   The boiling point of R-114 is:

    a. 32.0 degrees F   b. 48.2 degrees F

    c. 38.4 degrees F   d. 200 degrees F

14.   What type of compressor does the R-114 work with most efficiently?

    a. adsorption type   b. absorption type

    c. compression type   d. rotary vane

15. What is the boiling point for R-115?

    a. 32.8 degrees F

    b. 38.4 degrees F

    c. -41.4 degrees F

    d. -37.7 degrees F

16. Where is R-115 most frequently used?

    a. in high-discharge temperatures in reciprocating compressors

    b. in household refrigerators and automobile A/C

    c. in commercial freezers

    d. in industrial applications

17. Freon 502 is a(n) _____ mixture composed of 48.8 percent Freon-22 and 51.2 percent Freon-115.

    a. azeotropic

    b. asbestos-based

    c. ready-mixed

    d. latex-based

18. Freon refrigerants are colorless and almost _____.

    a. odorless

    b. impure

    c. tasteless

    d. flammable

19. Chemically, Freon refrigerants are inert, _____, and stable up to temperatures far beyond conditions found in actual operation.

    a. colorless

    b. odorless

    c. thermal

    d. dangerous

20. How is Freon harmful if released into the atmosphere?

    a. it depletes the ozone layer around the earth

    b. it is highly toxic to vegetation

    c. it is highly toxic to people

    d. it is harmful to wildlife

21. What are determining factors involved with compressor displacement?

    a. compressor type

    b. compression ratio

    c. refrigerating effect and specific volume of the refrigerant vapor

    d. amount of ultraviolet rays radiated

22. Freon is nonflammable and _____ under conditions in which appreciable quantities contact flame or hot metal surfaces.

    a. noncompressible   b. nonpoisonous

    c. noncombustible   d. noncolorable

23. How many pounds of R-114 refrigerant must be expanded per minute to produce refrigeration?

    a. 4.64   b. 2.237

    c. 2.887   d. 3.916

24. The amount of work required to produce a given refrigerating effect increases as the temperature level to which the heat is pumped from the cold body is _____.

    a. lowered   b. increased

    c. decreased   d. none of these

25. Volumes of liquid required to circulate for a given refrigerant effect should be _____.

    a. reduced   b. elevated

    c. low   d. high

26. One of the requirements of an ideal refrigerant is that it must be _____.

    a. toxic   b. nontoxic

    c. easily vaporized   d. easily compressed

27. What is another name for Freon R-12?

    a. sulfur dioxide   b. carbon dioxide

    c. ammonia   d. Carrene

28. Freon-12 has a relatively _____ volume displacement.

    a. high   b. low

    c. medium   d. extreme

29. In dealing with condensers, it is _____ condensing pressure that must be kept in mind for leakage and other problems.

    a. high   b. low

    c. extreme   d. minuscule

30. The freezing point of a refrigerant should be _____ any temperature that might be encountered in the system.

    a. close to   b. above

    c. below   d. around

31. High condensing pressure has a tendency to _____ leakage on the low side as well as on the high side when pressure builds up during idle periods.

    a. reduce   b. decrease

    c. increase   d. not allow

32. The freezing point of a refrigerant should be _____ any temperature that might be encountered in the system.

    a. around   b. near

    c. above   d. below

33. The evaporator pressure is the _____ side in the system.

    a. high   b. low

    c. very high   d. very low

34. The critical temperature of a refrigerant is the _____ temperature at which it can be condensed to a liquid, regardless of a higher pressure.

    a. highest   b. lowest

    c. best   d. nominal

35. Except for carbon dioxide and ethane, all common refrigerants have satisfactorily high _____ temperatures.

    a. evaporator   b. condenser

    c. critical   d. compressor

36. Dupont's brand name for its widely used refrigerant R 134a is _____.

    a. CFC-12   b. HFC-1

    c. Autosmart   d. Suva

37. Refrigerants with high Btu refrigerating effects are not always _____.

    a. available   b. free of contamination

    c. desirable   d. used

38. A low specific heat of the liquid is _____ in a refrigerant.

    a. unwanted   b. preferred

    c. desirable   d. undesirable

39. In a perfect system _____ Btu is the maximum refrigeration obtainable per Btu of energy used to operate the refrigeration system.

    a. 5.74   b. 6.00

    c. 7.33   d. 9.01

40. The amount of work required to produce a given refrigerating effect increases as the temperature level to which the heat is pumped from the cold body is _____.

    a. increased   b. decreased

    c. moderated   d. slowly changed

41. Which two of the older refrigerants can cause suffocation when in vapor form?

    a. sulfur dioxide and ammonia

    b. carbon dioxide and ammonia

    c. sulfur dioxide and carbon dioxide

    d. none of these

42. Remember that liquid refrigerant will refrigerate or remove heat from anything it meets when released from a container. In the case of contact with refrigerant, the affected area should be treated as if it has been _____ or frostbitten.

    a. frozen   b. scraped

    c. scarred   d. burned

43. Refrigerant cylinders should be stored in a dry, sheltered, and _____ area.

    a. well-ventilated   b. enclosed

    c. open   d. none of the above

44. Refrigerant cylinders should not be exposed to continuous dampness, salt water, or salt _____.

    a. nuggets   b. spray

    c. crystals   d. air

45. There are three basic properties that lubricants must possess. Which of the following is not one of them:

    a. viscosity   b. chemical stability

    c. lubricity   d. liquidity

46. Lubricity refers to a lubricant's ability to protect metal surfaces from _____.

    a. abrasion   b. wear

    c. disintegrating   d. contamination

47. The blended refrigerant R-134a is a long-term HFC alternative with similar properties to _____.

    a. R-114   b. R-501

    c. R-12   d. R-22

48. Automotive A/C systems may use polyalkaline glycols, which are abbreviated as:

    a. HFCs   b. alkalines

    c. PAGs   d. POEs

49. Retrofitting equipment with a substitute for R-12 is sometimes _____ because there are a number of considerations to be examined before undertaking the task.

    a. difficult   b. easy

    c. demanding   d. expensive

50. Another blended refrigerant that can be used to substitute for R-12 is _____.

    a. R-22   b. R-401A

    c. R-401B   d. R-402B

TRUE-FALSE EXAM

1. The EPA listing is updated when additional refrigerant reclaimers are approved.

   True   False

2. An EPA reclaimer program is voluntary, and the certification program is operated by the ARI.

   True   False

3. Freon-11 is also employed as brine for low-temperature applications.

   True   False

4. Freon R-12 is the most widely used refrigerant.

   True   False

5.   The automotive industry has replaced R-12 in car air conditioners with R-134a.

   True   False

6.   In 1987 the Montreal Protocol established requirements that began the worldwide phaseout of ozone-depleting chlorofluorocarbons (CFCs).

   True   False

7.   The Clean Air Act does not allow any refrigerant to be vented into the atmosphere during installation.

   True   False

8.   After 2020 the servicing of R-22-based systems will rely on recycled refrigerants.

   True   False

9.   Existing units using R-22 can continue to be serviced with R-22.

   True   False

10.   One of the substitutes for R-22 is R-410A.

   True   False

11.   The Clean Air Act outlines specific refrigerant containment and management practices for HVAC manufacturers, distributors, dealers, and technicians.

   True   False

12.   The handling and recycling of refrigerants used in motor-vehicle air-conditioning systems are governed under section 609 of the Clean Air Act.

   True   False

13.   If a leak is discovered in commercial equipment, owners or operators have 30 days to obtain a repair.

   True   False

14.   The technician who repairs or recharges equipment must be EPA- or ARI-certified as a reclaimer.

   True   False

15.   It is still legal under certain circumstances to vent refrigerant into the atmosphere.

   True   False

MULTIPLE-CHOICE ANSWERS

1. C	11. D	21. C	31. C	41. A
2. D	12. A	22. C	32. D	42. A
3. A	13. C	23. A	33. B	43. A
4. A	14. D	24. B	34. A	44. B
5. B	15. C	25. C	35. C	45. D
6. C	16. B	26. B	36. D	46. B
7. A	17. A	27. D	37. C	47. C
8. D	18. A	28. B	38. C	48. C
9. B	19. C	29. A	39. A	49. A
10. D	20. A	30. C	40. A	50. B

TRUE-FALSE ANSWERS

1. T	5. T	9. T	13. T
2. T	6. T	10. T	14. T
3. T	7. T	11. T	15. F
4. T	8. T	12. T	

—NOTES—

Chapter 13
HEATING CIRCUITS

Hot-air furnaces are self-contained and self-enclosed units. They are usually centrally located within a building or house. Their purpose is to make sure the temperature of the interior of the structure is maintained at a comfortable level throughout. The design of the furnace is determined by the type of fuel used to fire it. Cool air enters the furnace and is heated as it comes in contact with the hot metal heating surfaces. As the air becomes warmer, it also becomes lighter, which causes it to rise. The warmer, lighter air continues to rise until it is either discharged directly into a room, as in the pipeless gravity system, or is carried through a duct system to warm air outlets located at some distance from the furnace.

After the hot air loses its heat, it becomes cooler and heavier. Its increased weight causes it to fall back to the furnace, where it is reheated and repeats the cycle. This is a very simplified description of the operating principles involved in hot-air heating. And it is especially typical of those involved in gravity heating systems. The forced-air system relies on a blower to make sure the air is delivered to its intended location. The blower also causes the return air to move back to the furnace faster than with the gravity system.

With the addition of a blower to the system, there must be some way of turning the blower on when needed to move the air and to turn it off when the room has reached the desired temperature. Thus, electrical controls are needed to control the blower action.

BASIC GAS-FURNACE OPERATION

The gas furnace is the simplest to operate and understand. Therefore, we will use it here to look at a typical heating system. This type of natural-gas furnace is used to heat millions of homes in the United States.

Figure 13.1 shows a simple circuit needed to control a furnace with a blower. Note the location of the blower switch and the limit switch. The transformer provides low voltage for control of the gas solenoid. If the limit switch opens (it is shown in a closed position), there is no power to the transformer and the gas solenoid cannot energize. This is a safety precaution, because the limit switch will open if the furnace gets too hot. When the thermostat closes, it provides 24 volts to the gas solenoid, which energizes and turns on the gas. The gas is ignited by the pilot light and provides heat to the plenum of the furnace. When the air in the

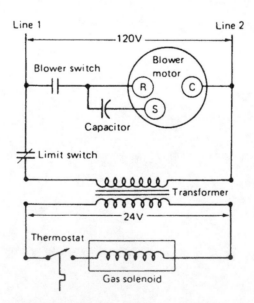

FIGURE 13.1   Simple one-stage furnace control system.

plenum reaches 120 degrees F, the fan switch closes and the fan starts. The fan switch provides the necessary 120 volts to the fan motor for it to operate.

Once the room has heated up to the desired thermostat setting, the thermostat opens. When it opens, the gas solenoid is de-energized, and the spring action of the solenoid causes it to close off the gas supply, thereby turning off the source of heat.

When the plenum on top of the furnace reaches 90 degrees F, the blower switch opens and turns off the blower. As the room cools down, causing the thermostat to once again close, the cycle starts over again. The gas solenoid opens to let in the gas, and the pilot light ignites it. The heat causes the temperature to rise in the plenum above the limit switch's setting, and the switch closes to start the blower. Once the thermostat setting has been reached, it opens and causes the gas solenoid to turn off the gas supply. The blower continues to run until the temperature in the plenum reaches 90 degrees F, and it turns off the blower by opening. This cycle is repeated over and over again to keep the room or house at a desired temperature.

BASIC ELECTRIC-HEATING SYSTEM

Electric-fired heat is the only heat produced almost as fast as the thermostat calls for it. It is almost instantaneous. There are no heat exchangers to warm up. The heating elements start producing heat the moment the thermostat calls for it. A number of types of electric-fired furnaces are available. They can be bought in 5- to 35-kilowatt sizes. The outside looks almost the same as the gas-fired furnace. The heating elements are located where the heat exchangers would normally be located. Since they draw high amperage, they need electrical controls that can take the high currents.

The operating principle is simple. The temperature selector on the thermostat is set for the desired temperature. When the temperature in the room falls below this setting, the thermostat calls for heat and causes the first heating circuit in the furnace to be turned on. There is generally a delay of about 15 seconds before the furnace blower starts. This prevents the blower from circulating cool air in the winter. After about 30 seconds, the second heating circuit is turned on. The other circuits are turned on one by one in a timed sequence.

When the temperature reaches the desired level, the thermostat opens. After a short time, the first heating circuit is shut off. The others are shut off one by one in a timed sequence. The blower continues to operate until the air temperature in the furnace drops below a specified temperature.

Basic Operation

Figure 13.2 shows that the electrical-heating system has a few more controls than the basic gas-fired furnace. The low-resistance element used for heating draws a lot of current, so the main contacts have to be of sufficient size to handle it.

The thermostat closes and completes the circuit to the heating sequencer coil. The sequencer coil heats the bimetal strip that causes the main contacts to close. Once the main contacts are closed, the heating element is in the circuit and across the 240-volt line. The auxiliary contacts will close at the same time as the main contacts. When the auxiliary contacts close, they complete the low-voltage circuit to the fan relay. The furnace fan will be turned on at this time.

Once the thermostat setting has been reached, it opens. This allows the heating sequencer coil to cool down slowly. Thus, the main contacts do not open immediately to remove the heating element from the line. So the furnace continues to produce heat after the thermostat setting has been reached. The bimetal cools down in about 2 minutes. Once it cools, it opens the main and auxiliary contacts, which removes the heating element from the line and also stops the fan motor. After the room cools down below the thermostat setting, the thermostat closes and starts the sequence all over again.

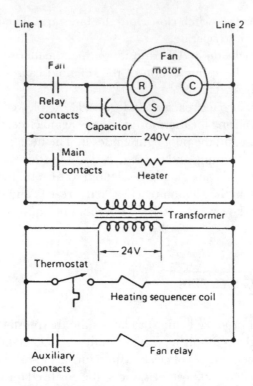

FIGURE 13.2 Ladder diagram for a
hot-air furnace.

LADDER DIAGRAMS

Electrical schematics are used to make it simple to trace the circuits of various devices. Some of these can appear complicated, but they are usually very simple when you start at the beginning and wind up at the end. The beginning is one side of the power line and the end is the other side of the line. What happens in between is that a number of switches are used to make sure the device turns on or off when it is supposed to cool, freeze, or heat.

The ladder diagram makes it easier to see how these devices are wired. It consists of two wires drawn parallel and representing the main power source. Along each side you find connections. By simply looking from left to right, you are able to trace the required power for the device. Symbols are used to represent the devices. There is usually a legend on the side of the diagram to tell you, for example, that CC means compressor contactor, EFR means evaporator fan relay, and HR means heating relay (see Figure 13.3).

Take a look at the thermostat in Figure 13.3. The location of the switch determines whether the evaporator fan-relay coil, the compressor-contactor coil, or the heating-relay coil is energized. Once the coil of the EFR is energized by having the thermostat turned to make contact with the desired point (G), it closes the points in the relay, and the evaporator fan motor starts to move. This means that the low voltage (24 volts) has energized the relay. The relay energizes and closes the EFR contacts located in the high-voltage (240 volts) circuit. If the thermostat is turned to W or the heating position, it will cause the heating-relay coil to be energized when the thermostat switch closes and demands heat. The energized heating-relay coil causes the HR contacts to close, which in turn places the heating element across the 240-volt line, and it begins to heat up. Note that the HR contacts are in parallel with the evaporator fan-relay contacts. Thus, the evaporator fan will operate when either the heating relay or the evaporator fan relay is energized.

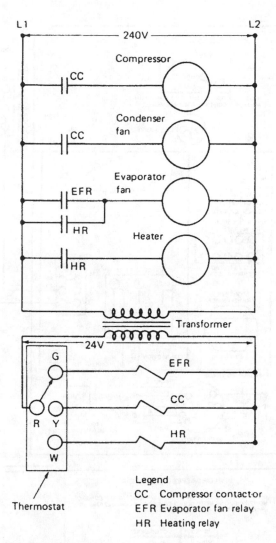

FIGURE 13.3 Ladder diagram for a heat-and-cool installation.

MANUFACTURER'S DIAGRAMS

Figure 13.4 shows how the manufacturer represents the location of the various furnace devices. The solid lines indicate the line voltage to be installed. The dotted lines are the low voltage to be installed when the furnace is put into service.

The motor is four speed. It has various colored leads to represent the speeds. You may have to change the speed of the motor to move the air to a given location. Most motors come from the factory with a medium-high speed selected. The speed is usually easily changed by removing a lead from one point and placing it on another where the proper color is located. In the schematic shown in Figure 13.5, the fan motor has a white lead connected to one side of the 120-volt line (neutral), and the red and black leads are switched by the indoor blower relay to black for the cooling speed and red for the heating speed. It takes a faster fan motor to push cold air than hot air because cold air is heavier.

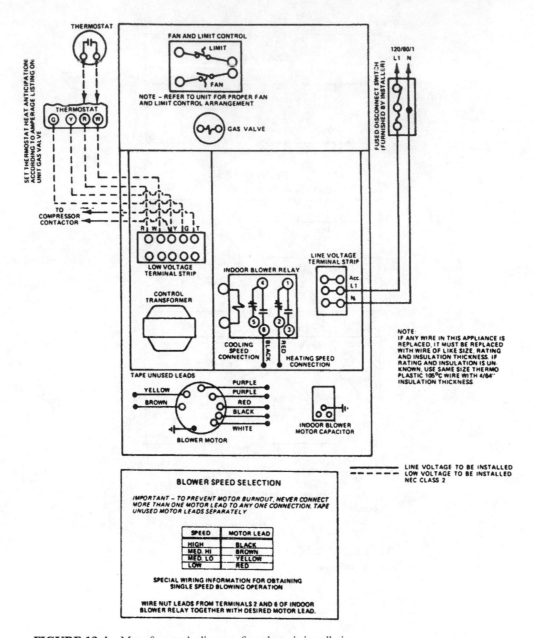

FIGURE 13.4 Manufacturer's diagram for a hot-air installation.

In Figure 13.5, the contacts on the thermostat are labeled R, W, Y, G, R, and W. They are used to place the thermostat in the circuit. It can be switched from W to Y manually by moving the heat-cool switch on the thermostat to the cool position.

Notice in Figure 13.5 that the indoor blower-relay coil is in the circuit all the time when the auto-on switch on the thermostat is located at the on position. The schematic also shows that the cool position has been selected manually, and the thermostat contacts will complete the circuit when it moves from W1 to Y1.

In Figure 13.4, note that the low-voltage terminal strip has a T on it. This is the common side of the low voltage from the transformer. In Figure 13.5, the T is the common side of the low-voltage transformer secondary. In Figure 13.4, the T terminal is connected to the compressor contactor by a wire run from the terminal to the contactor. Note that the other wire to the contactor runs from Y on the terminal strip. Now go back to Figure 13.5, where the Y and T terminals are shown as connection points for the compressor con-

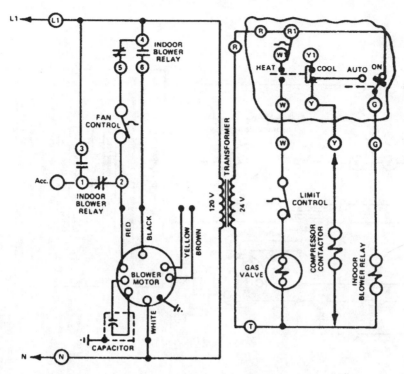

FIGURE 13.5   Schematic for a hot-air installation

tactor. Are you able to relate the schematic to the actual device? The gas valve is wired by attaching wire T of the terminal strip to one side of the solenoid and running a wire from the limit switch to the other side of the solenoid.

Figure 13.6 shows how the wiring diagram comes from the factory. It is usually located inside the cover for the cold-air return. In most instances it is glued to the cover so that it is handy for the person working on the furnace whenever there is a problem after installation.

FIELD WIRING

The installation of a new furnace requires you to follow a factory diagram furnished in a booklet that accompanies the unit. The wiring to be done in the field is represented by the dotted lines in Figure 13.7. All electrical connections should be made in accordance with the National Electrical Code and any local codes or ordinances that might apply.

WARNING: The unit cabinet must have an uninterrupted or unbroken electrical ground to minimize personal injury if an electrical fault should occur. This may consist of electrical wire or approved conduit when installed in accordance with existing electrical codes.

Low-Voltage Wiring

Make the field low-voltage connections at the low-voltage terminal strip shown in Figure 13.7. Set the thermostat heat anticipator at 0.60 ampere (or whatever is called for by the manufacturer). If additional controls are connected in the thermostat circuit, their amperage draw must be added to this setting. Failure to make the setting will result in improper operation of the thermostat.

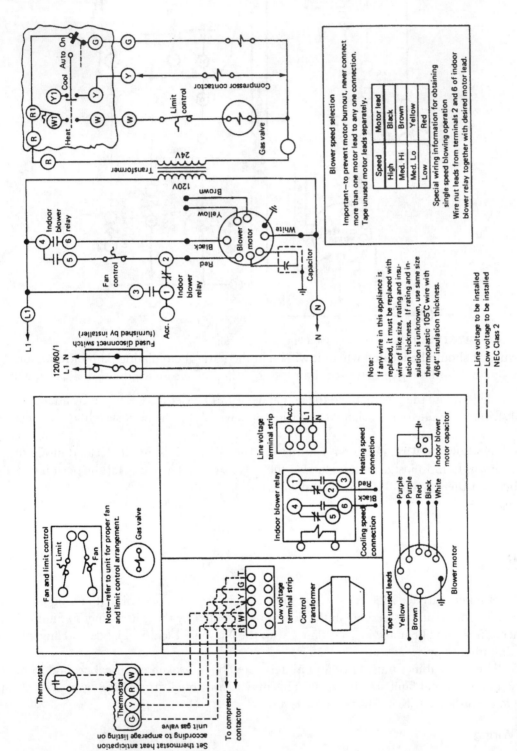

FIGURE 13.6   Complete instruction page packaged with a hot-air furnace.

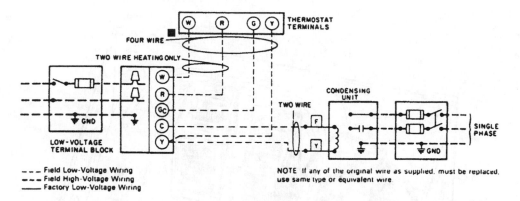

FIGURE 13.7 Heating and cooling application wiring diagram. *(Carrier)*

With the addition of an automatic vent damper, the anticipator setting would then be 0.12 ampere. The anticipator resistor is in series with whatever is in the circuit and is to be controlled by the thermostat. The more devices that are controlled by the thermostat, the more current will be drawn from the transformer to energize them. As the current demand increases, the current through the anticipator is also increased. As you remember from previous studies, a series circuit has the same current through each component in the circuit.

Thermostat Location

The room thermostat should be located where it will be in the natural circulating path of room air. Avoid locations where the thermostat is exposed to cold-air infiltration, drafts from windows, doors or other openings leading to the outside, air currents from warm- or cold-air registers, or to exposure where the natural circulation of the air is cut off, such as behind doors and above or below mantels or shelves. Also keep the thermostat out of direct sunlight.

The thermostat should not be exposed to heat from nearby fireplaces, radios, televisions, lamps, or rays from the sun. Nor should the thermostat be mounted on a wall containing pipes or warm-air ducts or a flue or vent that could affect its operation and prevent it from properly controlling the room temperature. Any hole in the plaster or panel through which the wires pass from the thermostat should be adequately sealed with suitable material to prevent drafts from affecting the thermostat.

Printed Circuit-Board Control Center

Newer hot-air furnaces feature a printed circuit control. The board shown in Figure 13.8 is such that it is easy for the technician installing the furnace to hook it up properly the first time. The markings are designed to make it easy to connect the furnace for accessories if needed. Figures 13.9 and Figure 13.10 show the factory-furnished schematic. See if you can trace the schematic and locate the various points on the printed circuit boards.

HEAT PUMPS

The heat pump is a heat multiplier. It takes warm air and makes it hot air. This is done by compressing the air and increasing its temperature. Heat pumps have received more attention since the fuel embargo of 1974. Energy conservation has become a more important concern for everyone. If a device can be made to take heat from the air and heat a home or commercial building, it is very useful to many people.

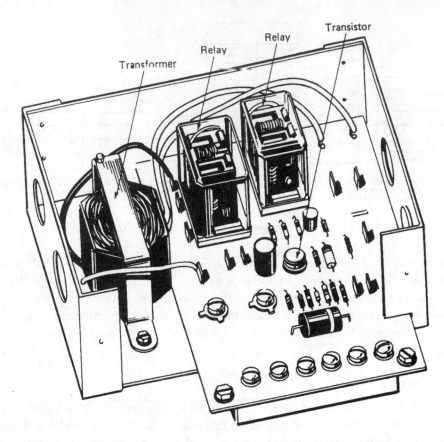

FIGURE 13.8   Printed circuit control center: heat-and-cool models. *(Carrier)*

The heat pump can take the heat generated by a refrigeration unit and use it to heat a house or room. Most take the heat from outside the home and move it indoors (see Figure 13.11). This unit can be used to air-condition the house in the summer and heat it in the winter by taking the heat from the outside air and moving it inside.

On mild-temperature heating days, the heat pump handles all heating needs. When the outdoor temperature reaches the balance point of the home—that is, when the heat loss is equal to the heat-pump heating capacity—the two-stage indoor thermostat activates the furnace (a secondary heat source, in most cases electric heating elements). As soon as the furnace is turned on, a heat relay de-energizes the heat pump.

When the second-stage (furnace) need is satisfied and the plenum temperature has cooled to below 90 and 100 degrees F, the heat-pump relay turns the heat pump back on and controls the conditioned space until the second-stage operation is required again. Figure 13.12 shows the heat-pump unit. The optional electric-heat unit shown in Figure 13.13 is added in geographic locations where needed. This particular unit can provide 23,000 to 56,000 Btus per hour (Btuh) and up to 112,700 Btuh with the addition of electric heat.

If the outdoor temperature drops below the setting of the low-temperature compressor monitor, the control shuts off the heat pump completely, and the furnace handles all the heating needs.

During the defrost cycle, the heat pump switches from heating to cooling. To prevent cool air from being circulated in the house when heating is needed, the control automatically turns on the furnace to compensate for the heat-pump defrost cycle (see Figure 13.14). When supply air temperature climbs above 110 to 120 degrees F, the defrost-limit control turns off the furnace and keeps indoor air from getting too warm.

If, after a defrost cycle, the air downstream of the coil rises above 115 degrees F, the closing point of the heat-pump relay, the compressor will stop until the heat exchanger has cooled down to 90 to 100 degrees F, as it does during normal cycling operation between furnace and heat pump.

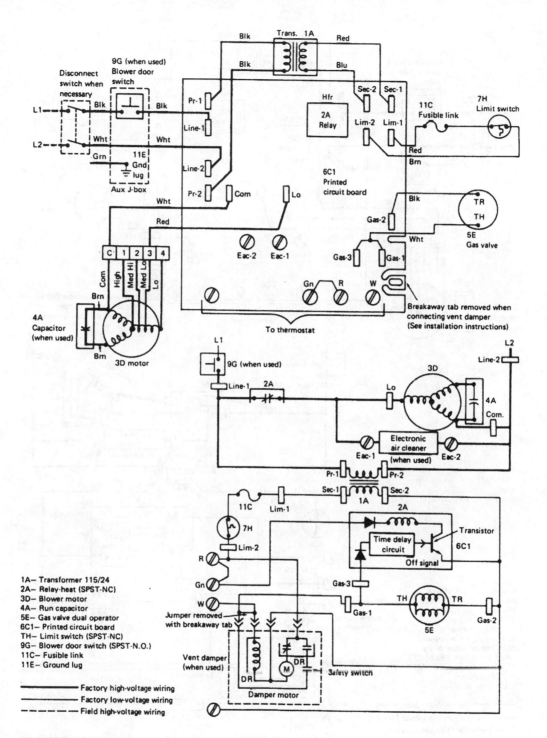

FIGURE 13.9 Wiring diagram: heating. *(Carrier)*

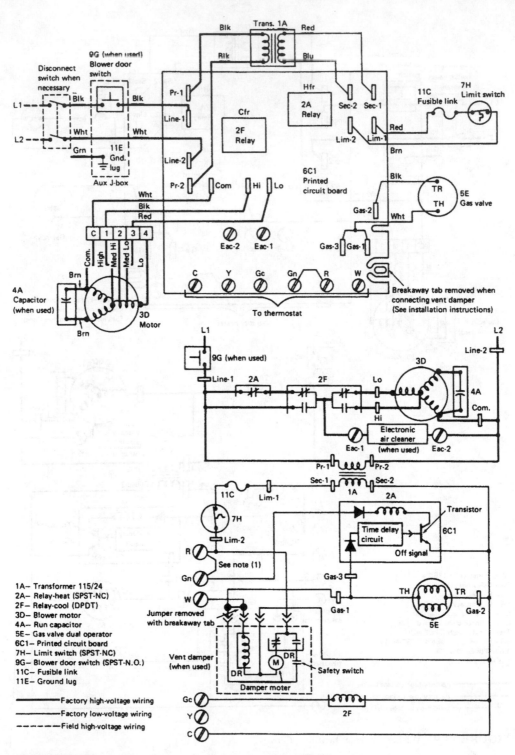

FIGURE 13.10 Wiring diagram: heating and cooling. *(Carrier)*

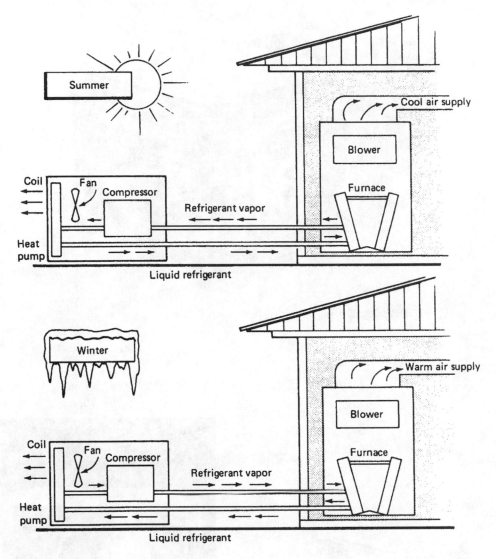

FIGURE 13.11   Basic operation of a heat pump.

During summer cooling, the heat pump works as a normal split system, using the furnace blower as the primary air mover (see Figure 13.15).

In a straight heat-pump/supplementary-electric-heater application, at least one outdoor thermostat is required to cycle the heaters as the outdoor temperature drops. In the system shown here, the indoor thermostat controls the supplemental heat source (furnace). The outdoor thermostat is not required.

Since the furnace is serving as the secondary heat source, the system does not require the home rewiring usually associated with supplemental electric strip heating.

Special Requirements of Heat-Pump Systems

The installation, maintenance, and operating efficiency of the heat-pump system are like those of no other comfort system. A heat-pump system requires the same air quantity for heating and cooling. Because of this, the air-moving capability of an existing furnace is extremely important. It should be carefully checked before a heat pump is added. Heating and load calculations must be accurate. System design and installation must be precise and according to the manufacturer's suggestions.

FIGURE 13.12   Single-package heat pump.

FIGURE 13.13   Optional elecrical
heat for a heat pump.

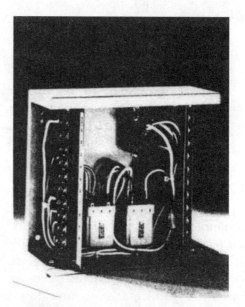

FIGURE 13.14   Control box for an add-
on type of heat pump.

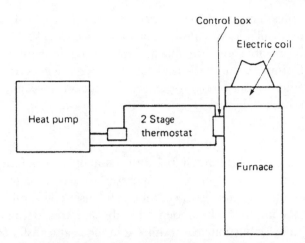

FIGURE 13.15   Heat pump with a two-stage
thermostat and control box mounted on the furnace.

The air-distribution system and diffuser location are equally important. Supply ducts must be properly sized and insulated. Adequate return air is also required. Heating-supply air is cooler than with other systems. This is quite noticeable to homeowners accustomed to gas or oil heat. This makes diffuser location and system balancing critical.

Heat-Pump Combinations

There are four ways to describe the heat-pump methods of transporting heat into the house:

- Air-to-air: the most common method and the type of system previously described

- Air-to-water: this method uses two different types of heat exchangers: warmed refrigerant flows through pipes to a heat exchanger in the boiler, and heated water flows into radiators located within the heated space

- Water-to-water: this type uses two water-to-refrigerant heat exchangers; heat is taken from the water source (well water, lakes, or the sea) and passed on by the refrigerant to the water used for heating; the reverse takes place in the cooling system

- Water-to-air: well water furnishes the heat by warming the refrigerant in the heat-exchanger coil; the refrigerant, compressed, flows to the top of the unit, where a fan blows air past the heat exchanger

Each type of heat pump has its advantages and disadvantages. The electrical connections and controls are used to do the job properly. Before attempting to work on this type of equipment, make sure you have a complete schematic of the electrical wiring and know all the component parts of the system.

HIGH-EFFICIENCY FURNACES

Newer furnaces (those made since 1981) have been designed with efficiencies of up to 97 percent, as compared to older types with efficiencies in the 60 percent range. The Lennox Pulse™ is one example of the types available.

The G-14-series pulse-combustion up-flow gas furnace provides efficiency of up to 97 percent. Eight models for natural gas and LPG are available with input capacities of 40,000, 60,000, 80,000, and 100,000 Btuh. The units operate on the pulse-combustion principle and do not require a pilot burner, main burners, conventional flue, or chimney. Compact, standard-sized cabinet design, with side or bottom return-air entry, permits installation in a basement, utility room, or closet. Evaporator coils may be added, as well as electronic air cleaners and power humidifiers (see Figure 13.16).

Operation

The high-efficiency furnaces achieve that level of fuel conversion by using a unique heat-exchanger design. It features a finned cast iron combustion chamber, temperature-resistant steel tailpipe, aluminized steel exhaust-de-coupler section, and a finned stainless steel tube condenser coil similar to an air-conditioner coil. Moisture from the products of combustion is condensed in the coil, thus wringing almost every usable Btu out of the gas. Since most of the combustion heat is utilized in the heat transfer from the coil, flue vent temperatures are as low as 100 F to 130 degrees F, allowing for the use of 2-inch-diameter polyvinyl chloride (PVC) pipe. The furnace is vented through a side wall or roof or to the top of an existing chimney with up

FIGURE 13.16   Lennox Pulse Furnace. *(Lennox)*

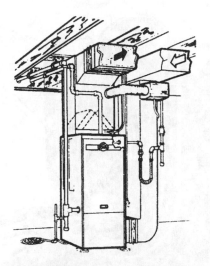

FIGURE 13.17 Basement installation of the Pulse® with cooling coil and automatic humidifier. Note the floor drain for condensate. *(Lennox)*

to 25 feet of PVC pipe and four 90-degree elbows. Condensate created in the coil may be disposed of in an indoor drain (see Figure 13.17). The condensate is not harmful to standard household plumbing and can be drained into city sewers and septic tanks without damage.

The furnace has no pilot light or burners. An automotive-type spark plug is used for ignition on the initial cycle only, saving gas and electrical energy. Due to the pulse-combustion principle, the use of atmospheric gas burners is eliminated, with the combustion process confined to the heat-exchanger combustion chamber. The sealed combustion system virtually eliminates the loss of conditioned air due to combustion and stack dilution. Combustion air from the outside is piped to the furnace with the same type of PVC pipe used for exhaust gases.

Electrical Controls

The furnace is equipped with a standard-type redundant gas valve in series with a gas-expansion tank, gas-intake flapper valve, and air-intake flapper valve. Also factory installed are a purge blower, spark plug igniter and flame sensor with solid-state control circuit board. The standard equipment includes a fan and limit control, a 30-VA transformer, blower cooling relay, flexible gas-line connector, and four isolation mounting pads, as well as a base insulation pad, condensate drip leg, and cleanable air filter. Flue vent/air intake line, roof- or wall-termination installation kits, LPG conversion kits, and thermostats are available as accessories and must be ordered extra, or you can use the existing ones when replacing a unit.

The printed circuit board is replaceable as a unit when there is a malfunction of one of the components. It uses a multivibrator transistorized circuit to generate the high voltages needed for the spark plug. The spark plug gets very little use except to start the combustion process. It has a long life expectancy. The spark gap is 0.115 inch, and the ground electrode is adjusted to 45 degrees (see Figure 13.20).

Sequence of Operation

On a demand for heat, the room thermostat initiates the purge blower operation for a prepurge cycle of 34 seconds, followed by energizing of the ignition and opening of the gas valve. As ignition occurs, the flame sensor senses proof of ignition and de-energizes the spark igniter and purge blower (see Figure 13.18). The furnace blower operation is initiated 30 to 45 seconds after combustion ignition. When the thermostat is

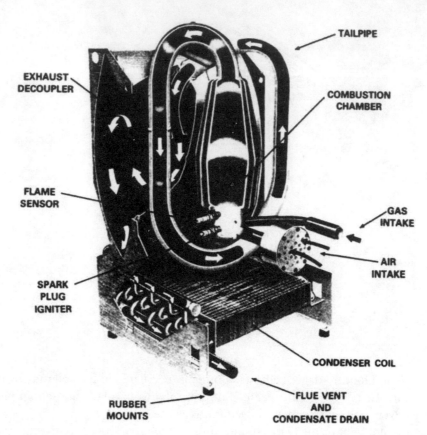

FIGURE 13-18 Cutaway view of the Pulse® furnace combustion chamber. *(Lennox)*

satisfied, the gas valve closes, and the purge blower is re-energized for a postpurge cycle of 34 seconds. The furnace blower remains in operation until the preset temperature setting (90 degrees F) of the fan control is reached. Should the loss of flame occur before the thermostat is satisfied, flame sensor controls will initiate three to five attempts at reignition before locking out the unit operation. In addition, loss of either combustion intake air or flue exhaust will automatically shut down the system.

Combustion Process

The process of pulse combustion begins as gas and air are introduced into the sealed combustion chamber with the spark-plug igniter. Spark from the plug ignites the gas/air mixture, which in turn causes a positive pressure buildup that closes the gas and air inlets. This pressure relieves itself by forcing the products of combustion out of the combustion chamber through the tailpipe into the heat-exchanger exhaust decoupler and on into the heat-exchanger coil. As the combustion chamber empties, its pressure becomes negative, drawing in air and gas for ignition of the next pulse. At the same instant, part of the pressure pulse is reflected back from the tailpipe at the top of the combustion chamber. The flame remnants of the previous pulse of combustion ignite the new gas/air mixture in the chamber, continuing the cycle.

Once combustion is started, it feeds on itself, allowing the purge blower and spark igniter to be turned off. Each pulse of gas/air mixture is ignited at the rate of 60 to 70 times per second, producing one-fourth to one-half Btu per pulse of combustion. Almost complete combustion occurs with each pulse. The force of these series of ignitions creates great turbulence, which forces the products of combination through the entire heat-exchanger assembly, resulting in maximum heart transfer (see Figure 13.18).

Startup procedures of the GSR-14Q series of Pulse furnaces, as well as maintenance and repair parts, are shown in Figure 13.19A-C.

ELECTRICAL

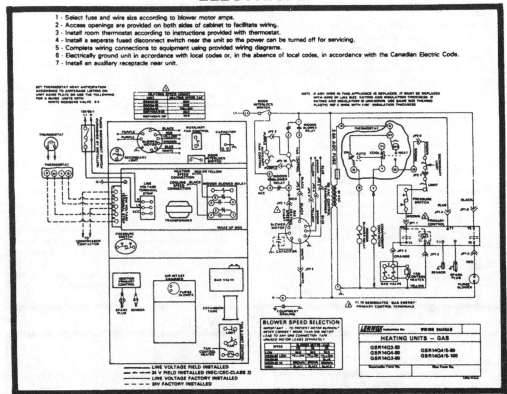

1 - Select fuse and wire size according to blower motor amps.
2 - Access openings are provided on both sides of cabinet to facilitate wiring.
3 - Install room thermostat according to instructions provided with thermostat.
4 - Install a separate fused disconnect switch near the unit so the power can be turned off for servicing.
5 - Complete wiring connections to equipment using provided wiring diagrams.
6 - Electrically ground unit in accordance with local codes or, in the absence of local codes, in accordance with the Canadian Electric Code.
7 - Install an auxiliary receptacle near unit.

START-UP/ADJUSTMENTS

START-UP

This unit is equipped with a direct spark ignition system with flame rectification. Once combustion has started, the purge blower and spark ignitor are turned off. To place furnace in operation:

1 - With thermostat set below room temperature make sure gas valve knob is in off position and wait 5 minutes.

2 - Turn manual knob of gas valve counterclockwise to ON position. Turn power on and set thermostat above room temperature. Unit will go into prepurge for approximately 30 seconds and then ignite.

3 - If the unit does not light on the first attempt, it will attempt four more ignitions before locking out.

4 - If lockout occurs, turn thermostat off and then back on.

To shut off furnace:

1 - Set thermostat to lowest temperature and turn power supply to furnace off.

2 - Turn manual knob of gas valve off.

FAILURE TO OPERATE

If unit fails to operate, check the following:

1 - Is thermostat calling for heat?
2 - Is main disconnect switch closed?
3 - Is there a blown fuse?
4 - Is filter dirty or plugged? Dirty or plugged filters will cause unit to go off on limit control.
5 - Is gas turned on at meter?
6 - Is manual main shut-off valve open?
7 - Is internal manual shut-off open?
8 - Are intake and exhaust pipes clogged?
9 - Is primary control locked out? (Turn thermostat off and then back on.)
10 - Is unit locked out on secondary limit? (Secondary limit is manually reset.)

GSR14 SERIES UNITS

LENNOX *Industries Inc.*

FIGURE 13.19A Electrical startup adjustments for the GSR-14Q furnaces. *(Lennox)*

ADJUSTMENTS (CONT.)

GAS FLOW

To check proper gas flow to combustion chamber, determine BTU input from the appliance rating plate. Divide this input rating by the BTU per cubic foot of available gas. Result is the number of cubic feet per hour required. Determine the flow of gas through gas meter for 2 minutes and multiply by 30 to get the hourly flow of gas to burner.

GAS PRESSURE

1 - Check gas line pressure with unit firing at maximum rate. Normal natural gas inlet line pressure should be 7.0 in. (178 mm) w.c. Normal line pressure for LP gas is 11 in. (280 mm) w.c.

   IMPORTANT - Minimum gas supply pressure is listed on unit rating plate for normal input. Operation below minimum pressure may cause nuisance lockouts.

2 - After line pressure is checked and adjusted, check regulator pressure. Correct manifold pressure (unit running) is specified on nameplate. To measure, connect gauge to pressure tap in elbow below expansion tank.

HEAT ANTICIPATOR SETTINGS

Units with White Rodgers gas valves — 0.9

FAN/LIMIT CONTROL

Limit Control — Factory set: No adjustment necessary
Fan Control — Factory set: ON — No adjustment necessary
OFF — 90°F (32°C)

TEMPERATURE RISE AND EXTERNAL STATIC PRESSURE

Check temperature rise and external static pressure. If necessary, adjust blower speed to maintain temperature rise and external static pressure within range shown on unit rating plate.

ELECTRICAL

1 - Check all wiring for loose connections.
2 - Check for correct voltage at unit (unit operating).
3 - Check amp-draw on blower motor.
   Motor Nameplate_____Actual_____

NOTE - Do not secure electrical conduit directly to duct work or structure.

BLOWER SPEEDS

Multi-tap drive motors are wired for different heating and cooling speeds. Speed may be changed by simply interchanging motor connections at indoor blower relay and fan control. Refer to speed selection chart on unit wiring diagram.

CAUTION - To prevent motor burnout, never connect more than one (1) motor lead to any one connection. Tape unused motor leads separately.

MAINTENANCE

NOTE - Disconnect power before servicing.

ANNUAL MAINTENANCE

At the beginning of each heating season, system should be checked by qualified serviceman as follows:

A - Blower
1 - Check and clean blower wheel.
2 - Check motor lubrication.
   Always relubricate motor according to manufacturer's lubrication instructions on each motor. If no instructions are provided, use the following as a guide:
   a - Motors Without Oiling Ports — Prelubricated and sealed. No further lubrication required.
   b - Direct Drive Motors with Oiling Ports — Prelubricated for an extended period of operation. For extended bearing life, relubricate with a few drops of SAE No. 10 non-detergent oil once every two years. It may be necessary to remove blower assembly for access to oiling ports.

B - Electrical
1 - Check all wiring for loose connections.
2 - check for correct voltage at unit (unit operating).
3 - Check amp-draw on blower motor.
   Motor nameplate _____ Actual _____
4 - Check to see that heat tape (if applicable) is operating.

C - Filters
1 - Filters must be cleaned or replaced when dirty to assure proper furnace operation.
2 - Reusable foam filters supplied with GSR14 can be washed with water and mild detergent. They should be sprayed with filter handicoater when dry prior to reinstallation. Filter handicoater is RP products coating No. 481 and is available as Lennox part No. P-8-5069.
3 - If replacement is necessary, order Lennox part No. P-8-7831 for 20 X 25 inch (508 X 635 mm) filter.

D - Intake and Exhaust Lines
Check intake and exhaust PVC lines and all connections for tightness and make sure there is no blockage. Also check condensate line for free flow during operation.

E - Insulation
Outdoor piping insulation should be inspected yearly for deterioration. If necessary, replace with same materials.

FIGURE 13.19B Electrical startup adjustments and maintenance for the GSR-14Q furnaces. *(Lennox)*

REPAIR PARTS LIST

The following repair parts are available through independent Lennox dealers. When ordering parts, include the complete furnace model number listed on unit rating plate. Example: GSR14Q3-50-1.

CABINET PARTS
Blower access panel
Control access panel
Upper vestibule panel
Lower vestibule panel
Control box cover

CONTROL PANEL PARTS
Transformer
Indoor blower relay
Low voltage terminal strip
High voltage terminal strip

BLOWER PARTS
Blower wheel
Motor
Motor mounting frame
Motor capacitor
Blower housing cut-off plate
Blower housing

HEATING PARTS
Heat exchanger assembly
Gas orifice
Gas valve
Gas decoupler
Gas flapper valve
Purge blower
Air intake flapper valve
Primary control board
Ignition lead
Flame sensor lead
Flame sensor
Primary fan and limit control
Secondary limit control
Auxiliary fan control
Differential pressure switch
Door interlock switch
Air filter

FIGURE 13.19C   Repair list for the GSR-14Q furnaces. *(Lennox)*

Troubleshooting The Pulse™ Furnace

Troubleshooting procedures for the Pulse™ furnaces are shown in Figures 13.20 A-C. Figure 13.21 shows the circuitry for the G-14Q series of furnaces. Note the difference in electrical circuitry for the G-14 and GSR-14. Blower-speed color-coded wires are also indicated for the different units. The 40, 60, 80, and 100 qualification after G-14Q indicates whether it is a 40,000, 60,000, 80,000, or 100,000 Btuh unit. Thermostat heat anticipation is also given for the Robertshaw valve and the Rodgers valve. This type of electrical diagram is usually glued to the cabinet so that it is with the unit whenever there is a need for troubleshooting.

The troubleshooting flow chart is typical of those furnished with newer equipment in the technical manuals the dealers provide. After locating the exact symptoms, check with the other part of Figure 13.20 to find how to use the multimeter to check out all the circuitry to see if the exact cause of the problem can be determined.

TROUBLESHOOTING

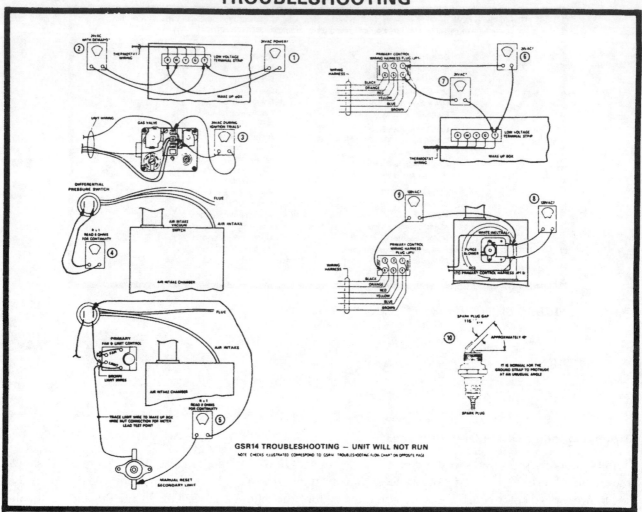

FIGURE 13.20A   Troubleshooting the GSR-14Q furnace with a meter. *(Lennox)*

TROUBLESHOOTING (CONT.)

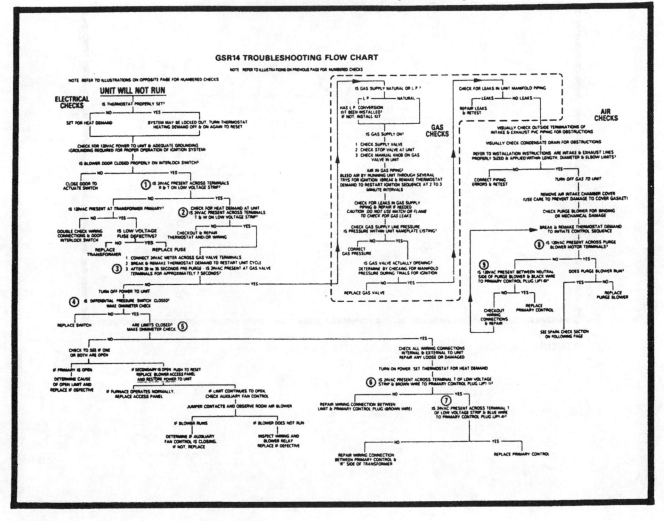

FIGURE 13.20B Troubleshooting the GSR-14Q furnace with a meter. *(Lennox)*

TROUBLESHOOTING (CONT.)

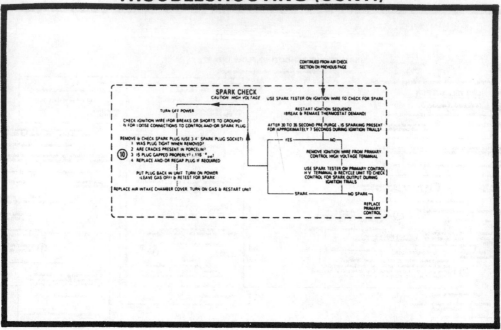

FIGURE 13.20C Troubleshooting the GSR-14Q furnace with a meter. *(Lennox)*

ELECTRICAL

1 - Select fuse and wire size according to blower motor amps.
2 - Access holes are provided on both sides of cabinet to facilitate wiring.
3 - Install room thermostat according to instruction provided with thermostat.
4 - Install a separate fused disconnect switch near the unit so the power can be turned off for servicing.

5 - Complete wiring connections to equipment using provided wiring diagrams.
6 - Electrically ground unit in accordance with local codes, or in the absence of local codes in accordance with the CSA Standards.
7 - Seal unused electrical openings with snap-plugs provided.

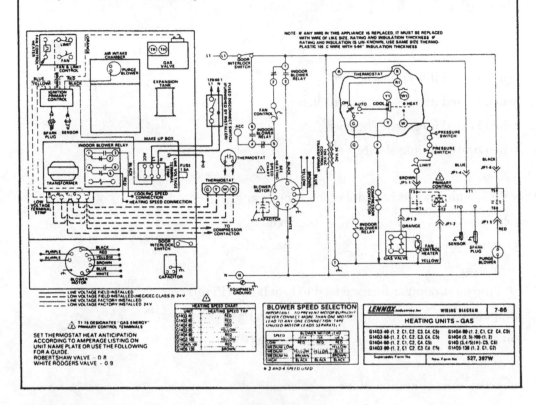

FIGURE 13.21 Electrical wiring for the G-14Q series of furnaces. *(Lennox)*

MULTIPLE-CHOICE EXAM

1. Hot air furnaces are self-contained and self-_____ units.
   a. healing   b. enclosed
   c. cleaning   d. stoking

2. Why does hot air rise?
   a. when heated it becomes lighter
   b. when heated it gains oxygen and floats
   c. when heated it becomes heavier
   d. when heated it sinks slowly

3. After the hot air loses its heat, it becomes cooler and _____.
   a. heavier   b. lighter
   c. moisture laden   d. freezes

4. What is the most popular furnace used to heat houses in this country?
   a. propane gas   b. natural-gas
   c. butane gas   d. coal gas

5. What controls the operation of the gas solenoid?
   a. ignitor switch   b. printed circuit switch
   c. blower switch   d. thermostat

6. How hot does the air in the plenum have to be before the fan switch closes and the fan starts?
   a. 950 degrees F   b. 1,000 degrees F
   c. 1,200 degrees F   d. 1,250 degrees F

7. What is the voltage required for the fan motor to operate?
   a. 240   b. 120
   c. 12   d. 24

8. How low does the temperature in the plenum have to drop to cause the fan to stop?
   a. 100 degrees F   b. 90 degrees F
   c. 85 degrees F   d. 75 degrees F

9. Why do electric-fired furnaces need large amounts of current to operate?

    a. the heating elements draw high current

    b. the fans draw a lot of current

    c. the heat exchanger is very inefficient

    d. none of the above

10. How fast does the electric-fired furnace begin producing heat?

    a. five minutes   b. instantly

    c. slowly   d. none of the above

11. Why aren't all the heating elements turned on at the same time?

    a. to prevent an overload on the electrical circuits in the house

    b. to prevent a switching problem

    c. to prevent excess heat being generated

    d. to prevent fuse blowing

12. What voltage does the heating elements in an electric-fired furnace use?

    a. 240   b. 120

    c. 440   d. 880

13. Why are ladder diagrams easy to read?

    a. They are often used for printed circuit boards

    b. They have everything labeled

    c. They make it easy to wire the circuits in production

    d. They look like a ladder and devices and switches are connected like rungs in a ladder

14. What does EFR stand for on a ladder diagram.

    a. evaporator function relay   b. evaporator fail-safe relay

    c. evaporator fan relay   d. none of the above

    e. all of the above

15. What does CC on a ladder diagram represent?

    a. compressor cleaner   b. compressor control

    c. compressor contactor   d. condenser control

16. On furnace manufacturers' diagrams what do the dotted lines represent?

    a. high voltage lines

    b. control lines

    c. low voltage line to be furnished at installation

    d. power lines for fans and controls

17. At what speed are fans' electric motors set as they come from the factory?

    a. medium   b. medium-high

    c. high   d. low

18. In electrical wiring for air-conditioners or refrigerators what does the white wire represent in the three wire plugs?

    a. ground or neutral   b. hot side

    c. low voltage   d. high voltage

19. The gas valve is wired by having low voltage wires from the _____ connected so that one wire is attached to one side of the solenoid and the other from the limit-switch to the other side of the solenoid.

    a. transformer   b. solenoid

    c. gas valve   d. limit-switch

20. Where is the wiring diagram for a furnace located as it comes from the factory?

    a. inside the cover for the cold-air return

    b. in the plenum

    c. in the hot air ductwork

    d. near the electronic controls

21. All electrical connections should comply with the _____ electrical code.

    a. county   b. local

    c. international   d. national

22. What happens when the heat anticipator is not adjusted and set to factory specs?

    a. improper use of the fans

    b. improper operation of the solenoid

    c. improper operation of the gas valve

    d. improper operation of the thermostat

23. The room _____ should be located where it will be in the natural circulating path of room air.

 a. thermostat   b. fan

 c. ventilator   d. heat diffuser

24. The series circuit has _____ amount of current in all parts of the circuit.

 a. the same   b. a less

 c. a greater   d. a lower

25. Why should the furnace cabinet have an un-interrupted or unbroken electrical ground?

 a. to minimize personal injury

 b. to maximize personal injury

 c. to protect the equipment from dust

 d. to protect the blower mechanism from damage

26. Which of the following is not a desirable place to locate a thermostat?

 a. behind doors   b. near a cold-air register

 c. near drafts from windows   d. in the sunlight

 e. all of the above   f. none of the above

27. The heat pump is a heat _____.

 a. generator   b. manipulator

 c. regular pump   d. multiplier

28. What does the heat pump do to air to heat it?

 a. it adds refrigeration to the air   b. it decompresses the air

 c. it compresses the air   d. it contaminates the air

29. The heat pump can take the heat generated by a refrigeration unit and use it to _____ a house or room.

 a. decontaminate   b. cool

 c. heat   d. dehumidify

30. How is the heat pump system augmented to produce heating in extremely cold weather?

 a. electrical heat is switched on and heat pump is deactivated

 b. gas heat is switched on and heat pump is also activated

 c. propane heat is substituted for the heat pump

 d. sunshine is relied upon for supplemental heat

31. During the defrost cycle, the heat pump switches from heating to _____.

    a. idle

    b. cooling

    c. ventilating

    d. freezing

32. In a straight heat pump/supplementary electric heater application, at least one outdoor _____ is required to cycle the heaters as the outdoor temperature drops.

    a. compressor

    b. transformer

    c. thermostat

    d. fan

33. In a heat pump installation the supply ducts must be properly sized and _____.

    a. insulated

    b. installed

    c. leaded

    d. sealed

34. Which of the following is NOT one of the ways to describe the heat-pump method of transporting heat into the house?

    a. air-to-air

    b. air-to-water

    c. water-to-water

    d. water-to-air

    e. man-to-man

35. Newer furnaces have been designed with efficiencies up to _____ percent.

    a. 55

    b. 85

    c. 80

    d. 97

36. High efficiency furnaces achieve a higher level of fuel conversion by using a _____ heat-exchanger design.

    a. unique

    b. different

    c. slower

    d. faster

37. How are exhaust gases from the Pulse furnace conducted to the outside?

    a. by using a sheet metal stack

    b. by using a PVC pipe

    c. by using a rubberized chimney

    d. by using lead pipe

38. The Pulse furnace has no pilot light or _____.

    a. burners   b. chimney problems

    c. open flame   d. none of the above

39. In the Pulse furnace, the pre-purge cycle lasts _____ seconds.

    a. 33   b. 34

    c. 35   d. 36

40. In the Pulse furnace, the furnace blower operation is initiated in 30 to _____ seconds after combustion ignition.

    a. 60   b. 35

    c. 45   d. 50

41. In the Pulse furnace, each pulse of gas/air mixture is ignited at the rate of 60 to _____ times per second.

    a. 65   b. 70

    c. 75   d. 80

TRUE-FALSE EXAM

1. Hot air furnaces are self-contained and self-enclosed units.

    True   False

2. After hot air loses its heat it becomes cooler and heavier.

    True   False

3. The gas furnace is the simplest to operate and understand.

    True   False

4. When the air in the plenum reaches 90 degrees F, the fan switch closes and the blower fan starts.

    True   False

5. A transformer is needed to provide low voltage for control circuits.

    True   False

6.   The fan switch provides the necessary 120 volts to the fan motor for it to operate.
True   False

7.   Once the room has heated up to the desired thermostat setting, the thermostat closes.
True   False

8.   Electric-fired heat is the only heat produced almost as fast as the thermostat calls for it.
True   False

9.   The outside of an electric-fired furnace looks almost the same as a gas-fired furnace.
True   False

10.   After the gas has been turned off, the blower continues to operate until the temperature in the plenum hits a pre-set point.
True   False

11.   The low-resistance heating element used in an electric-fired furnace draws a lot of current.
True   False

12.   In the electric-fired furnace, when the auxiliary contacts open they remove the heating element from the line and stop the fan motor.
True   False

13.   The ladder diagram makes it easier to see how devices connected to the line are wired.
True   False

14.   The EFR evaporator fan relay also controls the blower fan.
True   False

15.   The evaporator fan will operate when either the heating relay or the evaporator fan relay is energized.
True   False

16.   The wiring diagram is usually located inside the cover for the cold air return when it comes from the factory.
True   False

17.   All electrical connections made in the field should be done in accordance with the American Electrical Code.
True   False

18. The room thermostat should be located where it will be in the natural circulating path of room air.

True   False

19. The thermostat should not be exposed to heat from nearby fireplaces, radios, televisions, lamps, or rays from the sun.

True   False

20. Newer hot-air furnaces feature printed circuit contro is.

True   False

21. The heat pump is a heat multiplier.

True   False

22. The heat pump can take the heat generated by a refrigeration unit and use it to heat a room or whole house.

True   False

23. On mild temperature days, the heat pump handles all heating needs.

True   False

24. When the second-stage furnace is satisfied and the plenum temperature has cooled to below 90 degrees and 100 degrees F, the heat pump relay turns the heat pump on and controls the conditioned space until the second-stage operation is required again.

True   False

25. During the defrost cycle, the heat pump switches from heating to cooling.

True   False

26. If after a defrost cycle, the air downstream of the coil gcts above 115 degree F, the closing point of the heat-pump relay, the compressor will stop until the heat exchanger has cooled down to 90 to 100 degrees as it does during normal cycling operation.

True   False

27. During the summer cooling season the heat pump works as a split system.

True   False

28. The installation, maintenance, and operating efficiency of the heat-pump system are like those of all other comfort systems.

True   False

29. In the heat-pump system the air distribution system and diffuser locations are not too important.
     True   False

30. There are four ways to describe the heat-pump methods of transporting heat into the house.
     True   False

31. Newer furnaces have been designed with efficiencies of up to 97 percent.
     True   False

32. High efficiency furnaces achieve a high level of fuel conversion by using a unique heat-exchanger design.
     True   False

33. The Pulse furnace uses a spark plug for an igniter.
     True   False

34. The Pulse furnace uses a pilot light.
     True   False

35. The Pulse furnace uses PVC for connections to the outside air.
     True   False

MULTIPLE-CHOICE ANSWERS

1.	B	10.	B	18.	A	26.	E	34.	E
2.	A	11.	A	19.	A	27.	D	35.	D
3.	A	12.	A	20.	A	28.	C	36.	A
4.	B	13.	D	21.	D	29.	C	37.	B
5.	D	14.	C	22.	D	30.	A	38.	A
6.	C	15.	C	23.	A	31.	B	39.	B
7.	B	16.	C	24.	A	32.	C	40.	C
8.	B	17.	B	25.	A	33.	A	41.	B
9.	A								

TRUE-FALSE ANSWERS

1.	T	10.	T	19.	T	28.	F
2.	T	11.	T	20.	T	29.	F
3.	T	12.	T	21.	T	30.	T
4.	F	13.	T	22.	T	31.	T
5.	T	14.	F	23.	T	32.	T
6.	T	15.	T	24.	T	33.	T
7.	F	16.	T	25.	T	34.	F
8.	T	17.	T	26.	T	35.	T
9.	T	18.	T	27.	T		

—NOTES—

Chapter 14
SAFETY ON THE JOB

HVAC Licensing Study Guide

This chapter includes useful safety information. The facts that a HVAC mechanic must know have been gleaned from years of HVAC experience. The apprentice (or even the journeyman) should find these tips useful.

BASIC SAFETY CONSIDERATIONS

Improper installation, adjustment, alteration, service, maintenance, or use of air-conditioning and refrigeration equipment can cause explosion, fire, electrical shock, or other conditions that may cause personal injury or property damage. Consult a qualified installer, a service agency, or a distributor branch for information or assistance. The qualified installer or agency must use factory-authorized kits or accessories when modifying any product. Refer to the individual instructions packaged with the kits or accessories when installing. You will notice that most codes tell you to observe the manufacturer's instructions when installing.

Here are some essential guidelines:

- Follow all safety codes

- Wear safety glasses and work gloves

- Use quenching cloths for brazing operations

- Have fire extinguishers available

- Read the instructions thoroughly and follow all warnings or cautions attached to the unit

- Consult local building codes and the National Electrical Code (NEC) for special requirements

It is important to recognize safety information. This is the safety-alert symbol: ⚠. When you see this symbol on the unit and also in installation instructions or manuals, be alert to the potential for personal injury.

Understand the signal words:

- DANGER

- WARNING

- CAUTION

- NOTE

These words are used with the safety-alert symbol. DANGER identifies the most serious hazards, which will result in severe personal injury or death. WARNING signifies hazards that could result in personal injury or death. CAUTION is used to identify unsafe practices that may result in minor personal injury or equipment and property damage. NOTE is used to highlight suggestions that will result in enchanced reliability or operation.

When working with electricity, it is very important to make sure that you do not touch the "hot" wire (either black or red in a three-wire cable) and ground. There is usually a bare wire (it goes to ground) and a white wire, which is neutral or also a ground. If you touch the white or the bare wire and either the black or red wire, you will be shocked with 120 volts in most instances. If you make contact with the black and red wires, you will probably receive a shock with 240 volts pushing enough current through your body to cause your heart to start fibrillating or stop. Take adequate precautions when working around 120 or 240 volts.

Sometimes the circuits you will be working with will be powered with 240 volts. You must be aware of and take appropriate safety precautions to prevent electrical shock hazards. Be sure to wear safety glasses while working.

In many situations the equipment will have electric motors for fans. Needless to say, the motors may have rotating shafts with belts and pulleys. Keep clear of rotating motor shafts and equipment powered by them. Fan blades can be very sharp and can inflict a great deal of personal damage.

Some compressors use a start-and-run capacitor for proper starting and operating. These electrolytic capacitors are capable of storing a charge sufficient to kill a human being. Be careful not to get across the two terminals of the capacitor.

You can discharge the capacitor by shorting across the terminals. Use a screwdriver with an insulated handle. The discharge produces a loud pop noise and may cause pitting of the screwdriver blade. The arc produced by the discharge can cause fires or other damage. Remember, a capacitor must be shorted five times in order to completely discharge it. You can still get a shock after the initial discharge surge.

With the split-phase type of motor it is suggested that you not wear loose-fitting clothing. Keep away from the rotating end of the motor shaft. Also keep your hands away from the shaft, belt, and pulley.

When working with pressure controls, make sure safety precautions are taken to protect your body and eyes. Wear safety glasses and gloves when appropriate. Make sure you use eye protection and gloves when working with refrigerant. When working with heating equipment, it is very possible to get burned by the heat exchanger, exhaust pipes, and other parts of the system. Wear proper protection.

SAFETY AND THE COOLING TOWER

Most areas of the tower must be inspected for safe working and operating conditions. A number of items should be inspected yearly and repaired immediately in order to guarantee the safety of maintenance personnel.

Scale remover contains acid and can cause skin irritation. Avoid contact with your eyes, skin, and clothing. In case of contact, flush the skin or eyes with plenty of water for at least 15 minutes. If the eyes are affected, get medical attention. Keep scale remover and other chemicals out of the reach of children. Do not drain the spent solution to the roof or to a septic tank. Always drain the solution in an environmentally safe way, not to the storm sewer. Safety is all-important. All chemicals, especially acids, should be treated with great respect and handled with care. Rubber gloves, acid-proof coveralls, and safety goggles should be worn when working with chemicals. Cleaning a system through the tower, although easier and faster than some of the other methods, presents one unique hazard—wind drift.

Wind drift, even with the tower fan off, is a definite possibility. Wind drift will carry tiny droplets of acid that can burn eyes and skin. These acid droplets will also damage automobile finishes and buildings. Should cleaning solution contact any part of the person, it should be washed off immediately with soap and water. Using forethought and reasonable precaution can prevent grief and expense.

SOLVENTS AND DETERGENTS

There are several uses for solvents and detergents in the ordinary maintenance schedule of air-cooled tin-coil condensers, evaporator coils, permanent-type air filters, and fan blades. In most instances, a high-pressure spray washer is used to clean the equipment with detergent, then a high-pressure spray rinse is used to clean the unit being scrubbed. The pump is usually rated at 2 gallons/minute at 500 pounds/inch^2 of pressure. The main function is to remove dirt and grease from fans and cooling surfaces. It takes about 10 to 15 minutes for the cleaning solution to do its job. It is then rinsed with clean water.

Using the dipping method cleans permanent-type filters. Prepare a cleaning solution of one part detergent to one part water. Use this solution as a bath in which the filters may be immersed briefly. After dipping, set the filter aside for 10 to 15 minutes. Flush with a stream of water. If running water is not available, good results may be obtained by brisk agitation in a tank filled with fresh water. When draining the solution used for cleaning purposes, be sure to follow the local codes on the use of storm sewers for disposal purposes. Proper disposal of the spent solution is critical for legal operation of this type of air-conditioning unit.

Another more recent requirement is the use of asbestos in the construction of the tower fill. If discovered when inspections are conducted, make sure it is replaced with the latest materials. The older towers can become more efficient with newer fill of more modern design.

Cooling-tower manufacturers design their units for a given performance standard and for conditions such as the type of chiller used, ambient temperatures, location, and specifications. As a system ages, it may lose efficiency. The cleanliness of the tower and its components are crucial to the success of the system. An unattended cold-water temperature will rise. This will send warmer water to the chiller. When the chiller kicks out on high head pressure, the system may shut down. Precautions should be taken to prevent shutdown from occurring.

HANDLING CYLINDERS

Refrigeration and air-conditioning servicepersons must be able to handle compressed gases. Accidents occur when compressed gases are not handled properly.

Oxygen or acetylene must never be used to pressurize a refrigeration system. Oxygen will explode when it comes in contact with oil. Acetylene will explode under pressure, except when properly dissolved in acetone as used in commercial acetylene cylinders.

Dry nitrogen or dry carbon dioxide are suitable gases for pressurizing refrigeration or air-conditioning systems for leak tests or system cleaning. However, the following specific restrictions must be observed:

- Nitrogen (N_2): commercial cylinders contain pressure in excess of 2,000 pounds/inch^2 at normal room temperature

- Carbon dioxide (CO_2): commercial cylinders contain pressures in excess of 800 pounds/inch^2 at normal room temperature

Cylinders should be handled carefully. Do not drop them or bump them. Keep cylinders in a vertical position and securely fastened to prevent them from tipping over. Do not heat the cylinder with a torch, or other open flame. If heat is necessary to withdraw gas from the cylinder, apply it by immersing the lower portion of the cylinder in warm water. Never heat a cylinder to a temperature over 110 degrees F (43 degrees C).

PRESSURIZING

Pressure-testing or cleaning refrigeration and air-conditioning systems can be dangerous! Extreme caution must be used in the selection and use of pressurizing equipment. Follow these procedures:

1. Never attempt to pressurize a system without first installing an appropriate pressure-regulating valve on the nitrogen or carbon dioxide cylinder discharge. This regulating valve should be equipped with two functioning pressure gauges. One gauge indicates cylinder pressure. The other gauge indicates discharge of downstream pressure.

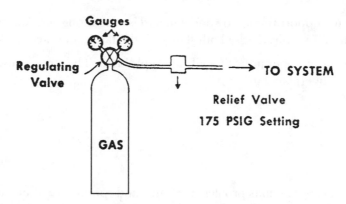

FIGURE 14.1 Pressurizing set-up for charging refrigeration systems.

2. Always install a pressure-relief valve or frangible disk type of pressure-relief device in the pressure-supply line. This device should have a discharge port of at least 1/2 inch national pipe thread (NPT) size. This valve or frangible-disk device should be set to release at 175 psg.

A system can be pressurized up to a maximum of 150 psig for leak testing or purging (see Figure 14.1).

ELECTRICAL SAFETY

Many Tecumseh single-phase compressors are installed in systems requiring off-cycle crankcase heating. This is designed to prevent refrigerant accumulation in the compressor housing. The power is on at all times. Even if the compressor is not running, power is applied to the compressor housing where the heating element is located.

Another popular system uses a run capacitor that is always connected to the compressor motor windings, even when the compressor is not running. Other devices are energized when the compressor is not running. That means that there is electrical power applied to the unit when the compressor is not running. This calls for an awareness of the situation and the proper safety procedures.

Be safe. Before you attempt to service any refrigeration system, make sure that the main circuit breaker is open and all power is off.

FIBERGLASS FANS

Fiberglass fans require care similar to steel fans except as noted below:

1. Be sure the impeller is not striking the inlet or housing and is rotating in the proper direction. Fiberglass may break on impact or fail quickly due to stress caused by improper rotation.

2. Do not operate a fiberglass fan in an abrasive atmosphere. Abrasives will erode resin surfaces of the FRP material and destroy the corrosion resistance of the fan.

3. Do not operate a fiberglass fan in temperatures above 150 degrees F (65 degrees C) without a specific resin and a fan design approved by the fan manufacturer.

4. Never attempt to support the fan by one flange. Use mounting brackets and appropriate vibration isolators if required. Use both flanges if the fan is mounted in ductwork.

5. Never allow the fan to operate with a vibration problem. The stress caused by vibration will quickly destroy a fiberglass fan.

LIFTING

Lifting heavy objects can cause serious problems. Strains and sprains are often caused by improper lifting methods. Figure 14.2 indicates the right and wrong way to lift heavy objects. In this case, an evaporator coil is shown.

To avoid injury, learn to lift the safe way. Bend your knees, keep your back erect, and lift gradually with your leg muscles.

The material you are lifting may slip from your hands and injure your feet. To prevent foot injuries, wear the proper steel-toed shoes.

HANDLING REFRIGERANTS SAFELY

One of the requirements of an ideal refrigerant is that it must be nontoxic. In reality, however, all gases (with the exception of pure air) are more or less toxic or asphyxiating. It is therefore important that wherever gases or highly volatile liquids are used, adequate ventilation be provided, because even nontoxic gases in air produce a suffocating effect.

Vaporized refrigerants (especially ammonia and sulfur dioxide) bring about irritation and congestion of the lungs and bronchial organs, accompanied by violent coughing, vomiting, and, when breathed in

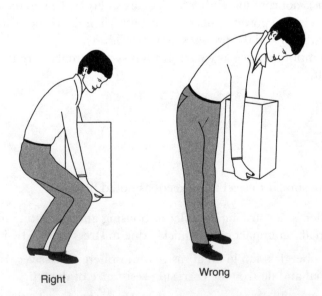

Right   Wrong

FIGURE 14.2   Safety first. Lift with the legs not the back.
(Tecumseh)

sufficient quantity, suffocation. It is of the utmost importance, therefore, that the serviceman subjected to a refrigerant gas find access to fresh air at frequent intervals to clear his lungs. When engaged in the repair of ammonia and sulfur dioxide machines, approved gas masks and goggles should be used. Carrene, Freon (R-12), and carbon dioxide fumes are not irritating and can be inhaled in considerable concentrations for short periods without serious consequences.

It should be remembered that liquid refrigerant would refrigerate or remove heat from anything it meets when released from a container. In the case of contact with refrigerant, the affected or injured area should be treated as if it has been frozen or frostbitten.

Refrigerant cylinders should be stored in a dry, sheltered, and well-ventilated area. The cylinders should be placed in a horizontal position, if possible, and held by blocks or saddles to prevent rolling. It is of utmost importance to handle refrigerant cylinders with care and to observe the following precautions:

- Never drop the cylinders or permit them to strike each other violently

- Never use a lifting magnet or a sling (rope or chain) when handling cylinders; a crane may be used when a safe cradle or platform is provided to hold the cylinders

- Caps provided for valve protection should be kept on the cylinders at all times except when the cylinders are actually in use

- Never overfill the cylinders; whenever refrigerant is discharged from or into a cylinder, weigh the cylinder and record the weight of the refrigerant remaining in it

- Never mix gases in a cylinder

- Never use cylinders for rollers, supports, or for any purpose other than to carry gas

- Never tamper with the safety devices in valves or on the cylinders

- Open the cylinder valves slowly; never use wrenches or tools except those provided or approved by the gas manufacturer

- Make sure that the threads on regulators or other unions are the same as those on the cylinder-valve outlets.; never force a connection that does not fit

- Regulators and gauges provided for use with one gas must not be used on cylinders containing a different gas

- Never attempt to repair or alter the cylinders or valves

- Never store the cylinders near highly flammable substances (such as oil, gasoline, or waste)

- Cylinders should not be exposed to continuous dampness, salt water, or salt spray

- Store full and empty cylinders apart to avoid confusion

- Protect the cylinders from any object that will produce a cut or other abrasion on the surface of the metal

MULTIPLE-CHOICE EXAM

1.   Which of the following is not considered a safety hazard?

   a. ignoring safety codes

   b. not wearing safety glasses

   c. not wearing steel-toed shoes

   d. having a fire extinguisher available

2.   Which of the following is not a recommended safety practice?

   a. following all safety codes

   b. wiring up the compressor with the power on

   c. reading instructions thoroughly and following them

   d. using a quenching cloth for brazing operations

3.   What is the symbol for safety alert?

   a. a triangle with an exclamation mark enclosed

   b. a circle with period enclosed

   c. a triangle with a flame in it

   d. a rectangle with the word "caution" written inside it

4.   Which of the following is not a signal word?

   a. danger   b. warning

   c. help   d. caution

5.   The word "danger" identifies the most serious hazards, which could result in personal injury or
   _____.

   a. harm   b. amputation

   c. death   d. hospitalization

6.   When working with electricity, it is important not to touch the _____ wire?

   a. low-voltage   b. gray

   c. green   d. hot

7. If you make contact with a white wire and a black wire, what voltage would you be aware of immediately?

   a. 120   b. 240

   c. 12   d. 24

8. If you make contact with a white wire and a red wire, what voltage would you come across?

   a. 120   b. 240

   c. 12   d. 24

9. If you are wiring a piece of equipment and and happen to get across the red and black wires in the power line, what would be the voltage present?

   a. 120   b. 240

   c. 12   d. 24

10. While working with the thermostat circuit, you make contact with the two wires coming from the transformer secondary. What is the voltage you receive this time?

    a. 200   b. 100

    c. 12   d. 24

11. The thermostat is in series with which component in a furnace circuit?

    a. transformer   b. push button

    c. gas valve   d. blower fan

12. The following two wires in a three-wire cable are ground wires:

    a. red and black wires   b. red and white wires

    c. bare and white wires d. black and bare wires

13. What type of capacitor is used as a start capacitor with a compressor motor?

    a. mica   b. ceramic

    c. electrolytic   d. paper

14. What type of capacitor is used as a run capacitor with some motors?

    a. mica   b. ceramic

    c. electrolytic   d. silver plated

15. With which motor is it recommended not to wear loose-fitting clothing?

    a. permanent-split capacitor   b. split-phase capacitor

    c. start capacitor   d. shaded-pole capacitor

16. How many times should you short across an electrolytic capacitor to prevent it from retaining a charge capable of doing physical harm?

    a. two   b. four

    c. five   d. six

17. When working with heating equipment (such as a hot-air furnace), it is very possible to get burned by the _____ exchanger.

    a. humidity   b. cold

    c. heat   d. root

18. In working with cooling towers, what is the main safety hazard to be watchful for:

    a. chemicals   b. scale

    c. water   d. refrigerant

19. In working with cooling towers, scale _____ contains acid and can cause skin irritation.

    a. residue   b. coater

    c. reactor   d. remover

20. Do not drain the spent scale-remover solution to the roof or to a(n) _____ tank.

    a. above-ground   b. septic

    c. disposal   d. underground

21. Wind drift can carry tiny droplets of _____ that can burn eyes and skin.

    a. salt water   b. acid

    c. calcium chloride   d. soap

22. There are several uses for solvents and _____ in the ordinary maintenance schedule of cooling towers as well as air-cooled tin-coil condensers, evaporator coils, permanent-type filters, and fan blades.

    a. detergents   b. soaps

    c. steel wool   d. sandpaper

23. What method of cleaning is used in cleaning permanent-type filters?

    a. spraying   b. dipping

    c. dripping   d. hosing

24. Cooling tower manufacturers design their units for:

    a. maximum efficiency no matter how they affect the environment

    b. minimum efficiency and maximum cooling

    c. a given performance standard

    d. fitting any location

25. Refrigeration and air-conditioning mechanics must be able to handle _____ gases.

    a. hydrolyzed   b. oxygenated

    c. compressed   d. acetylenized

26. Dry nitrogen and dry _____ dioxide are suitable gases for pressurizing refrigeration or air-conditioning systems for leak tests.

    a. ferric   b. carbon

    c. sodium   d. calcium

27. Which of the following gases should never be used to pressurize a refrigeration system?

    a. nitrogen   b. acetylene

    c. hydrogen   d. chlorine

28. Commercial cylinders of carbon dioxide contain pressures in excess of _____ pound/inch^2 at normal room temperature.

    a. 100   b. 200

    c. 600   d. 800

29. Never heat a gas-filled cylinder to a temperature over _____ degrees F.

    a. 110   b. 220

    c. 120   d. 230

30. For leak testing and purging, a system can be pressurized up to a maximum of _____ psig.

    a. 150   b. 200

    c. 250   d. 300

TRUE-FALSE EXAM

1.   Fiberglass fans require care similar to steel fans.

    True   False

2.   Do not operate a fiberglass fan in an abrasive atmosphere.

    True   False

3.   Never attempt to support a fiberglass fan by one flange.

    True   False

4.   Never allow a fiberglass fan to operate with a vibration problem.

    True   False

5.   You can operate a fiberglass fan above 150 degrees F.

    True   False

6.   Lifting heavy objects, such as an evaporator coil, can cause serious injury.

    True   False

7.   Before you attempt to service any refrigeration system, make sure that the main circuit breaker is open and all power is off.

    True   False

8.   To avoid injury when lifting heavy objects, lift the safe way. Bend your knees.

    True   False

9.   Improper installation, adjustment, alteration, service, maintenance, or use of refrigeration equipment can cause an explosion. Fire, electrical shock, or other dangerous conditions may be encountered.

    True   False

10.   Quenching clothes are necessary when brazing.

    True   False

11.   One of the requirements of an ideal refrigerant is that it must be toxic.

    True   False

12. Refrigerants such as sulfur dioxide and ammonia can bring about serious illness.

True   False

13. Vaporized refrigerants in enclosed areas can cause irritation and congestion of the lungs and bronchial organs.

True   False

14. Carrene is another name for Freon R-12.

True   False

15. A gas mask and goggles should be worn when working with ammonia as a refrigerant.

True   False

16. When the human body has been directly exposed to a refrigerant, the victim should treat the resulting area as if it were frozen or frostbitten.

True   False

17. It is not necessary to store refrigerant cylinders in a dry, sheltered, and well-ventilated area.

True   False

18. Never drop cylinders or permit them to strike each other violently.

True   False

19. Never attempt to repair or alter the cylinders or valves.

True   False

20. Never use a lifting magnet or a sling when handling cylinders.

True   False

21. Caps are provided for valve protection on gas-filled cylinders.

True   False

22. Never mix gases in a cylinder.

True   False

23. Never overfill cylinders.

True   False

24. Make sure that the threads on regulators or other unions are the same as those on the cylinder-valve outlets.

    True   False

25. Regulators and gauges provided for use with one gas can be used on cylinders containing a different gas.

    True   False

26. When storing cylinders, it is best to keep them in a vertical position.

    True   False

27. A crane may be used to lift cylinders if a safe cradle is provided to hold them.

    True   False

28. It is necessary to open the cylinder valves slowly.

    True   False

29. Never attempt to repair or alter cylinders or valves.

    True   False

30. Store full and empty cylinders apart to avoid confusion.

    True   False

31. Never store cylinders near highly flammable substances such as oil, gasoline, or waste.

    True   False

32. Cylinders should not be exposed to continuous dampness, salt water, or salt spray.

    True   False

33. When working with refrigeration equipment, wear safety glasses and work gloves.

    True   False

34. When installing new refrigeration equipment, be sure to consult all local, state, and national codes for their requirements.

    True   False

35. Some compressors use a start-and-run capacitor for proper operation.

    True   False

MULTIPLE-CHOICE ANSWERS

1.	D	7.	A	13.	C	19.	D	25.	C
2.	B	8.	A	14.	C	20.	B	26.	B
3.	A	9.	B	15.	B	21.	B	27.	B
4.	C	10.	D	16.	C	22.	A	28.	D
5.	C	11.	C	17.	C	23.	B	29.	A
6.	D	12.	C	18.	A	24.	C	30.	A

TRUE-FALSE ANSWERS

1.	T	10.	T	19.	T	28.	T
2.	T	11.	F	20.	T	29.	T
3.	T	12.	T	21.	T	30.	T
4.	T	13.	T	22.	T	31.	T
5.	F	14.	T	23.	T	32.	T
6.	T	15.	T	24.	T	33.	T
7.	T	16.	T	25.	F	34.	T
8.	T	17.	F	26.	F	35.	T
9.	T	18.	T	27.	T		

—NOTES—

Chapter 15

TRADE ASSOCIATIONS AND CODES

There are a number of codes available to builders, contractors, home owners, and municipalities. They have been prepared by the *International Code Council (ICC)* to protect the public by making certain standards and codes into law.

Tradespeople are expected to "know" the codes. However, there are many codes covering, in some cases, the same topics or subjects. Therefore, it is to the advantage of the air-conditioning and refrigeration mechanic to become familiar with what is expected of him/her when working in this field.

The following is a listing of some of the codes that have refrigeration and air-conditioning installation and repair as their concern. Topics covered in the code are listed for you to be able to grasp the meaning of the requirements for the work you do for a living.

CODES

International Mechanical Code

Ventilation
Exhaust systems
Duct systems
Combustion air
Chimneys and vents
Specific appliances
Fireplaces, solid fuel-burning equipment
Boilers
Water heaters and pressure vessels
Refrigeration
Hydronic piping
Fuel oil piping and storage
Solar systems

International Fire Code and International Fuel Gas Code

Gas-piping installations
Chimneys and vents
Specific appliances
Gaseous hydrogen systems
Referenced standards

International Building Code
International Energy Conservation Code

The International Codes are designed and promulgated to be adopted by reference and ordinance. Local jurisdictions adopt the codes and add their own suggestions.

National Electrical Code (NEC)®

All aspects of installation and design of electrical equipment and systems are included. This is the national guide to the safe installation of electrical wiring and equipment. It is designed primarily for use by trained electrical people and is necessarily terse in its wording.

The latest information on the code (usually revised and updated every three years) can be found on the Internet.

TRADE ASSOCIATIONS

American National Standards Institute

1819 L Street NW
Washington, DC   20036

www.ansi.org

American Society for Testing and Materials

100 Barr Harbor Drive
W. Conshohocken, PA   19428-2959

www.astm.org

American Society of Heating, Refrigerating and Air-Conditioning Engineers, Inc.

1791 Tullie Circle
Atlanta, GA   30329-2305

www.ashrae.org

American Society of Safety Engineers

1800 East Oakton
Des Plaines, IL   60018

www.asse.org

Building and Fire Research Laboratory, National Institute of Standards and Technology

Building 226, Room 6216
Gaithersburg, MD   20899

www.bfrl.nist.gov

Building Research Establishment

Bucknalls Lane
Garston Watford, WD2 7 JR, England

www.bre.co.uk

Building Officials and Code Administrators International

4051 W Flossmoor Road
Country Club Hills, IL   50478-5794

www.bocai.org

International Code Council

5203 Leesburg Pike, Suite 600
Falls Church, VA   22041

www.intlcode.org

International Organization for Standardization

1, rue de Varembe
Case postale 56 CH 1211
Geneva 20 Switzerland

www.iso.ch

National Institute of Standards and Technology

100 Bureau Drive, Stop 3460
Gaithersburg, MD   20899-3460

www.nist.gov

Southern Building Code Congress International

900 Montclair Road
Birmingham, AL   35213-1206

www.sbcci.org

Underwriters' Laboratories, Inc.

333 Pfingsten Road
Northbrook, IL   60062-2096

www.ul.com

WEB SITE SOURCES

Architects, contractors, and manufacturers
www.construction.com

Association acronyms
www.acronymfinder.com

Code-related web sites
www.codesourcepc.com

HVAC industry links
www.hvacportal.com

Internet Public Library general reference site
www.ipl.org

Construction waste
www.thebluebook.com

Building products directory
www.virtual-engineer.net

Global codes and standards
www.plumbinglinks.com

GLOSSARY

HVAC Licensing Study Guide

absolute humidity

The weight of water vapor per unit volume; grains per cubic foot; or grams per cubic meter.

absolute pressure

The sum of gauge pressure and atmospheric pressure. Thus, for example, if the gauge pressure is 154 psi (pounds per square inch), the absolute pressure will be 154 + 14.7, or 168.7 psi.

absolute zero

A temperature equal to -459.6 degrees F or -273 degrees C. At this temperature the volume of an ideal gas maintained at a constant pressure becomes zero.

absorption

The action of a material in extracting one or more substances present in the atmosphere or a mixture of gases or liquids accompanied by physical change, chemical change, or both.

acceleration

The time rate of change of velocity. It is the derivative of velocity with respect to time.

accumulator

A shell placed in a suction line for separating the liquid entrained in the suction gas. A storage tank at the evaporator exit or suction line used to prevent floodbacks to the compressor.

acrolein

A warning agent often used with methyl chloride to call attention to the escape of refrigerant. The material has a compelling, pungent odor and causes irritation of the throat and eyes. Acrolein reacts with sulfur dioxide to form a sludge.

ACR tube

A copper tube usually hard drawn and sold to the trade cleaned and sealed with nitrogen inside to prevent oxidation. Identified by its actual outside diameter (OD).

activated alumina

A form of aluminum oxide (Al_2O_3) that absorbs moisture readily and is used as a drying agent.

adiabatic

Referring to a change in gas conditions where no heat is added or removed except in the form of work.

adiabatic process

Any thermodynamic process taking place in a closed system without the addition or removal of heat.

adsorbent

A sorbent that changes physically, chemically, or both during the sorption process.

aeration

Exposing a substance or area to air circulation.

agitation

A condition in which a device causes circulation in a tank containing fluid.

air, ambient

Generally speaking, the air surrounding an object.

air changes

A method of expressing the amount of air leakage into or out of a building or room in terms of the number of building volumes or room volumes exchanged per unit of time.

air circulation

Natural or imparted motion of air.

air cleaner

A device designed for the purpose of removing airborne impurities such as dust, gases, vapors, fumes, and smoke. An air cleaner includes air washers, air filters, electrostatic precipitors and charcoal filters.

air conditioner

An assembly of equipment for the control of at least the first three items enumerated in the definition of air conditioning.

air conditioner, room

A factory-made assembly designed as a unit for mounting in a window, through a wall, or as a console. It is designed for free delivery of conditioned air to an enclosed space without ducts.

air conditioning

The simultaneous control of all, or at least the first three, of the following factors affecting the physical and chemical conditions of the atmosphere within a structure—temperature, humidity, motion, distribution, dust, bacteria, odors, toxic gases, and ionization—most of which affect human health or comfort.

air-conditioning system, central fan

A mechanical indirect system of heating, ventilating, or air conditioning in which the air is treated or handled by equipment located outside the rooms served, usually at a central location and conveyed to and from the rooms by means of a fan and a system of distributing ducts.

air-conditioning system, year round

An air-conditioning system that ventilates, heats, and humidifies in winter, and cools and dehumidifies in summer to provide the desired degree of air motion and cleanliness.

air-conditioning unit

A piece of equipment designed as a specific air-treating combination, consisting of a means for ventilation, air circulation, air cleaning, and heat transfer with a control means for maintaining temperature and humidity within prescribed limits.

air cooler

A factory-assembled unit including elements, whereby the temperature of air passing through the unit is reduced.

air cooler, spray type

A forced-circulation air cooler, wherein the coil surface capacity is augmented by a liquid spray during the period of operation.

air cooling

A reduction in air temperature due to the removal of heat as a result of contact with a medium held at a temperature lower than that of the air.

air diffuser

A circular, square, or rectangular air-distribution outlet, generally located in the ceiling, and comprised of deflecting members discharging supply air in various directions and planes, arranged to promote mixing of primary air with secondary room air.

air, dry

In psychrometry, air unmixed with or containing no water vapor.

air infiltration

The in-leakage of air through cracks, crevices, doors, windows, or other openings caused by wind pressure or temperature difference.

air, recirculated

Return air passed through the conditioner before being again supplied to the conditioned space.

air, return

Air returned from conditioned or refrigerated space.

air, saturated

Moist air in which the partial pressure of the water vapor is equal to the vapor pressure of water at the existing temperature. This occurs when dry air and saturated water vapor coexist at the same dry-bulb temperature.

air, standard

Air with a density of 0.075 lb/ft^3 and an absolute viscosity of 1.22×10 lb mass/ft-s. This is substantially equivalent to dry air at 70 degrees F and 29.92 in. Hg barometer.

air washer

An enclosure in which air is forced through a spray of water in order to cleanse, humidify, or precool the air.

ambient temperature

The temperature of the medium surrounding an object. In a domestic system having an air-cooled condenser, it is the temperature of the air entering the condenser.

ammonia machine

An abbreviation for a compression-refrigerating machine using ammonia as a refrigerant. Similarly, Freon machine, sulfur dioxide machine, and so forth.

ampere

Unit used to measure electrical current. It is equal to 1 coulomb of electrons flowing past a point in 1 second. A coulomb is 6.28×10^{18} electrons.

analyzer

A device used in the high side of an absorption system for increasing the concentration of vapor entering the rectifier or condenser.

anemometer

An instrument for measuring the velocity of air in motion.

antifreeze, liquid

A substance added to the refrigerant to prevent formation of ice crystals at the expansion valve. Antifreeze agents in general do not prevent corrosion due to moisture. The use of a liquid should be a temporary measure where large quantities of water are involved, unless a drier is used to reduce the moisture content. Ice crystals may form when moisture is present below the corrosion limits, and in such instances, a suitable noncorrosive antifreeze liquid is often of value. Materials such as alcohol are corrosive and, if used, should be allowed to remain in the machine for a limited time only.

atmospheric condenser

A condenser operated with water that is exposed to the atmosphere.

atmospheric pressure

The pressure exerted by the atmosphere in all directions as indicated by a barometer. Standard atmospheric pressure is considered to be 14.695 psi (pounds per square inch), which is equivalent to 29.92 in. Hg (inches of mercury).

atomize

To reduce to a fine spray.

automatic air conditioning

An air-conditioning system that regulates itself to maintain a definite set of conditions by means of automatic controls and valves usually responsive to temperature or pressure.

automatic expansion valve

A pressure-actuated device that regulates the flow of refrigerant from the liquid line into the evaporator to maintain a constant evaporator pressure.

baffle

A partition used to divert the flow of air or a fluid.

balanced pressure

The same pressure in a system or container that exists outside the system or container.

barometer

An instrument for measuring atmospheric pressure.

blast heater

A set of heat-transfer coils or sections used to heat air that is drawn or forced through it by a fan.

bleeder

A pipe sometimes attached to a condenser to bleed off liquid refrigerant parallel to the main flow.

boiler

A closed vessel in which liquid is heated or vaporized.

boiler horsepower

The equivalent evaporation of 34.5 lb of water per hour at 212 degrees F, which is equal to a heat output of $970.3 \times 34.5 = 33,475$ Btu.

boiling point

The temperature at which a liquid is vaporized upon the addition of heat, dependent on the refrigerant and the absolute pressure at the surface of the liquid and vapor.

bore

The inside diameter of a cylinder.

bourdon tube

Tube of elastic metal bent into circular shape that is found inside a pressure gauge.

brine

Any liquid cooled by a refrigerant and used for transmission of heat without a change in its state.

brine system

A system whereby brine cooled by a refrigerating system is circulated through pipes to the point where the refrigeration is needed.

British thermal unit (Btu)

The amount of heat required to raise the temperature of 1 lb of water 1 degree F. It is also the measure of the amount of heat removed in cooling 1 lb of water 1 degree F and is so used as a measure of refrigerating effect.

butane

A hydrocarbon, flammable refrigerant used to a limited extent in small units.

calcium chloride

A chemical having the formula $CaCl_2$, which, in granular form, is used as a drier. This material is soluble in water, and in the presence of large quantities of moisture may dissolve and plug up the drier unit or even pass into the system beyond the drier.

calcium sulfate

A solid chemical of the formula $CaSO_4$, which may be used as a drying agent.

calibration

The process of dividing and numbering the scale of an instrument; also of correcting and determining the error of an existing scale.

calorie

Heat required to raise the temperature of 1 g of water 1 degree C (actually, from 4 to 5 degrees C). Mean calorie is equal to one-hundredth part of the heat required to raise 1 g of water from 0 to 100 degrees C.

capacitor

An electrical device that has the ability to store an electrical charge. It is used to start motors, among other purposes.

capacity, refrigerating

The ability of a refrigerating system, or part thereof, to remove heat. Expressed as a rate of heat removal, it is usually measured in Btu/h or tons/24 h.

capacity reducer

In a compressor, a device, such as a clearance pocket, movable cylinder head, or suction bypass, by which compressor capacity can be adjusted without otherwise changing the operating conditions.

capillarity

The action by which the surface of a liquid in contact with a solid (as in a slender tube) is raised or lowered.

capillary tube

In refrigeration practice, a tube of small internal diameter used as a liquid refrigerant-flow control or expansion device between high and low sides; also used to transmit pressure from the sensitive bulb of some temperature controls to the operating element.

carbon dioxide ice

Compressed solid CO_2; dry ice.

celsius

A thermometric system in which the freezing point of water is called 0 degrees C and its boiling point 100 degrees C at normal pressure. This system is used in the scientific community for research work and also by most European countries and Canada. This book has the Celsius value of each Fahrenheit temperature in parenthesis.

centrifugal compressor

A compressor employing centrifugal force for compression.

centrifuge

A device for separating liquids of different densities by centrifugal action.

change of air

Introduction of new, cleansed, or recirculated air to a conditioned space, measured by the number of complete changes per unit of time.

change of state

Change from one state to another, as from a liquid to a solid, from a liquid to a gas, and so forth.

charge

The amount of refrigerant in a system.

chimney effect

The tendency of air or gas in a duct or other vertical passage to rise when heated due to its lower density compared with that of the surrounding air or gas. In buildings, the tendency toward displacement, caused by the difference in temperature, of internal heated air by unheated outside air due to the difference in density of outside and inside air.

clearance

Space in a cylinder not occupied by a piston at the end of the compression stroke or volume of gas remaining in a cylinder at the same point, measured in percentage of piston displacement.

coefficient of expansion

The fractional increase in length or volume of a material per degree rise in temperature.

coefficient of performance (heat pump)

Ratio of heating effect produced to the energy supplied, each expressed in the same thermal units.

coil

Any heating or cooling element made of pipe or tubing connected in series.

cold storage

A trade or process of preserving perishables on a large scale by refrigeration.

comfort chart

A chart showing effective temperatures with dry-bulb temperatures and humidities (and sometimes air motion) by which the effects of various air conditions on human comfort may be compared.

compression system

A refrigerating system in which the pressure-imposing element is mechanically operated.

compressor

That part of a mechanical refrigerating system which receives the refrigerant vapor at low pressure and compresses it into a smaller volume at higher pressure.

compressor, centrifugal

A nonpositive displacement compressor that depends on centrifugal effect, at least in part, for pressure rise.

compressor displacement

Compressor volume in cubic inches found by multiplying piston area by stroke by the number of cylinders.

$$\text{Displacement in cubic feet per minute} = \frac{\pi \times r^2 \times L \times \text{rpm} \times n}{1728}$$

compressor, open-type

A compressor with a shaft or other moving part, extending through a casing, to be driven by an outside source of power, thus requiring a stuffing box, shaft seals, or equivalent rubbing contact between a fixed and moving part.

compressor, reciprocating

A positive-displacement compressor with a piston or pistons moving in a straight line but alternately in opposite directions.

compressor, rotary

One in which compression is attained in a cylinder by rotation of a positive-displacement member.

compressor booster

A compressor for very low pressures, usually discharging into the suction line of another compressor.

condenser

A heat-transfer device that receives high-pressure vapor at temperatures above that of the cooling medium, such as air or water, to which the condenser passes latent heat from the refrigerant, causing the refrigerant vapor to liquefy.

condensing

The process of giving up latent heat of vaporization in order to liquefy a vapor.

condensing unit

A specific refrigerating machine combination, for a given refrigerant, consisting of one or more power-driven compressors, condensers, liquid receivers (when required), and the regularly furnished accessories.

condensing unit, sealed

A mechanical condensing unit in which the compressor and compressor motor are enclosed in the same housing, with no external shaft or shaft seal, and the compressor motor is operating in the refrigerant atmosphere.

conduction, thermal

Passage of heat from one point to another by transmission of molecular energy from particle to particle through a conductor.

conductivity, thermal

The ability of a material to pass heat from one point to another, generally expressed in terms of Btu per hour per square foot of material per inch of thickness per degree temperature difference.

conductor, electrical

A material that will pass an electric current as part of an electrical system.

connecting rod

A device connecting the piston to a crank and used to change rotating motion into reciprocating motion, or vice versa, as from a rotating crankshaft to a reciprocating piston.

constant-pressure valve

A valve of the throttling type, responsive to pressure, located in the suction line of an evaporator to maintain a desired constant pressure in the evaporator higher than the main suction-line pressure.

constant-temperature valve

A valve of the throttling type, responsive to the temperature of a thermostatic bulb. This valve is located in the suction line of an evaporator to reduce the refrigerating effect on the coil to just maintain a desired minimum temperature.

control

Any device for regulation of a system or component in normal operation either manual or automatic. If automatic, the implication is that it is responsive to changes of temperature, pressure, or any other property whose magnitude is to be regulated.

control, high-pressure

A pressure-responsive device (usually an electric switch) actuated directly by the refrigerant-vapor pressure on the high side of a refrigerating system (usually compressor-head pressure).

control, low-pressure

An electric switch, responsive to pressure, connected into the low-pressure part of a refrigerating system (usually closes at high pressure and opens at low pressure).

control, temperature

An electric switch or relay that is responsive to the temperature change of a thermostatic bulb or element.

convection

The circulatory motion that occurs in a fluid at a nonuniform temperature, owing to the variation of its density and the action of gravity.

convection, forced

Convection resulting from forced circulation of a fluid as by a fan, jet, or pump.

cooling tower, water

An enclosed device for evaporative cooling water by contact with air.

cooling unit

A specific air-treating combination consisting of a means for air circulation and cooling within prescribed temperature limits.

cooling water

Water used for condensation of refrigerant. Condenser water.

copper plating

Formation of a film of copper, usually on compressor walls, pistons, or discharge valves caused by moisture in a methyl chloride system.

corrosive

Having a chemically destructive effect on metals (occasionally on other materials).

counter-flow

In the heat exchange between two fluids, the opposite direction of flow, the coldest portion of one meeting the coldest portion of the other.

critical pressure

The vapor pressure corresponding to the critical temperature.

critical temperature

The temperature above which a vapor cannot be liquefied, regardless of pressure.

critical velocity

The velocity above which fluid flow is turbulent.

cryohydrate

An eutectic brine mixture of water and any salt mixed in proportions to give the lowest freezing temperature.

cycle

A complete course of operation of working fluid back to a starting point measured in thermodynamic terms. Also used in general for any repeated process in a system.

cycle, defrosting

That portion of a refrigeration operation which permits the cooling unit to defrost.

cycle, refrigeration

A complete course of operation of a refrigerant back to the starting point measured in thermodynamic terms. Also used in general for any repeated process for any system.

Dalton's law of partial pressure

Each constituent of a mixture of gases behaves thermodynamically as if it alone occupied the space. The sum of the individual pressures of the constituents equals the total pressure of the mixture.

defrosting

The removal of accumulated ice from a cooling unit.

degree day

A unit based on temperature difference and time used to specify the nominal heating load in winter. For one day there exist as many degree-days as there are degrees Fahrenheit difference in temperature between the average outside air temperature, taken over a 24-h period, and a temperature of 65 degrees F.

dehumidifier

An air cooler used for lowering the moisture content of the air passing through it. An absorption or adsorption device for removing moisture from the air.

dehumidify

To remove water vapor from the atmosphere or to remove water or liquid from stored goods.

dehydrator

A device used to remove moisture from the refrigerant.

density

The mass or weight per unit of volume.

dew point, air

The temperature at which a specified sample of air, with no moisture added or removed, is completely saturated. The temperature at which the air, on being cooled, gives up moisture or dew.

differential (of a control)

The difference between the cut-in and cut-out temperature. A valve that opens at one pressure and closes at another. This allows a system to adjust itself with a minimum of overcorrection.

direct connected

Driver and driven, as motor and compressor, positively connected in line to operate at the same speed.

direct expansion

A system in which the evaporator is located in the material or space refrigerated or in the air-circulating passages communicating with such space.

discharge gas

Hot, high-pressure vapor refrigerant, which has just left the compressor.

displacement, actual

The volume of gas at the compressor inlet actually moved in a given time.

displacement, theoretical

The total volume displaced by all the pistons of a compressor for every stroke during a definite interval (usually measured in cubic feet per minute).

drier

Synonymous with dehydrator.

dry-type evaporator

An evaporator of the continuous-tube type where the refrigerant from a pressure-reducing device is fed into one end and the suction line is connected to the outlet end.

duct

A passageway made of sheet metal or other suitable material, not necessarily leaktight, used for conveying air or other gas at low pressure.

dust

An air suspension (aerosol) of solid particles of earthy material, as differentiated from smoke.

economizer

A reservoir or chamber wherein energy or material from a process is reclaimed for a further useful purpose.

efficiency, mechanical

The ratio of the output of a machine to the input in equivalent units.

efficiency, volumetric

The ratio of the volume of gas actually pumped by a compressor or pump to the theoretical displacement of the compressor.

ejector

A device that utilizes static pressure to build up a high fluid velocity in a restricted area to obtain a lower static pressure at that point so that fluid from another source may be drawn in.

element, bimetallic

An element formed of two metals having different coefficients of thermal expansion, such as used in temperature-indicating and -controlling devices.

emulsion

A relatively stable suspension of small, but not colloidal, particles of a substance in a liquid.

engine

Prime mover; device for transforming fuel or heat energy into mechanical energy.

enthalpy

The total heat content of a substance, compared to a standard value 32 degrees F or 0 degrees C for water vapor. The ratio of the heat added to a substance to the absolute temperature at which it is added.

equalizer

A piping arrangement to maintain a common liquid level or pressure between two or more chambers.

eutectic solution

A solution of such concentration as to have a constant freezing point at the lowest freezing temperature for the solution.

evaporative condenser

A refrigerant condenser utilizing the evaporation of water by air at the condenser surface as a means of dissipating heat.

evaporative cooling

The process of cooling by means of the evaporation of water in air.

evaporator

A device in which the refrigerant evaporates while absorbing heat.

expansion valve, automatic

A device that regulates the flow of refrigerant from the liquid line into the evaporator to maintain a constant evaporator pressure.

expansion valve, thermostatic

A device that regulates the flow of refrigerant into an evaporator so as to maintain an evaporation temperature in a definite relationship to the temperature of a thermostatic bulb.

extended surface

The evaporator or condenser surface that is not a primary surface. Fins or other surfaces that transmit heat from or to a primary surface, which is part of the refrigerant container.

external equalizer

In a thermostatic expansion valve, a tube connection from the chamber containing the pressure-actuated element of the valve to the outlet of the evaporator coil. A device to compensate for excessive pressure drop through the coil.

fahrenheit

A thermometric system in which 32 degrees F denotes the freezing point of water and 212 degrees F the boiling point under normal pressure.

fan

An air-moving device comprising a wheel, or blade, and housing or orifice plate.

fan, centrifugal

A fan rotor or wheel within a scroll-type housing and including driving-mechanism supports for either belt-drive or direct connection.

fan, propeller

A propeller or disk-type wheel within a mounting ring or plate and including driving-mechanism supports for either belt-drive or direct connection.

fan, tube-axial

A disk-type wheel within a cylinder, a set of air-guide vanes located either before or after the wheel, and driving-mechanism supports for either belt-drive or direct connection.

filter

A device to remove solid material from a fluid by a straining action.

flammability

The ability of a material to burn.

flare fitting

A type of connector for soft tubing that involves the flaring of the tube to provide a mechanical seal.

flash gas

The gas resulting from the instantaneous evaporation of the refrigerant in a pressure-reducing device to cool the refrigerant to the evaporation temperature obtained at the reduced pressure.

float valve

Valve actuated by a float immersed in a liquid container.

flooded system

A system in which the refrigerant enters into a header from a pressure-reducing valve and the evaporator maintains a liquid level. Opposed to dry evaporator.

fluid

A gas or liquid.

foaming

Formation of a foam or froth of oil refrigerant due to rapid boiling out of the refrigerant dissolved in the oil when the pressure is suddenly reduced. This occurs when the compressor operates; and if large quantities of refrigerant have been dissolved, large quantities of oil may "boil" out and be carried through the refrigerant lines.

freezeup

Failure of a refrigeration unit to operate normally due to formation of ice at the expansion valve. The valve may be frozen closed or open, causing improper refrigeration in either case.

freezing point

The temperature at which a liquid will solidify upon the removal of heat.

Freon-12

The common name for dichlorodifluoromethane (CCl_2F_2).

frostback

The flooding of liquid from an evaporator into the suction line, accompanied by frost formation on the suction line in most cases.

furnace

That part of a boiler or warm-up heating plant in which combustion takes place. Also a complete heating unit for transferring heat from fuel being burned to the air supplied to a heating system.

fusible plug

A safety plug used in vessels containing refrigerant. The plug is designed to melt at high temperatures (usually about 165 degrees F) to prevent excessive pressure from bursting the vessel.

gauge

An instrument used for measuring various pressures or liquid levels. (Sometimes spelled gage).

gas

The vapor state of a material.

generator

A basic component of any absorption-refrigeration system.

gravity, specific

The density of a standard material usually compared to that of water or air.

grille

A perforated or louvered covering for an air passage, usually installed in a sidewall, ceiling, or floor.

halide torch

A leak tester generally using alcohol and burning with a blue flame; when the sampling tube draws in halocarbon refrigerant vapor, the color of the flame changes to bright green. Gas given off by the burning halocarbon is phosgene, a deadly gas used in World War I in Europe against Allied troops (can be deadly if breathed in a closed or confined area).

halogen

An element from the halogen group that consists of chlorine, fluorine, bromine, and iodine. Two halogens may be present in chlorofluorocarbon refrigerants.

heat

Basic form of energy that may be partially converted into other forms and into which all other forms may be entirely converted.

heat of fusion

Latent heat involved in changing between the solid and the liquid states.

heat, sensible

Heat that is associated with a change in temperature; specific heat exchange of temperature, in contrast to a heat interchange in which a change of state (latent heat) occurs.

heat, specific

The ratio of the quantity of heat required to raise the temperature of a given mass of any substance 1 degree to the quantity required to raise the temperature of an equal mass of a standard substance (usually water at 59 degrees F) 1 degree.

heat of vaporization

Latent heat involved in the change between liquid and vapor states.

heat pump

A refrigerating system employed to transfer heat into a space or substance. The condenser provides the heat, while the evaporator is arranged to pick up heat from air, water, and so forth. By shifting the flow of the refrigerant, a heat-pump system may also be used to cool the space.

heating system

Any of the several heating methods usually termed according to the method used in its generation, such as steam heating, warm-air heating, and the like.

heating system, electric

Heating produced by the rise of temperature caused by the passage of an electric current through a conductor having a high resistance to the current flow. Residence electric-heating systems generally consist of one or several resistance units installed in a frame or casing, the degree of heating being thermostatically controlled.

heating system, steam

A heating system in which heat is transferred from a boiler or other source to the heating units by steam at, above, or below atmospheric pressure.

heating system, vacuum

A two-pipe steam heating system equipped with the necessary accessory apparatus to permit operating the system below atmospheric pressure.

heating system, warm-air

A warm-air heating plant consisting of a heating unit (fuel burning furnace) enclosed in a casing from which the heated air is distributed to various rooms of the building through ducts.

hermetically sealed unit

A refrigerating unit containing the motor and compressor in a sealed container.

high-pressure cutout

A control device connected into the high-pressure part of a refrigerating system to stop the machine when the pressure becomes excessive.

high side

That part of the refrigerating system containing the high-pressure refrigerant. Also the term used to refer to the condensing unit consisting of the motor, compressor, condenser, and receiver mounted on a single base.

high-side float valve

A float valve that floats in high-pressure liquid. Opens on an increase in liquid level.

hold over

In an evaporator, the ability to stay cold after heat removal from the evaporator stops.

horsepower

A unit of power. Work done at the rate of 33,000 ft-lb/min, or 550 ft-lb/s.

humidifier

A device to add moisture to the air.

humidify

To add water vapor to the atmosphere; to add water vapor or moisture to any material.

humidistat

A control device actuated by changes in humidity and used for automatic control of relative humidity.

humidity, absolute

The definite amount of water contained in a definite quantity of air (usually measured in grains of water per pound or per cubic foot of air).

humidity, relative

The ratio of the water-vapor pressure of air compared to the vapor pressure it would have if saturated at its dry-bulb temperature. Very near to the ratio of the amount of moisture contained in air compared to what it could hold at the existing temperature.

humidity, specific

The weight of vapor associated with 1 lb of dry air; also termed *humidity ratio*.

hydrocarbons

A series of chemicals of similar chemical nature, ranging from methane (the main constituent of natural gas) through butane, octane, and so forth, to heavy lubricating oils. All are more or less flammable. Butane and isobutane have been used to a limited extent as refrigerants.

hydrolysis

Reaction of a material, such as Freon-12 or methyl chloride, with water. Acid materials in general are formed.

hydrostatic pressure

The pressure due to liquid in a container that contains no gas space.

hygrometer

An instrument used to measure moisture in the air.

hygroscope

See humidistat.

ice-melting equivalent

The amount of heat (144 Btu) absorbed by 1 lb of ice at 32 degrees F in liquefying to water at 32 degrees F.

indirect cooling system

See brine system.

infiltration

The leakage of air into a building or space.

insulation

A material of low heat conductivity.

irritant refrigerant

Any refrigerant that has an irritating effect on the eyes, nose, throat, or lungs.

isobutane

A hydrocarbon refrigerant used to a limited extent. It is flammable.

kilowatt

Unit of electrical power equal to 1,000 W, or 1.34 hp, approximately.

lag of temperature control

The delay in action of a temperature-responsive element due to the time required for the temperature of the element to reach the surrounding temperature.

latent heat

The quantity of heat that may be added to a substance during a change of state without causing a temperature change.

latent heat of evaporation

The quantity of heat required changing 1 lb of liquid into a vapor with no change in temperature. Reversible.

leak detector

A device used to detect refrigerant leaks in a refrigerating system.

liquid

The state of a material in which its top surface in a vessel will become horizontal. Distinguished from solid or vapor forms.

liquid line

The tube or pipe that carries the refrigerant liquid from the condenser or receiver of a refrigerating system to a pressure-reducing device.

liquid receiver

That part of the condensing unit that stores the liquid refrigerant.

load

The required rate of heat removal.

low-pressure control

An electric switch and pressure-responsive element connected into the suction side of a refrigerating unit to control the operation of the system.

low side

That part of a refrigerating system which normally operates under low pressure, as opposed to the high side. Also used to refer to the evaporator.

low-side float

A valve operated by the low-pressure liquid, which opens at a low level and closes at a high level.

main

A pipe or duct for distributing to or collecting conditioned air from various branches.

manometer

A U-shaped liquid-filled tube for measuring pressure differences.

mechanical efficiency

The ratio of work done by a machine to the work done on it or energy used by it.

mechanical equivalent of heat

An energy-conversion ratio of 778.18 ft-lb = 1 Btu.

methyl chloride

A refrigerant having the chemical formula CH_3Cl.

micron (μ)

A unit of length, the thousandth part of 1 mm or the millionth part of a meter.

Mollier chart

A graphical representation of thermal properties of fluids, with total heat and entropy as coordinates.

motor

A device for transforming electrical energy into mechanical energy.

motor capacitor

A device designed to improve the starting ability of single-phase induction motors.

noncondensables

Foreign gases mixed with a refrigerant, which cannot be condensed into liquid form at the temperatures and pressures at which the refrigerant condenses.

oil trap

A device to separate oil from the high-pressure vapor from the compressor. Usually contains a float valve to return the oil to the compressor crankcase.

output

Net refrigeration produced by the system.

ozone

The O_3 form of oxygen, sometimes used in air conditioning or cold-storage rooms to eliminate odors; can be toxic in concentrations of 0.5 ppm and over.

packing

The stuffing around a shaft to prevent fluid leakage between the shaft and parts around the shaft.

packless valve

A valve that does not use packing to prevent leaks around the valve stem. Flexible material is usually used to seal against leaks and still permit valve movement.

performance factor

The ratio of the heat moved by a refrigerating system to heat equivalent of the energy used. Varies with conditions.

phosphorous pentoxide

An efficient drier material that becomes gummy reacting with moisture and hence is not used alone as a drying agent.

pour point, oil

The temperature below which the oil surface will not change when the oil container is tilted.

power

The rate of doing work measured in horsepower, watts, kilowatts, and so forth.

power factor, electrical devices

The ratio of watts to volt-amperes in an alternating current circuit.

pressure

The force exerted per unit of area.

pressure drop

Loss in pressure, as from one end of a refrigerant line to the other, due to friction, static head, and the like.

pressure gauge

See gauge.

pressure-relief valve

A valve or rupture member designed to relieve excessive pressure automatically.

psychrometric chart

A chart used to determine the specific volume, heat content, dew point, relative humidity, absolute humidity, and wet- and dry-bulb temperatures, knowing any two independent items of those mentioned.

purging

The act of blowing out refrigerant gas from a refrigerant containing vessel usually for the purpose of removing noncondensables.

pyrometer

An instrument for the measurement of high temperatures.

radiation

The passage of heat from one object to another without warming the space between. The heat is passed by wave motion similar to light.

refrigerant

The medium of heat transfer in a refrigerating system that picks up heat by evaporating at a low temperature and gives up heat by condensing at a higher temperature.

refrigerating system

A combination of parts in which a refrigerant is circulated for the purpose of extracting heat.

relative humidity

The ratio of the water-vapor pressure of air compared to the vapor pressure it would have if saturated at its dry-bulb temperature. Very nearly the ratio of the amount of moisture contained in air compared to what it could hold at the existing temperature.

relief valve

A valve designed to open at excessively high pressures to allow the refrigerant to escape.

resistance, electrical

The opposition to electric-current flow, measured in ohms.

resistance, thermal

The reciprocal of thermal conductivity.

room cooler

A cooling element for a room. In air conditioning, a device for conditioning small volumes of air for comfort.

rotary compressor

A compressor in which compression is attained in a cylinder by rotation of a semiradial member.

running time

Usually indicates percent of time a refrigerant compressor operates.

saturated vapor

Vapor not superheated but of 100 percent quality, that is, containing no unvaporized liquid.

seal, shaft

A mechanical system of parts for preventing gas leakage between a rotating shaft and a stationary crankcase.

sealed unit

See hermetically sealed unit.

shell and tube

Pertaining to heat exchangers in which a coil of tubing or pipe is contained in a shell or container. The pipe is provided with openings to allow the passage of a fluid through it, while the shell is also provided with an inlet and outlet for a fluid flow.

silica gel

A drier material having the formula SiO_2.

sludge

A decomposition product formed in a refrigerant due to impurities in the oil or due to moisture. Sludges may be gummy or hard.

soda lime

A material used for removing moisture. Not recommended for refrigeration use.

solenoid valve

A valve opened by a magnetic effect of an electric current through a solenoid coil.

solid

The state of matter in which a force can be exerted in a downward direction only when not confined. As distinguished from fluids.

solubility

The ability of one material to enter into solution with another.

solution

The homogeneous mixture of two or more materials.

specific gravity

The weight of a volume of a material compared to the weight of the same volume of water.

specific heat

The quantity of heat required to raise the temperature of a definite mass of a material a definite amount compared to that required to raise the temperature of the same mass of water the same amount. May be expressed as Btu/pound/degrees Fahrenheit.

specific volume

The volume of a definite weight of a material. Usually expressed in cubic feet per pound. The reciprocal of density.

spray pond

An arrangement for lowering the temperature of water by evaporative cooling of the water in contact with outside air. The water to be cooled is sprayed by nozzles into the space above a body of previously cooled water and allowed to fall by gravity into it.

steam

Water in the vapor phase.

steam trap

A device for allowing the passage of condensate, or air and condensate, and preventing the passage of steam.

subcooled

Cooled below the condensing temperature corresponding to the existing pressure.

sublimation

The change from a solid to a vapor state without an intermediate liquid state.

suction line

The tube or pipe that carries refrigerant vapor from the evaporator to the compressor inlet.

suction pressure

Pressure on the suction side of the compressor.

superheater

A heat exchanger used on flooded evaporators, wherein hot liquid on its way to enter the evaporator is cooled by supplying heat to dry and superheat the wet vapor leaving the evaporator.

sweating

Condensation of moisture from the air on surfaces below the dew-point temperature.

system

A heating or refrigerating scheme or machine, usually confined to those parts in contact with the heating or refrigerating medium.

temperature

Heat level or pressure. The thermal state of a body with respect to its ability to pick up heat from or pass heat to another body.

thermal conductivity

The ability of a material to conduct heat from one point to another. Indicated in terms of Btu/per hour per square foot per inches of thickness per degrees Fahrenheit.

thermocouple

A device consisting of two electrical conductors having two junctions—one at a point whose temperature is to be measured, and the other at a known temperature. The temperature between the two junctions is determined by the material characteristics and the electrical potential setup.

thermodynamics

The science of the mechanics of heat.

thermometer

A device for indicating temperature.

thermostat

A temperature-actuated switch.

ton of refrigeration

Refrigeration equivalent to the melting of 1 ton of ice per 24 h, 288,000 Btu/day, 12,000 Btu/h, or 200 Btu/min.

total heat

The total heat added to a refrigerant above an arbitrary starting point to bring it to a given set of conditions (usually expressed in Btu/pound). For instance, in a super-heated gas, the combined heat added to the liquid necessary to raise its temperature from an arbitrary starting point to the evaporation temperature to complete evaporation, and to raise the temperature to the final temperature where the gas is superheated.

total pressure

In fluid flow, the sum of static pressure and velocity pressure.

turbulent flow

Fluid flow in which the fluid moves transversely as well as in the direction of the tube or pipe axis, as opposed to streamline or viscous flow.

unit heater

A direct-heating, factory-made, encased assembly including a heating element, fan, motor, and directional outlet.

unit system

A system that can be removed from the user's premises without disconnecting refrigerant-containing parts, water connection, or fixed electrical connections.

unloader

A device in a compressor for equalizing high- and low-side pressures when the compressor stops and for a brief period after it starts so as to decrease the starting load on the motor.

vacuum

A pressure below atmospheric, usually measured in inches of mercury below atmospheric pressure.

valve

In refrigeration, a device for regulation of a liquid, air, or gas.

vapor

A gas, particularly one near to equilibrium with the liquid phase of the substance, which does not follow the gas laws. Frequently used instead of gas for a refrigerant and, in general, for any gas below the critical temperature.

viscosity

The property of a fluid to resist flow or change of shape.

water cooler

Evaporator for cooling water in an indirect refrigerating system.

wax

A material that may separate when oil/refrigerant mixtures are cooled. Wax may plug the expansion valve and reduce heat transfer of the coil.

wet-bulb depression

Difference between dry- and wet-bulb temperatures.

wet compression

A system of refrigeration in which some liquid refrigerant is mixed with vapor entering the compressor so as to cause discharge vapors from the compressor to tend to be saturated rather than superheated.

xylene

A flammable solvent, similar to kerosene, used for dissolving or loosening sludges, and for cleaning compressors and lines.

zero, absolute, of pressure

The pressure existing in a vessel that is entirely empty. The lowest possible pressure. Perfect vacuum.

zero, absolute, of temperature

The temperature at which a body has no heat in it (- 459.6 degrees F or - 273.1 degrees C).

zone, comfort (average)

The range of effective temperature during which the majority of adults feel comfortable.

DECODING WORDS

HVAC Licensing Study Guide

Most individuals will not look at the glossary until they find a word that is not in their vocabulary. One of the things about learning a new language, which incidentally you are doing by studying this trade, is discovering new word meanings. Many of the words you use in the trade are used elsewhere in the English language.

By decoding the words you would normally encounter in your studies, you can develop a better understanding of the word's meaning and why it was applied to this particular situation. Now, while you are looking at the Glossary for your hint or answer, also read the meaning of the word as used in the trade. The practice will help you identify equipment and talk with others, both inside and outside the trade.

INSTRUCTIONS

In the first column is a jumbled-up word concerning air conditioning and/or refrigeration. In the blank column write out the decoded word. When finished, check the Answer Key to see how you did.

Hint: Check the Glossary before checking the Answer Key.

1. oerz _____

2. eubt _____

3. mentabit _____

4. moor _____

5. oncditinore _____

6. ira _____

7. tnui _____

8. reoolc _____

9. ifdfsuer _____

10. rnutre _____

11. quidil _____

12. perame _____

13. dardstan _____

14. rhawse _____

15. eeeattparum _____

16. mmaaion _____

17. susererp _____

18. avvle _____

19. ffabel _____

20. eeelrbd _____

21. rbolie _____

22. oreb _____

23. erinb _____

24. orielac _____

25. reerduc _____

26. ssilecu _____

27. rgehac _____

28. loic _____

29. ssorrepmoc _____

30. neymihc _____

31. aryrot _____

32. reensdonc _____

33. loontcr _____

34. lemarht _____

35. veisorroc _____

36. darethyorc _____

37. reryd _____

38. utds _____

39. uctd _____

40. ortejec _____

41. inegne _____

42. opytner _____

43. poraevatro   _____

44. riflte   _____

45. eithnerhfa   _____

46. acesurf   _____

47. orpellerp   _____

48. liaxa   _____

49. halsf   _____

50. enacruf   _____

51. noref   _____

52. riellg   _____

53. liedah   _____

54. thea   _____

55. usniof   _____

56. outtuc   _____

57. mauucv   _____

58. ppmu   _____

59. metsys   _____

60. owperorseh   _____

61. iiierhfdmu   _____

62. carhydbonsro   _____

63. drohysisly   _____

64. rometyhgre   _____

65. talinsunoi   _____

66. buisoteen   _____

67. attwoilk   _____

68. torcteed   _____

69. teemranmo _____

70. rnoicm _____

71. romille _____

72. prat _____

73. oonez _____

74. sslekcap _____

75. xideopten _____

76. eowrp _____

77. chrompsyetric _____

78. gingrup _____

79. eteropymr _____

80. rantefriger _____

81. vtealeri _____

82. ravpo _____

83. fatsh _____

84. ludges _____

85. noitsolu _____

86. meast _____

87. eaterpersuh _____

88. gineatow _____

89. ttthersoma _____

90. lentturub _____

91. reloaund _____

92. cosivsity _____

93. yxleen _____

94. luteosba _____

95. wflo   _____

96. ragavee   _____

97. secsmopnior   _____

98. animula   _____

99. mocoulephter   _____

100. eatrepush   _____

DECODING ANSWER KEY

1.	zero	44.	filter
2.	tube	45.	Fahrenheit
3.	ambient	46.	surface
4.	room	47.	propeller
5.	conditioner	48.	axial
6.	air	49.	flash
7.	unit	50.	furnace
8.	cooler	51.	Freon
9.	diffuser	52.	grille
10.	return	53.	halide
11.	liquid	54.	heat
12.	ampere	55.	fusion
13.	standard	56.	cutout
14.	washer	57.	vacuum
15.	temperature	58.	pump
16.	ammonia	59.	system
17.	pressure	60.	horsepower
18.	valve	61.	humidifier
19.	baffle	62.	hydrocarbons
20.	bleeder	63.	hydroysis
21.	boiler	64.	hygrometer
22.	bore	65.	insulation
23.	brine	66.	isobutene
24.	calorie	67.	kilowatt
25.	reducer	68.	detector
26.	Celsius	69.	manometer
27.	charge	70.	micron
28.	coil	71.	mollier
29.	compressor	72.	trap
30.	chimney	73.	ozone
31.	rotary	74.	packless
32.	condenser	75.	pentoxide
33.	control	76.	power
34.	thermal	77.	psychrometrc
35.	corrosive	78.	purging
36.	crohydrate	79.	pyrometer
37.	drier	80.	refrigerant
38.	dust	81.	relative
39.	duct	82.	vapor
40.	ejector	83.	shaft
41.	engine	84.	sludge
42.	entropy	85.	solution
43.	evaporator	86.	steam

(continued)

87. superheater
88. sweating
89. thermostat
90. turbulent
91. unloader
92. viscosity
93. xylene

94. absolute
95. flow
96. average
97. compression
98. alumina
99. thermocouple
100. superheat

APPENDICES

HVAC Licensing Study Guide

Appendix 1
Professional Organizations

This Table provides a listing of professional organizations, a description of the organization, and contact information.

Table A1-1 Professional Organizations

Organization	Description	Contact Information
Air Movement and Control Association International, Inc. (AMCA)	The Air Movement and Control Association International, Inc. is a not-for-profit international association of the world's manufacturers of related air system equipment - primarily, but not limited to: fans, louvers, dampers, air curtains, airflow measurement stations, acoustic attenuators, and other air system components for the industrial, commercial and residential markets-	**Website:** http://www.amca.org **Address:** 30 West University Drive, Arlington Heights, IL 60004-1893 **Phone:** 847-394-0150 **Fax:** 847-253-0088 **Email:** amca@amca.org
Air-Conditioning Contractors of America (ACCA)		**Website:** http://www.acca.org
American Institute of Architects (AIA)		**Website:** http://www.aia.org **Address:** 1735 New York Ave. NW, Washington, DC 20006 **Phone:** 800-AIA-3837 **Fax:** 202-626-7547 **Email:** infocentral@aia.org
American National Standards Institute (ANSI)	ANSI is a private, non-profit organization (501(c)3-that administers and coordinates the U.S. voluntary standardization and conformity assessment system- The Institute's mission is to enhance both the global competitiveness of U.S. business and the U.S. quality of life by promoting and facilitating voluntary consensus standards and conformity assessment systems, and safeguarding their integrity.	**Website:** http://www.ansi.org **Address:** 25 West 43rd Street, 4th Floor, New York , NY 10036 **Phone:** 212 642 4900 **Fax:** 212 398 0023 **Email:** info@ansi.org
HVAC Excellence	HVAC Excellence, a not for profit organization, establishes standards of excellence within the Heating, Ventilation, Air Conditioning and Refrigeration (HVAC&R-Industry. The two primary responsibilities of HVAC Excellence are Technician Competency Certification and Programmatic Accreditation. HVAC Excellence has achieved national acceptance and is involved in cooperative efforts with the National Skiffs Standard Board, the Manufactures Skills Standards Council,	**Website:** http://www.hvacexcellence.org **Address:** Box 491, Mount Prospect, IL 60056-0491 **Phone:** 800-394-5268 **Fax:** 800-546-3726

(continued)

Table A1-1 Professional Organizations *(continued)*

	VTECHS and various other industry and educational organizations.	
Heating, Refrigeration, and Air Conditioning Institute of Canada (HRAI)	The Heating, Refrigeration and Air Conditioning Institute of Canada (HRAE), founded in 1968 is a non-profit trade association of manufacturers, wholesalers and contractors in the Canadian heating, ventilation, air conditioning and refrigeration (HVACR-industry. HRAI member companies provide products and services for indoor comfort and essential refrigeration processes.	**Website:** http://www.hrai.ca **Address:** 5045 Orbitor Drive, Building 11, Suite 300, Mississauga , ON L4W 4Y4 Canada **Phone:** 800-267-2231; 905-602-4700 **Fax:** 905-602-1197 **Email:** hraimail@hrai.ca
International Institute of Ammonia Refrigeration (IIAR)	IIAR is an international association serving those who use ammonia refrigeration technology. Your membership in HAR is an investment that pays dividends for your company and for the ammonia refrigeration industry.	**Website:** http://www.iiar.org **Address:** 1110 North Globe Road, Suite 250, Arlington , VA 22201 **Phone:** 703-312-4200 **Fax:** 703-312-0065 **Email:** iiar@naii.org
National Fire Protection Association (NFPA)	The mission of the International nonprofit NFPA is to reduce the worldwide burden of fire and other hazards on the quality of life by providing and advocating scientifically-based consensus codes and standards, research, training and education. NFPA membership totals more than 75,000 individuals from around the world and more than 80 national trade and professional organizations.	**Website:** http://www.nfpa.org **Address:** 1 Batterymarch Park, Quincy , MA 02269-9101 **Phone:** 617- 770-3000 **Fax:** 617- 770-0700
BOCA International, Inc. See ICC Code	Founded in 1915, Building Officials and Code Administrators International, Inc., is a nonprofit membership association, comprised of more than 16,000 members who span the building community, from code enforcement officials to materials manufacturers. We are dedicated to preserving the public health, safety and welfare In the environment through the effective, efficient use and enforcement of Model Codes. BOCA provides a unique opportunity for any individual to join and derive the benefits of membership. Our members are professionals who are directly or indirectly engaged in the construction and regulatory process. BOCA is the Original professional association representing the full spectrum of code enforcement disciplines and construction industry interests We are the premier publishers of model codes. If you are interested in the development, maintenance and enforcement of progressive and	**Website:** http://www.bocai.org **Address:** 4051 West Flossmoor Road, Country Club Hills , IL 60478 **Phone:** 708/799-2300 **Fax:** 708/799-4981

Table A1-1 Professional Organizations *(continued)*

	responsive building regulations, take a closer look at how membership in BOCA International can promote excellence in your profession.	
Cooling Tower Institute (CTI)	The CTI mission is to advocate and promote the use of environmentally responsible Evaporative Heat Transfer Systems (EHTS-for the benefit of the public by encouraging Education, Research. Standards Development and Verification, Government Relations, and Technical Information Exchange	**Website:** http://www.cti.org **Address:** 2611 FM 1960 West, Suite H-200, Houston , TX 77068-3730 **Phone:** 281.583.4087 **Fax:** 281.537.1721 **Email:** vmanser@cti.org
Federation of European Heating and Air-Conditioning Associations	In was in the early sixties that the idea of some form of co-operation between technical associations in the field of heating, ventilating and air-conditioning would be extremely useful was recognized by leading professionals who met each other in international meetings and conferences. On September the 27th 1963 representatives of technical associations of nine European countries met in the Hague, The Netherlands, at the invitation of the Dutch association TWL These associations were ATIC (Belgium), AICVF (France), VDI TGA (Germany-IHVE (United Kingdom), CARR (Italy), TWL (The Netherlands-Norsk WS (Norway), SWKI (Switzerland-and Swedish WS (Sweden). An impressive list of subjects about which further co-operation was essential, was brought for discussion. Ways of exchanging technical knowledge was at the top of the list, together with the promotion of European standardization and regulations, education and co-ordination of congress and conferences.	**Website:** http://www.rehva.com **Address:** Ravenstein 3 Brussels 1000 BE **Phone:** 1: 32-2-514.11.71

—NOTES—

Appendix 2

Table A2-1 Decimal and Millimeter Eqivalents of Fractional Parts of an Inch

Parts of Inch		Decimal	mm	Parts of Inch		Decimal	mm
1-32	1-64	.01563	.397	17-32	33-64	.51563	13.097
		.03125	.794			.53125	13.097
	3-64	.04688	1.191		35-64	.54688	13.890
1-16		.0625	1.587	9-16		.5625	14.287
3-32	5-64	.07813	1.984	19-32	37-64	.57813	14.684
		.09375	2.381			.59375	15.081
	7-64	.10938	2.778		39-64	.60938	15.478
1-8		.125	3.175	5-8		.625	15.875
5-32	9-64	.14063	3.572	21-32	41-64	.64063	16.272
		.15625	3.969			.65625	16.669
	11-64	.17188	4.366		43-64	.67188	17.065
3-16		.1875	4.762	11-16		.6875	17.462
7-32	13-64	.20313	5.159	23-32	45-64	.70313	17.859
		.21875	5.556			.71875	18.256
	15-64	.23438	5.953		47-64	.73438	18.653
1-4		.25	6.350	3-4		.75	19.050
9-32	17-64	.26563	6.747	25-32	49-64	.76563	19.447
		.28125	7.144			.78125	19.844
	19-64	.29688	7.541		51-64	.79688	20.240
5-16		.3125	7.937	13-16		.8125	20.637
11-32	21-64	.32813	8.334	27-32	53-64	.82813	21.034
		.34375	8.731			.84375	21.431
	23-64	.35938	9.128		55-64	.85938	21.828
3-8		.375	9.525	7-8		.875	22.225
13-32	25-64	.39063	9.922	29-32	57-64	.89063	22.622
		.40625	10.319			.90625	23.019
	27-64	.42188	10.716		59-64	.92188	23.415
7-16		.4375	11.113	15-16		.9375	23.812
15-32	29-64	.45313	11.509	31-32	61-64	.95313	24.209
		.46875	11.906			.96875	24.606
	31-64	.48438	12.303		63-64	.98438	25.003
1-2		.5	12.700	1		1.00000	25.400

(continued)

Table A-2 lists the pipe sizes and wall thicknesses currently established as standard, or specifically:

1. The traditional standard weight, extra strong, and double extra strong pipe.
2. The pipe wall thickness schedules listed in American Standard B36.10, which are applicable to carbon steel and alloys other than stainless steels.
3. The pipe wall thickness schedules listed in American Standard B36.19, which are applicable only to stainless steels.

Table A2-2 Commercial Pipe Sizes and Wall Thicknesses

Nominal Pipe Size	Outside Diam.	Nominal Wall Thickness For													
		Sched. 5*	Sched. 10*	Sched. 20	Sched. 30	Standard†	Sched. 40	Sched. 60	Extra Strong‡	Sched. 80	Sched. 100	Sched. 120	Sched. 140	Sched. 160	XX Strong
⅛	0.405	–	0.049	–	–	0.068	0.068	–	0.095	0.095	–	–	–	–	–
¼	0.540	–	0.065	–	–	0.068	0.086	–	0.119	0.119	–	–	–	–	–
⅜	0.675	–	0.065	–	–	0.091	0.091	–	0.126	0.126	–	–	–	–	–
½	0.840	–	0.083	–	–	0.109	0.109	–	0.147	0.147	–	–	–	0.187	0.294
¾	1.050	0.065	0.083	–	–	0.113	0.113	–	0.154	0.154	–	–	–	0.218	0.308
1	1.315	0.065	0.109	–	–	0.133	0.133	–	0.179	0.179	–	–	–	0.250	0.358
1¼	1.660	0.065	0.109	–	–	0.140	0.140	–	0.191	0.191	–	–	–	0.250	0.382
1½	1.900	0.065	0.109	–	–	0.145	0.145	–	0.200	0.200	–	–	–	0.281	0.400
2	2.375	0.065	0.109	–	–	0.154	0.154	–	0.218	0.218	–	–	–	0.343	0.436
2½	2.875	0.083	0.120	–	–	0.203	0.203	–	0.276	0.276	–	–	–	0.375	0.552
3	3.5	0.082	0.120	–	–	0.216	0.216	–	0.300	0.300	–	–	–	0.438	0.600
3½	4.0	0.083	0.120	–	–	0.226	0.226	–	0.318	0.318	–	–	–	–	–
4	4.5	0.083	0.120	–	–	0.237	0.237	–	0.337	0.337	–	0.438	–	0.531	0.674
5	5.563	0.109	0.134	–	–	0.258	0.258	–	0.375	0.375	–	0.500	–	0.625	0.750
6	6.625	0.109	0.134	–	–	0.280	0.280	0.406	0.432	0.432	–	0.562	–	0.718	0.864
8	8.625	0.109	0.148	0.250	0.277	0.322	0.322	0.406	0.500	0.500	0.593	0.718	0.812	0.906	0.875
10	10.75	0.134	0.165	0.250	0.307	0.365	0.365	0.500	0.500	0.593	0.713	0.843	1.000	1.125	–
12	12.75	0.156	0.180	0.250	0.330	0.375	0.406	0.562	0.500	0.687	0.843	1.000	1.125	1.312	–
14 O.D.	14.0	–	0.250	0.312	0.375	0.375	0.438	0.593	0.500	0.750	0.937	1.093	1.250	1.406	–
16 O.D.	16.0	–	0.250	0.312	0.375	0.375	0.500	0.656	0.500	0.843	1.031	1.218	1.438	1.593	–
18 O.D.	18.0	–	0.250	0.312	0.438	0.375	0.562	0.750	0.500	0.937	1.156	1.375	1.562	1.781	–
20 O.D.	20.0	–	0.250	0.375	0.500	0.375	0.593	0.812	0.500	1.031	1.281	1.500	1.750	1.968	–
22 O.D.	22.0	–	0.250	0.375	–	0.375	–	–	0.500	–	–	–	–	–	–
24 O.D.	24.0	–	0.250	0.375	0.562	0.375	0.687	0.968	0.500	1.218	1.531	1.812	2.062	2.343	–
26 O.D.	26.0	–	–	–	–	0.375	–	–	0.500	–	–	–	–	–	–
30 O.D.	30.0	–	0.312	0.500	0.625	0.375	–	–	0.500	–	–	–	–	–	–
34 O.D.	34.0	–	–	–	–	0.375	–	–	0.500	–	–	–	–	–	–
36 O.D.	36.0	–	–	–	–	0.375	–	–	0.500	–	–	–	–	–	–
42 O.D.	42.0	–	–	–	–	0.375	–	–	0.500	–	–	–	–	–	–

Appendix 3

Table A3-1 Weights and Measures

Commercial Weight

16 drams (dr.)	= 1 ounce (oz.)
16 ounces	= 1 pound (lb.)
2000 pounds	= 1 ton (T.)

Dry Measure

2 pints (pt.)	= 1 quart (qt.)
8 quarts	= 1 peck (pk.)
4 pecks	= 1 bushel (bu.)

Long Measure

12 inches (in.)	= 1 foot (ft.)
3 feet	= 1 yard (yd.)
16 ½ feet	= 1 rod (rd.)
320 rd. (5280 ft.)	= 1 mile (ml.)

Time Measure

60 seconds (sec.)	= 1 minute (min.)
60 minutes	= 1 hour (hr.)
24 hours	= 1 day (da.)
365 ¼ days	= 1 year (yr.)

Square Measure

144 square inches	= 1 square foot
9 square feet	= 1 square yard
30 ¼ sq. yards	= 1 square rod
272 ¼ sq. feet	= 1 square rod
40 square rods	= 1 square rood
4 square roods	= 1 square acre
43,560 square feet	= 1 acre
640 acres	= 1 square mile

Troy Weight

24 grains (gr.)	= 1 pennyweight (pwt.)
20 pennyw'ts	= 1 ounce (oz.)
12 ounces	= 1 pound (lb.)

Circular Measure

60 seconds (")	= 1 minute (')
60 minutes (')	= 1 degree (°)
360 degrees (°)	= 1 circle

Surveyors' Measure

7.92 inches (in.)	= 1 link (lk.)
25 links (16 ½ ft.)	= 1 rod (rd.)
4 rods (66 ft.)	= 1 chain (ch.)
80 chains	= 1 mile (ml.)
Gunter's chain	= 22 yards or 100 links
10 square chains	= 1 acre

Liquid Measure

4 gills (gi.)	= 1 pint (pt.)
2 pints	= 1 quart (qt.)
4 quarts	= 1 gallon (gal.)
31 ½ gallons	= 1 barrel (bbl.)

Cubic Measure

231 cubic inches	= 1 gallon
2150.4 cubic inches	= 1 bushel
1728 cubic inches	= 1 cubic foot
27 cubic feet	= 1 cubic yard
128 cubic feet	= 1 cord (wood)
24 3/4 cubic feet	= 1 perch (stone)

Table A3-2   Ratings and Test Limits for A/C Electrolytic Capacitors

Capacity Rating (microfarads)			110-Volt Ratings		125-Volt-Ratings		220-Volt Ratings	
Nominal	Limits	Average	Amps. at Rated Voltage, 60 Hz	Approx. Max. Watts	Amps. at Rated Voltage 60 Hz	Approx. Max. Watts	Amps. At Rated Voltage, 60 Hz	Approx. Max. Watts
	25- 30	27.5	1.04- 1.24	10.9	1.18- 1.41	14.1	2.07-2.49	43.8
	32- 36	34	1.33- 1.49	13.1	1.51- 1.70	17	2.65-2.99	52.6
	38- 42	40	1.56- 1.74	15.3	1.79- 1.98	19.8	3.15-3.48	61.2
	43- 48	45.5	1.78- 1.99	17.5	2.03- 2.26	22.6	3.57-3.98	70
50	53- 60	56.5	2.20- 2.49	21.9	2.50- 2.83	28.3	4.40-4.98	87.6
60	64- 72	68	2.65- 2.99	26.3	3.02- 3.39	33.9	5.31-5.97	118.2
65	70- 78	74	2.90- 3.23	28.4	3.30- 3.68	36.8	5.81-6.47	128.1
70	75- 84	79.5	3.11- 3.48	30.6	3.53- 3.96	39.6	6.22-6.97	138
80	86- 96	91	3.57- 3.98	35	4.05- 4.52	45.2	7.13-7.96	157.6
90	97-107	102	4.02- 4.44	39.1	4.57- 5.04	50.4	8.05-8.87	175.6
100	108-120	114	4.48- 4.98	43.8	5.09- 5.65	56.5	8.96-9.95	197
115	124-138	131	5.14- 5.72	50.3	5.84- 6.50	65		
135	145-162	154	6.01- 6.72	62.8	6.83- 7.63	85.8		
150	161-180	170	6.68- 7.46	69.8	7.59- 8.48	95.4		
175	189-210	200	7.84- 8.71	81.4	8.91- 9.90	111.4		
180	194-216	205	8.05- 8.96	83.8	9.14-10.18	114.5		
200	216-240	228	8.96- 9.95	93	10.18-11.31	127.2		
215	233-260	247	9.66-10.78	106.7	10.98-12.25	145.5		
225	243-270	257	10.08-11.20	110.9	11.45-12.72	151		
250	270-300	285	11.20-12.44	123.2	12.72-14.14	167.9		
300	324-360	342	13.44-14.93	147.8	15.27-16.96	201.4		
315	340-380	360	14.10-15.76	156				
350	378-420	399	15.68-17.42	172.5				
400	430-480	455	17.83-19.91	197.1				

Appendix 4

There are a number of distributors and manufacturers of refrigeration and air-conditioning equipment and refrigerants. The Internet is an ideal place to look for the latest tools, equipment, and refrigerant sources. The following are but just of the many distributors online.

An online catalog for refrigeration and A/C service tools is available at **www.mastercool.com**. Another online catalog is found at **www.yellowjacket.com**, which also has leak detectors, fin straighteners, and other equipment for the tradesman.

Mastercool, Inc.

1 Aspen Drive

Randolph, NJ 07869

www.mastercool.com

Ritchie Engineering Company, Inc.

10950 Hampshire Avenue S.

Bloomington, MN 55438-2623

www.yellowjacket.com

These are examples of the questions and answers available on the Web site for those interested in furthering their careers and being able to respond to customer questions and problems. (Used through the courtesy of Ritchie Engineering, Inc.)

FREQUENTLY ASKED QUESTIONS

Q.   **For what refrigerants are the R 60a, R 70a, and R 80a rated? See Figure A4.1 for recovery units.**

A.   They are tested by UL (Underwriter's Laboratories, Inc.) to ARI 740-98 and approved for medium pressure refrigerants R-12, R-134a, R-401C, R406A, R-500; medium high pressure refrigerants R-401A, R-409A, R-401B, R-412A, R-411A, R-407D, R-22, R411B, R-502, R-407C, R-402B, R-408A, R-509; and high pressure refrigerants R-407A, R-404A, R-402A, R-507, R407B, R-410A.

Q.   **Why should I purchase a Recovery System?**

A.   With the Yellow Jacket® name on a hose, you know you have got the genuine item for performance backed by more than 50 years. Now, you will also find the name on refrigerant recovery systems that are based on RRTI and RST proven designs. RRTI was one of the original recovery companies and helped DuPont design its original unit. With the purchase of RST in 1998, Ritchie Engineering combined Yellow Jacket standards of manufacturing and testing with the RST track record of tough reliability.

R-100 Recovery system
(A)

Refrigerant recovery cylinders
(B)

FIGURE A4.1   (A) Recovery units. (B) Recovery unit cylinders. (*Ritchie Engineering, available at www.yellowjacket.com*)

Q. **Is ARI (American Refrigeration Institute) the only testing agency?**

A. No. ARI is only a certifying agency that hires another agency to perform the actual testing. UL is also EPA (Enviromental Protection Agency) approved as a testing and certifying agency. Yellow Jacket Systems are UL tested for performance. Some Yellow Jacket Systems are tested to CSA, CUL, CE, and TUV safety standards, which go beyond the ARI performance standards.

Q. **Can I compare systems by comparing their ARI or UL ratings?**

A. Yes. ARI and UL test standards should be the same. And remember that manufacturers can change the conditions under which they test their own machines to give the appearance of enhanced performance. Only ARI and UL test results provide consistent benchmarks and controls on which to make objective comparisons.

Q. **How dependable are Yellow Jacket refrigerant recovery systems?**

A. Yellow Jacket recovery systems get pushed to the limits day in and day out in dirty conditions, on roof tops, and sometimes in freezing or high ambient temperatures. Yellow Jacket equipment has been tested at thousands of cycles, and is backed with the experience of units in the field since 1992.

Q. **Is pumping liquid the fastest way to move refrigerant?**

A. Yes, and the R 50a, R 60a, R 70a, R 80a, and R 100 monitor liquid flow at a rate safe for the compressor. In Yellow Jacket lab testing, over 80,000 lbs of virgin liquid R-22 were continuously and successfully pumped. That is over 2,500 standard 30-lb tanks, or the equivalent of refrigerant in over 25,000 typical central A/C systems.

Q. **What is the push/pull recovery method?**

A. Many technicians use this recovery method, particularly on large jobs. By switching the R 50a, R 60a, R 70a, R 80a, or R 100 discharge valve to "recover" during push/pull recovery, the condenser is bypassed, increasing the "push" pressure and speeding the recovery.

Q. **Why do the R 50a through 100 feature a built-in filter?**

A. Every recovery machine requires an in-line filter to protect the machine against the particles and "gunk" that can be found in failed refrigeration systems. For your convenience, the R 50a, R 60a, R 70a, R 80a, and R 100 series incorporate a built-in 200-mesh filter that you can clean, and if necessary, replace. The filter traps 150-micron particles and protects against the dirtiest systems to maximize service life. In case of a burn-out, an acid-core filter/drier is mandatory. The Yellow Jacket filter is built-in to prevent breaking off like some competitive units with external filters.

Q. **What is auto purge and how does it work?**

A. At the end of each cycle, several ounces of refrigerant can be left in the recovery machine to possibly contaminate the next job, or be illegally vented. Many competitive recovery machines require switching hoses, tuning the unit off and on, or other time-consuming procedures. The R 50a, R 60a, R 70a, R 80a, and R 100 can be quickly purged "on the fly" by simply closing the inlet valve and switching the discharge valve to purge. In a few seconds, all residual refrigerant is purged and you are finished. Purging is completed without switching off the recovery unit.

Q.   Can increased airflow benefit recovery cylinder pressure?

A.   Yes. For reliable performance in the high ambient temperatures, Yellow Jacket units are engineered with a larger condenser and more aggressive fan blade with a greater pitch. This allows the unit to run cooler and keeps the refrigerant cooler in the recovery cylinder.

Q.   Can I service a Yellow Jacket System in the field?

A.   Although Yellow Jacket Systems feature either a full one or optional two year warranty, there are times when a unit will need a tune-up. The service manual with every unit includes a wide variety of information such as tips to speed recovery, troubleshooting guides, and parts listings. On the side of every unit you will find hook-up instructions, a quick start guide, and simple tips for troubleshooting. And if ever in doubt, just call 1-800-769-8370 and ask for customer service.

All service and repair parts are readily available through your nearest Yellow Jacket wholesaler.

Q.   Are there any recovery systems certified for R-410A?

A.   ARI 740-98 has been written but not yet enacted by the EPA. The Yellow Jacket R 60a, R 70a, and R 80a series have also been UL tested and certified for high-pressure gases such as R-410A that are covered under ARI 740-98.

Q.   What features should I demand in a system to be used for R-410A?

A.   Look for the following three features as a minimum:

- High-volume airflow through an oversized condenser to keep the unit running cooler and help eliminate cut-outs in high ambient temperatures
- Single automatic internal high-pressure switch for simple operation
- Constant pressure regulator (CPR) valve rated to 600 psi for safety eliminating the need to monitor and regulate the unit during recovery.

Q.   What is a CPR valve?

A.   The CPR valve is the feature that makes the Yellow Jacket R 70a and R 80a the first truly automatic recovery systems for every refrigerant. The single 600 psi rated high-pressure switch covers all refrigerants and eliminates the need for a control panel with two selector switches for R-410A.

The CPR valve automatically reduces the pressure of the refrigerant being recovered. Regardless of which refrigerant, input is automatically regulated through a small orifice that allows refrigerant to flash into vapor for compression without slugging the compressor. Throttling is not required.

Under normal conditions, you could turn the machine to the "liquid" and "recover" settings. The machine will complete the job while you work elsewhere.

Q.   **With the Yellow Jacket R 70a design, do I have to manually reset a pressure switch between medium and high-pressure gasses?**

A.   No. Some competitive machines require you to choose between medium and high-pressure gas settings before recovery. You will see the switch on their control panels. With the Yellow Jacket R 70a, the single internal automatic high-pressure switch makes the choice for you. That is why only R 70a is truly automatic.

FREQUENTLY ASKED QUESTIONS ABOUT PUMPS

Q.   **How can I select the right pump cfm?**

A.   See Figure A4.2 for a sampling of pumps. The following guidelines are for domestic through commercial applications.

System (tons)	Pump (cfm)
1-10	1.5
10-15	2.0
15-30	4.0
30-45	6.0
45-60	8.0
60-90	12.0
90-130	18.0
Over 130	24.0

Large capacity supervac™ pumps 12, 18, 24 cfm

FIGURE A4.2   Vacuum pumps. *(Ritchie Engineering, available at www.yellowjacket.com)*

Q.   **Can I use a vacuum pump for recovery?**

A.   A vacuum pump removes water vapor and is not for refrigerant recovery. Connecting a vacuum pump to a pressurized line will damage the pump and vent refrigerant to the atmosphere, which is a crime.

Q.   **How much of a vacuum should I pull?**

A.   A properly evacuated system is at 2,500 microns or less. This is 1 inch and impossible to detect without an electronic vacuum gauge. For most refrigeration systems, American Society of Heating, Refrigerating, and Air-Conditioning Engineers (ASHRAE) recommends pulling vacuum to 1,000 microns or less. Most system manufacturers recommend pulling to an even lower number of microns.

Q.   **Do I connect an electronic vacuum gauge to the system or pump?**

A.   To monitor evacuation progress, connect it to the system with a vacuum/charge valve.

Q.   **Why does the gauge micron reading rise after the system is isolated from the vacuum pump?**

A.   This indicates that water molecules are still detaching from the system's interior surfaces. The rate of rise indicates the level of system contamination and if evacuation should continue.

Q.   **Why does frost form on the system exterior during evacuation?**

A.   Because ice has formed inside. Use a heat gun to thaw all spots. This helps molecules move off system walls more quickly toward the pump.

Q.   **How can I speed evacuation?**

A.   • Use clean vacuum pump oil. Milky oil is water saturated and limits pump efficiency.

   • Remove valve cores from both high and low fittings with a vacuum/charge valve tool to reduce time through this orifice by at least 20 percent.

   • Evacuate both high and low sides at the same time. Use short, 3/8 inch diameter and larger hoses.

   SuperEvac™ Systems can reduce evacuation time by over 50 to 60 percent. SuperEvac pumps are rated at 15 microns (or less) to pull a vacuum quickly. A large inlet allows you to connect a large diameter hose. With large oil capacity, SuperEvac pumps can remove more moisture from systems between oil changes.

Q.   **What hose construction is best for evacuation?**

A.   Stainless steel. There is no permeation and outgassing.

UV LEAK DETECTION TOOLS

53515 MACH IV FLEXIBLE UV LIGHT

The MACH IV flexible UV light has 4 TRUE UV LEDS delivering a brilliant fluorescent glow. Gets into tight areas easily.

53012 UV SWIVEL HEAD LIGHT

- 12V/100 WATT
- 180° Swivel Head Gets Into Tight Places
- Instant ON
- Heavy Duty Metal Construction
- Includes UV Enhancing Safety Glasses

53312 UV MINI LIGHT

- 12V/50 WATT
- High Intensity
- Compact & Lightweight
- Instant ON
- 16 ft. (5m) Cord
- Includes UV Enhancing Safety Glasses

RECHARGEABLE UV LIGHT

53411
- High Intensity 12V/50 WATT
- Cordless with Rechargeable Battery Cartridge and Charger
- Compact & Portable
- Includes UV Enhancing Safety Glasses

53412
- Same as 53411 (less Battery Charger and glasses)

53413 BATTERY CARTRIDGE

- 12V/50 WATT
- This powerful rechargeable battery cartridge holds a charge equal to 30 minutes of continuous use

BATTERY CHARGER

53414 110V/60 HZ
53414-220 220V/50-60 HZ

LIGHT PART#	LIGHT BULB PART#	LENS PART#
53012	53012-B	53012-L
53112	53012-B	53110-L
53312	53312-B	53312-L
53411	53312-B	53312-L
53515	53515-B	–

ACCESSORIES

53314 DYE REMOVER - 4 oz
53315 SERVICE LABELS - 25 per pack. Bright yellow label indicating that the system has been charged with UV Dye.
92398 UV ENHANCING SAFETY GLASSES This is a must for protection against ultraviolet light during leak detection.

53809 MINI DYE INJECTOR

"Cartridge Type" Dye Injector (10 Appl.) Concentrated Dye.

53223 "CARTRIDGE TYPE" UNIVERSAL DYE/OIL INJECTOR

Adding dye with the new 53223 is fast and easy, simply connect the injector to the low side of the A/C system and twist the handle to the next application line. The replaceable cartridge provides 25 applications of universal dye that is compatible with R12/22/502 and R134a systems. The injector hose comes complete with a R134a coupler and auto shut-off valve adapter for 1/4"FL systems.

"REFILLABLE" UNIVERSAL DYE/OIL INJECTOR

To inject the oil or dye, simply connect the injector to the low side of the A/C system and twist the handle until you reach the desired amount. The injector hose comes complete with a R134a coupler and auto shut-off valve adapter for 1/4"FL systems.

53123 REFILLABLE UNIVERSAL DYE INJECTOR with 2 oz Bottle of Concentrated A/C Dye (25 Appl.)
53123-A REFILLABLE UNIVERSAL DYE/OIL INJECTOR (without dye)
53134 REFILLABLE DYE INJECTOR with (R134a 13mm Connection) and 2 oz Bottle of Concentrated A/C Dye (25 Appl.)
53322 REFILLABLE DYE INJECTOR with (1/4" Auto Shut-off Valve Connection) and 2 oz Bottle of Concentrated A/C Dye (25 Appl.)

ULTRAVIOLET DYES

Standard Universal A/C Dyes

Pack 1/4 oz - 1 Application
92699 Standard Universal Dye Six 1/4 oz Packs (6 pcs)

8 oz Bottle - 32 Applications
92708 R12/22/502 Standard Universal Dye

32 oz Bottle - 128 Applications
92732 R12/22/502 Standard Universal Dye

Concentrated Universal Dyes

Replaceable Cartridge - 25 Applications
53825 Universal Dye Cartridge
Replaceable Cartridge - 10 Applications
53810 Concentrated Dye Cartridge
2 oz Bottle - 25 Applications
53625 Universal Dye

FIGURE A4.3 Leak scanners (flourescent type). *(Mastercool, available at www.yellowjacket.com)*

FREQUENTLY ASKED QUESTIONS ABOUT FLUORESCENT LEAK SCANNERS

Q.   **Does the ultra violet (UV) scanner light work better than an electronic leak detector?**

A.   No one detection system is better for all situations. But, with a UV lamp you can scan a system more quickly and moving air is never a problem. Solutions also leave a telltale mark at every leak site. Multiple leaks are found more quickly. (See Figure A4.3).

Q.   **How effective are new light emitting diode (LED) type UV lights?**

A.   LEDs are small, compact lights for use in close range. (See Figure A4.4). Most effective at 6-in. range. The model with two blue UV and three UV bulbs has a slightly greater range. Higher power Yellow Jacket lights are available.

Q.   **Can LED bulbs be replaced?**

A.   No. The average life is 110,000 hours.

Q   **Are RediBeam lamps as effective as the System II lamps?**

A.   The RediBeam lamp has slightly less power to provide lightweight portability. But with the patented reflector and filter technologies, the RediBeam 100-W bulb produces sufficient UV light for pinpointing leaks.

Q.   **Does the solution mix completely in the system?**

A.   Solutions are combinations of compatible refrigeration oil and fluorescent material designed to mix completely with the oil type in the system.

Universal UV system

FIGURE A4-4   LED-ultraviolet lights.
(Ritchie Engineering, available at www.yellow-jacket.com)

Q.   How are solutions different?

A.   Solutions are available with mineral, alkylbenzene, PAG, or polyol ester base stock to match oil in the system.

Q.   What is universal solution?

A.   It is made from polyol ester and mixes well with newer oils. It also works in mineral oil systems, but can be harder to see.

Q.   What is the lowest operating temperature?

A.   It is -40 degrees F for all solutions. Alkylbenzene in alkylbenzene systems to -100 degrees F.

Q.   Does solution stay in the system?

A.   Yes. When future leaks develop, just scan for the sources. In over six years of testing, the fluorescent color retained contrast. When the oil is changed in the system, scanner solution must be added to the new oil.

Q.   Is the solution safe?

A.   Solutions were tested for three years before introduction and have been performance proven in the field since 1989. Results have shown the solutions safe for technicians, the environment and all equipment when used as directed. Solutions are pure and do not contain lead, chromium, or chlorofluorocarbon (CFC) products.

   Presently, solutions are approved and used by major manufacturers of compressors, refrigerant, and equipment.

Q.   How do I determine oil type in a system?

A.   Many times the oil is known due to the type of refrigerant or equipment application. Systems should be marked with the kind of oil used. Always tag a system when oil type is changed.

Q.   In a system with a mix of mineral and alkylbenzene oil, which scanner solution should be used?

A.   Base your choice of solution on whatever oil is present in the larger quantity. If you do not know which oil is in greater quantity, assume it to be alkylbenzene.

Q. **How do I add solution to the system?**

A. In addition to adding solution using injectors or mist infuser, you have other possibilities.

   If you do not want to add more gas to the system, connect the injector between the high and low side allowing system pressure to do the job. Or, remove some oil from the system, then add solution to the oil and pump back in.

Q. **How is the solution different from visible colored dyes?**

A. Unlike colored dyes, Yellow Jacket fluorescent solutions mix completely with refrigerant and oil and do not settle out. Lubrication, cooling capacity, and unit life are not affected; and there is no threat to valves or plugging of filters. The solutions will also work in a system containing dytel.

Q. **How do I test the system?**

A. Put solution into a running system to be mixed with oil and carried throughout the system. Nitrogen charging for test purposes will not work since nitrogen will not carry the oil. To confirm solution in the system, shine the lamp into the system's sight glass. Another way is to connect a hose and a sight glass between the high and low sides, and monitor flow with the lamp. The most common reason for inadequate fluorescence is insufficient solution in the system.

Q. **Can you tell me more about bulbs?**

A. 115-V systems are sold with self-ballasted bulbs in the 150-W range. Bulbs operate in the 365 nanometer long range UV area and produce the light necessary to activate the fluorescing material in the solution. A filter on the front of the lamp allows only "B" band rays to come through. "B" band rays are not harmful.

Q. **What is the most effective way to perform an acid test?**

A. Scanner solution affects the color of the oil slightly. Use a two-step acid-test kit which factors out the solution in the oil, giving a reliable result.

Q. **Can fluorescent product be used in nonrefrigerant applications?**

A. Yes, in many applications.

ACCUPROBE™ LEAK DETECTOR WITH HEATED SENSOR TIP

This is the only tool you need for fast, easy, and certain leak detection. The heated sensor tip of the Yellow Jacket® Accuprobe™ leak detector positively identifies the leak source for all refrigerants. That includes CFCs such as R-12 and R-502, hydrochlorofluorocarbons (HCFCs) as small as 0.03 oz/year, and hydrofluorocarbon (HFC) leaks of 0.06 oz/year, even R-404A, and R-410A. See Figure A4.5.

Frequency of flashes in the tip and audible beeping increases the closer you get to the leak source. You zero-in and the exclusive Smart Alarm LED shows how big or small the leak is on a scale of 1 to 9. Maximum value helps you determine if the leak needs immediate repair.

Service life of the replaceable sensor is more than 300 h with minimal cleaning and no adjustments. Replaceable filters help keep out moisture and dust that can trigger false sensing and alarms.

Three sensitivity levels include ultrahigh to detect leaks that could be missed with other detection systems.

Additional features and benefits:

- Detects all HFCs including R-134a, R-404A, R-410A, and R407C and R-507; all HCFCs such as R-22 and all CFCs

  - Audible beeping can be muted

- Extended flexible probe for easy access in hard-to-reach areas

- Sensor not poisoned by large amount of refrigerant and does not need recalibration

  - Sensor failure report mode

- Weighs only about 15 oz for handling comfort and ease

- Carrying holster and hard, protective case included

- Bottle of nonozone depleting chemical included for use as a leak standard to verify proper functioning of sensor and electronic circuitry.

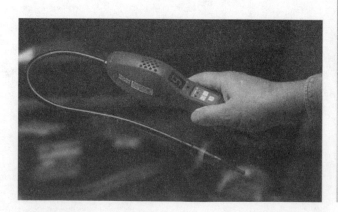
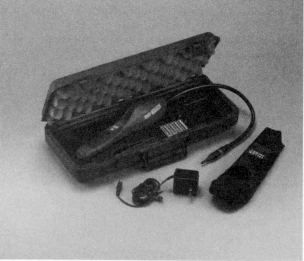

FIGURE A4.5   Accuprobe leak detector. *(Ritchie Engineering, available at www.yellowjacket.com)*

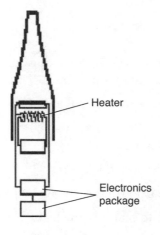

FIGURE A4.6   Heated sensor leak detector. *(Ritchie Engineering, available at www.yellowjacket.com)*

TECHNOLOGY COMPARISON–HEATED SENSOR OR NEGATIVE CORONA?

Heated Sensor Leak Detectors

When the heated sensing element is exposed to refrigerant, an electrochemical reaction changes the electrical resistance within the element, causing an alarm. The sensor is refrigerant specific with superior sensitivity to all HFCs, HCFCs, and CFCs and minimal chance of false alarms. When exposed to large amounts of refrigerant which could poison other systems, the heated sensor clears quickly and does not need recalibration before reuse. See Figure A4.6.

In the sensor of an old-style corona detector, high voltage applied to a pointed electrode creates a corona. When refrigerant breaks the corona arc, the degree of breakage generates the level of the alarm. This technology has good sensitivity to R-12 and R-22, but only fair for R-134a, and poor for R-41 OA, R-404A, and R-407C. Sensitivity decreases with exposure to dirt, oils, and water. And false alarms can be triggered by dust, dirt specks, soap bubbles, humidity, smoke, small variations in the electrode emission, high levels of hydrocarbon vapors, and other nonrefrigerant variables. See Figure A4.7.

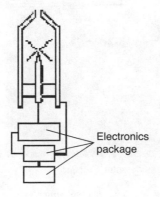

FIGURE A4.7   Negative corona leak detector. *(Ritchie Engineering, available at www.yellowjacket.com)*

TIPS FOR DETECTING SYSTEM LEAKS

1. Inspect entire A/C system for signs of oil leakage, corrosion cracks, or other damage. Follow the system in a continuous path so no potential leaks are missed.

2. Make sure there is enough refrigerant in a system (about 15 percent of system capacity or 50 psi/min) to generate pressure to detect leaks.

3. Check all service access port fittings. Check seals in caps.

4. Move detector probe at 1 inch/s within inch of suspected leak area.

5. Refrigerant is heavier than air, so position probe below test point.

6. Minimize air movement in area to make it easier to pinpoint the leak.

7. Verify an apparent leak by blowing air into the suspected leak.

8. When checking for evaporator leaks, check for gas in condensate drain tube.

9. Use heated sensor type detector for difficult-to-detect R-134a, R-410A, R-407C, and R-404A.

NEW COMBUSTIBLE GAS DETECTOR (with ultrasensitive, long life sensor)

- Hand-held precision equipment detects all hydrocarbon and other combustible gases including propane, methane, butane, industrial solvents, and more. See Figure A4.8.

- Sensitivity, bar graph, and beeping to signal how much and how close.

- Unit is preset at normal sensitivity, but you can switch to high or low. After warm-up you will hear a slow beeping. Frequency increases when a leak is detected until an alarm sounds when moving into high gas concentration. The illuminated bar graph indicates leak size.

Combustible gas detector

FIGURE A4.8   Gas detectors. *(Ritchie Engineering, available at www.yellowjacket.com)*

- If no leak is detected in an area you suspect, select high sensitivity. This will detect even low levels in the area to confirm your suspicions. Use low sensitivity as you move the tip over more defined areas, and you will be alerted when the tip encounters the concentration at the leak source.

- Ultrasensitive sensor detects less than 5 ppm methane and better than 2 ppm for propane. It performs equally well on a complete list of detectable gases including acetylene, butane, and isobutane.

- Automatic calibration and zeroing.

- Long-life sensor easily replaced after full service life.

Applications

- Gas lines/pipes
- Propane filling stations
- Gas heaters
- Combustion appliances
- Hydrocarbon refrigerant
- Heat exchangers
- Marine bilges
- Manholes
- Air quality
- Arson residue (accelerants)

FREQUENTLY ASKED QUESTIONS ABOUT FIXED MONITORING SYSTEMS

See Figure A4.9.

Q.   **Are calibrated leak testers available to confirm that the monitor is calibrated correctly?**

A.   The Yellow Jacket calibrated leak is a nonreactive mixture of R-134a or NH_3 and CO_2. The non-returnable cylinders contain 10 L of test gas. The cylinders require a reusable control valve and flow indicator. Test gases can be ordered for 100 ppm or 1,000 ppm mixtures.

Q.   **What refrigerants will the leak monitors detect?**

A.   Leak monitors will detect most CFC, HFC, and HCFCs such as R-11, R-12, R-13, R-22, R-113, R-123, R-134a, R-404A, R407C, R-410A, R-500, R-502, and R-507. Yellow Jacket also has leak monitors available for ammonia and hydrocarbon-based refrigerants.

Q.   **Can the leak monitor be calibrated for specific applications?**

A.   Yes, the Yellow Jacket leak monitor can be calibrated for specific applications. Contact customer service for your specific need.

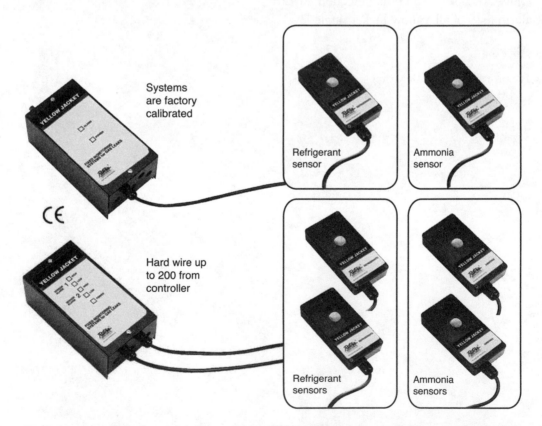

FIGURE A4.9   Fixed monitoring systems. *(Ritchie Engineering, available at www.yellow-jacket.com)*

Q. **If the unit goes into alarm, can it switch on the fan? Can it turn off the system at the same time?**

A. The leak monitor has a pair of dry, normally open/normally closed contacts that can handle 10 A at 115 V. When the sensor indicates a gas presence higher than the set point, it opens the closed contacts and closes the open contacts which will turn equipment on or off.

Q. **After a unit goes into alarm and the contacts close, what can it be connected to?**

A. The open contacts can shut the system down, call a phone number, turn on a fan, or emergency light.

Q. **How does the sensor work?**

A. When the sintered metal oxide surface within the sensor absorbs gas molecules, electrical resistance is reduced in the surface allowing electrons to flow more easily. The system controller reads this increase in conductivity and signals an alarm. Metal oxide technology is proven for stability and performance.

Q. **What is the detection sensitivity level of Yellow Jacket fixed monitors?**

A. The dual sensitivity system has a low alarm level of about 100 ppm and a high level of about 1,000 ppm for most CFC, HFC, and HCFC products. The high level for R-123 is an exception at about 300 ppm. Ammonia detection levels are about 100 ppm low and about 150 ppm high. The alarm level of all Yellow Jacket single-level systems is about 100 ppm.

Detection levels are preset at the factory to cover most situations. If necessary, however, you can order a custom level, or adjust the set point on site.

Q. **What gas concentration must be detected?**

A. It depends on the refrigerant. For a more thorough answer, terms established by U.S. agencies must first be understood:

- IDLH—immediately dangerous to life and health

- TWA—time weighted average concentration value over an 8-hour work day or a 40-hour work week (OSHA or NIOSH levels)

- STEL—short term exposure level measured over 15 minutes (NIOSH)

- Ceiling concentration—should not be exceeded in a working day (OSHA)

Obviously, the first consideration is IDLH. For most refrigerants, the IDLH is relatively high (e.g., R-12 is 15,000 ppm), and such a concentration would be unusual in a typical refrigerant leak situation. Leak detection, however, is still an immediate condition, so the STEL should be the next consideration, followed then by the 8-h TWA or ceiling concentration. R-22, for example, has a STEL of 1,250 ppm and a TWA of 1,000 ppm. (The TWA for most refrigerants is 1,000 ppm.)

The draft UL standard for leak monitors requires gas detection at 50 percent of the IDLH. In other words, R-12 with a IDLH value of 15,000 ppm must be detected at 7,500 ppm. As with most refrigerants, the TWA is 1,000 ppm.

All of the foregoing suggests that for most CFC, HFC and HCFCs, detection at 1,000 ppm provides a necessary safety margin for repair personnel. Ammonia with a significantly lower IDLH of 300 ppm and a TWA of 25 ppm requires detection at 150 ppm to comply with 50 percent IDLH requirements. R-123 has a TWA of 50 ppm and an IDLH of 1,000 ppm, therefore detection at 100 ppm provides a good margin of safety. A monitor with a detection threshold of about 100 ppm for any gas provides an early warning so that repairs can be made quickly. This can save refrigerant, money, and the environment.

Q. **How frequently should the system be calibrated?**

A. Factory calibration should be adequate for 5 to 8 years. Routine calibration is unnecessary when used with intended refrigerants. Yellow Jacket sensors can not be poisoned, show negligible drift, and are stable long term. You should, however, routinely check performance.

Q. **Can there be a false alarm?**

A. For monitoring mixtures, the semiconductor must be able to respond to molecularly similar gases. With such sensitivity, false alarms can be possible. Engineered features in Yellow Jacket monitors help minimize false alarms:

• The two-level system waits about 30 seconds until it is "certain" that gas is present before signalling.

• At about 100 to 1,000-ppm calibration level, false alarms are unlikely.

To prevent an unnecessary alarm, turn off the unit or disable the siren during maintenance involving refrigerants or solvents. Temperature, humidity, or transient gases may occasionally cause an alarm. If in doubt, check with the manufacturer.

Q. **What are alternative technologies for monitoring and detecting refrigeration gas leaks?**

A. Infrared technology is sensitive down to 1 ppm. This level is not normally required for refrigeration gases and is also very expensive compared to semiconductor technology. As an infrared beam passes through an air sample, each substance in the air absorbs the beam differently. Variations in the beam indicate the presence of a particular substance. The technique is very gas specific and in a room of mixed refrigerants, more than one system would be required. To get over this problem, newer models work on a broad band principle so they can see a range of gases. As a result they do not generally operate below 50 ppm and can experience false alarms.

Electrochemical cells can be used for ammonia. These cells are very accurate, but are expensive, and are normally used to detect low levels (less than 500 ppm), and perform for about two years.

With air sampling transport systems, tubing extends from the area(s) to be monitored back to a central controller/sensor.

Micropumps move air through the system, eliminating a number of on-site sensors, but there may be problems:

- Air in the area of concern is sampled at intervals rather than monitored continuously. This can slow the response to changing conditions.

- Dirt can be sucked into the tubes, blocking filters.

- Gases can be absorbed by the tube or leak out of the tube providing a concentration at the sensor lower than in the monitored area.

- Gases can leak into the tube in transit rather than the area monitored. The reading would be misleading.

NEW AND OLD TOOLS, OR CATALOG SHOPPING AND UPDATING

Refer to Figure A4.10 for the following tools and supplies.

The Mastercool Company is indicative of the supply house supplies provided for those working in the refrigeration and air-conditioning field. Some of the equipment you should be aware of as you continue to work in the field are shown in their catalog. A few of them are shown here as an example of some of the latest devices available to make your work day more efficient. A convenient way to categorize the tools you work with is shown in the following example of a listing of available tools. This listing may change in time as the requirements for handling new refrigerants are brought about by accrediting agencies and standards writers.

- *Leak detection* relies on electronic detectors as well as the older types that have been around for years. Ultraviolet rays have now been utilized to more accurately identify and locate leaks. There are various dyes and injectors that need examining for keeping up. The combustible gas leak detector should also be examined as gases other than refrigerants are encountered on the job.

- *Manifold/gauges and hoses* are another of the categories most often recognized as essential to the technician working in the field and in-house. Hoses can stand some examination since they have been constantly improving through the years. And, there is always the chance a hose will or has, ruptured, leaked or deteriorated. Newer hoses are usually designed for a longer life than previously.

- Another category for classifying devices, tools, and other equipment is the *recovery equipment*, now so necessary to keep within the letter of the law and protect the environment.

- *Vacuum pumps* now have the rotary vane to produce deep vacuums. There are a number of pumps, oils, and accessories that fall into this classification process.

- *Refrigerant scales* have certified scales and programmable scales. The charging program allows the user to program desired quantities, and before the charge is complete, an alarm will sound allowing ample time to turn off the refrigerant supply. There are now features such as pause/charge: empty/full tank that allow the user to know the amount of refrigerant left in the tank at any time. There is a repeat function that allows the user to charge to the previously stored amount. The scales are multilingual and have memory that allows programming for any number of vehicles or refrigerant applications.

- *Specialty hydraulic tools,* such as the tube expanding tool kit and the hydraulic flaring and swaging tool, are updated also. The new features are a hand-held hydraulic press that accurately flares and swages copper tubing. Once the die and adapter are secured in the fixture, a few pumps of the handle and you are done. This tool really takes the work out of swaging and flaring, especially on larger tube sizes. The kit includes dies and adapters for flaring and swaging copper tubing sizes from 1/4 to 7/8 inch.

- *Tube cutters* have carbide steel cutting wheels for cutting hard and soft copper, aluminum, brass, thin wall steel as well as stainless steel.

- *ing station,* a lightweight durable steel frame cart, contains all the necessary tools to conveniently charge the A/C system. No need for different charging cylinders have a rugged die cast electronic scale. Simply place the refrigerant cylinder on harge.

FIGURE A4.10 (A) Electronic leak detector. (B) Manifold and gauges. (C) Recovery unit. (D) Vacuum pumps. (E) Refrigerant scales. (F) Lasere thermometer. (G) Hydraulic tools. (H) Tube cutters. *(Mastercool, available at www.yellowjacket.com)*

- *The electronic tank heater blanket* speeds up recharge time. It also assures total discharge of refrigerant from 30 lb and 50 lb tanks of 125 degrees/55 degrees C and maximum pressure of 185 psi (R134a) and 170 psi for R-12. They are available for use with 120 or 240 V.

- *Air content analyzer:* when an A/C system leaks, refrigerant is lost and air enters the system. Your refrigerant recycler cannot tell the difference between refrigerant and air-it cycles both from partially filled systems. You end up with an unknown quantity of efficiency robbing air in your supply tank. Excess actual pressure in your supply tank indicates the pressure of air, also called "noncondensable gases" or NCGs. When you release the excess pressure, you are also releasing air. The result is purer refrigerant which will work more efficiently. This one can be left on the supply tank for regular monitoring or it can be removed to check all your tanks.

- *Thermometers, valve core tools, and accessories:* a valve core remover/installer controls refrigerant flow a 1/4 turn of the valve lever. Lever position also gives visual indication of whether the valve is opened or closed. The infrared thermometer with laser has a back-lit LCD display and an expanded temperature range of -20 to 500 degreees C or -4 to 932 degrees F. An alkaline battery furnishes power for up to 15 hours.